Lecture Notes in Mathematics

Edited by A. Dold and B. Ec

719

Categorical Topology

Proceedings of the International Conference,
Berlin, August 27th to September 2nd, 1978

Edited by H. Herrlich and G. Preuß

Springer-Verlag
Berlin Heidelberg New York 1979

Editors

Horst Herrlich
Universität Bremen
Fachbereich Mathematik
Achterstraße
D-2800 Bremen

Gerhard Preuß
Freie Universität Berlin
Institut für Mathematik I
Hüttenweg 9
D-1000 Berlin 33

AMS Subject Classifications (1970): 18-XX, 46 A 20, 46 E XX, 46 H XX, 46 M XX, 54-XX, 55 D XX

ISBN 3-540-09503-9 Springer-Verlag Berlin Heidelberg New York
ISBN 0-387-09503-9 Springer-Verlag New York Heidelberg Berlin

© by Springer-Verlag Berlin Heidelberg 1979
Printed in Germany

Printing and binding: Beltz Offsetdruck, Hemsbach/Bergstr.
2141/3140-543210

PREFACE

The International Conference on Categorical Topology at the Free University Berlin, from the 27th of August to the 2nd of September 1978, brought together fifty mathematicians from 16 countries of the world. The meeting was organized by the editors of these proceedings. Especially, we would like to thank Priv.-Doz. Dr. Thomas Marny (scientific secretary) and Mrs. Karla Lautenschläger (congress bureau) for their assistance. We are also grateful for the help of the members of the Categorical Topology Research Group in Berlin. We should not forget to thank the Deutsche Forschungsgemeinschaft, the Senator of Science and Research in Berlin and the Free University Berlin for financial support.
And last not least we thank the Vice-President of the Free University Berlin, Professor Heckelmann, for his willingness to welcome the participants of the conference.

The scientific program of the conference consisted of talks on various aspects of Categorical Topology. The subjects treated in this volume may roughly be classified as follows:

1) Internal problems in particular topological categories (Top, Unif, Conv, Near, Bitop):
   Bentley and Herrlich, Brandenburg, Brümmer, Frolík, Hajek and Mysior, Heldermann, Hoffmann, Hušek, Kannan, Kannan and Rajagopalan, Marny, Raghavan and Reilly, Salbany, Schwarz, Wattel.
2) Topological categories
   a) Function spaces and monoidal closed structures:
      Greve , Nel, Porst and Wischnewsky, Wyler.
   b) Connectedness-theories:
      Preuß, Salicrup and Vázquez.
3) Topological functors and related concepts (structure functors, (E,M)-functors, semi-topological functors, initial completions):

Börger, Harvey, Herrlich, Herrlich and Strecker, Nakagawa, Tholen, Wischnewsky.

4) Applications
   a) topological algebra:
      Smith-Thomas
   b) algebraic topology:
      Frei and Kleisli, Hardie, Heath and Kamps, Mac Donald
   c) functional analysis:
      Banaschewski, Cooper and Schachermayer, Mulvey, Nel, Semadeni and Zidenberg-Spirydonow
   d) statistic metric spaces
      Lüschow

5) Topoi:
   Johnstone, Linton and Paré.

Thus, the papers give a survey on the recent state of research in Categorical Topology and its applications. Finally, we would like to express our thanks to the Springer Verlag for publishing the proceedings of this conference in the Lecture Notes series.

H. Herrlich G. Preuß

Address list of authors and speakers

B. Banaschewski	Department of Mathematics, McMaster University, Hamilton, Ontario L8S 4K1, Canada
H.L. Bentley	Department of Mathematics, University of Toledo, Toledo, Ohio 43606, USA
R. Börger	Fachbereich Mathematik, Fernuniversität, Postfach 940, D-5800 Hagen, Fed. Rep. Germany
H. Brandenburg	Freie Universität Berlin, Institut für Mathematik I, Hüttenweg 9, D-1000 Berlin 33
G.C.L. Brümmer	Department of Mathematics, University of Cape Town, Private Bag, Rondebosch 7700, Republic of South Africa
J.B. Cooper	Institut für Mathematik, Universität Linz, A-4045 Linz-Auhof, Austria
A. Frei	Department of Mathematics, University of British Columbia, Vancouver, B.C., Canada
Z. Frolík	Matematický Ústav ČSAV, Zitná 25, 115 67 Praha 1, Czechoslovakia
G. Greve	Fachbereich Mathematik, Fernuniversität, Lützowstr. 125, D-5800 Hagen, Fed. Rep. Germany
D.W. Hajek	Department of Mathematics, University of Puerto Rico, Mayaguez, Puerto Rico 00708
K.A. Hardie	Department of Mathematics, University of Cape Town, Rondebosch 7700, Republic of South Africa
J.M. Harvey	Department of Mathematics, University of Rhodesia, P.O.Box MP 167, Salisbury, Rhodesia
P.R. Heath	Department of Mathematics, Memorial University of Newfoundland, St. John's, Newfoundland, Canada, A1B 3X7
N.C. Heldermann	Zentralblatt für Mathematik, Otto-Suhr-Allee 26, D-1000 Berlin 10

H. Herrlich	Universität Bremen, Fachsektion Mathematik, Achterstrasse 33, D-28 Bremen, Fed. Rep. Germany
R.-E. Hoffmann	Universität Bremen, Fachsektion Mathematik, Achterstrasse 33, D-28 Bremen, Fed. Rep. Germany
M. Hušek	Matematický Ústav, Karlovy University, Sokolovská 83, 18600 Praha 8 - Karlin, Czechoslovakia
P.T. Johnstone	Pure Mathematics, 16 Mill Lane, Cambridge, CB2 1SB, England
K.H. Kamps	Fachbereich Mathematik, Fernuniversität, Postfach 940, D-5800 Hagen, Fed. Rep. Germany
V. Kannan	Department of Mathematics, Hyderabad University, Nampally Station Road, Hyderabad - 500 001, India
H. Kleisli	Institut de Mathématiques, Faculté des Sciences, Université de Fribourg, CH-1700 Fribourg, Suisse
D. Leseberg	Freie Universität Berlin, Institut für Mathematik I, Hüttenweg 9, D-1000 Berlin 33
F.E.J. Linton	Department of Mathematics, Wesleyan University, Middletown, Connecticut 06457, USA
R.B. Lüschow	Department of Mathematics and Computation, Faculty of Science, Universidad Tecnica del Estado, Santiago, Chile
J.L. MacDonald	Department of Mathematics, University of British Columbia, Vancouver, B.C., Canada
T. Marny	Freie Universität Berlin, Institut für Mathematik I, Hüttenweg 9, D-1000 Berlin 33
E. Michael	Department of Mathematics, University of Washington, Seattle, Washington 98195, USA
C.J. Mulvey	Mathematics Division, University of Sussex, Falmer, Brighton, BN1 9QH, England

A. Mysior	Institute of Mathematics, University of Gdańsk, 80-952 Gdańsk, Wita Stwosza 57, Poland
R. Nakagawa	Department of Mathematics, University of Tsukuba, Ibaraki, Japan
L.D. Nel	Department of Mathematics, Carleton University, Ottawa, Ontario, Canada K1S 5B6
R. Paré	Mathematics Department, Dalhousie University, Halifax, New Scotland, Canada B3H 4H8
H.-E. Porst	Universität Bremen, Fachsektion Mathematik, Achterstraße 33, D-28 Bremen, Fed. Rep. Germany
G. Preuß	Freie Universität Berlin, Institut für Mathematik I, Hüttenweg 9, D-1000 Berlin 33
H. Pust	Freie Universität Berlin, Institut für Mathematik I, Hüttenweg 9, D-1000 Berlin 33
T.G. Raghavan	Department of Mathematics, University of Auckland, Private Bag, Auckland, New Zealand
M. Rajagopalan	Universidad de los Andes, Facultad de Ciencas, Departemento de Matematicas, Merida, Venezuela
I.L. Reilly	Department of Mathematics, University of Auckland, Private Bag, Auckland, New Zealand
M.D. Rice	Department of Mathematics, George Mason University, Fairfax, Virginia 22030, USA
S. Salbany	Department of Mathematics, University of Cape Town, Rondebosch 7700, Republic of South Africa
G. Salicrup	Instituto de Matemáticas U.N.A.M., Ciudad Universitaria, México 20, D.F.
W. Schachermayer	Institut für Mathematik, Universität Linz, A-4045 Linz-Auhof, Austria
F. Schwarz	Institut für Mathematik, Technische Universität, Welfengarten 1, D-3000 Hannover 1, Fed. Rep. Germany

Z. Semadeni	Instytut Matematyczny, Polskiej Akademii Nauk, ul. Sniadeckich 8, skr. poczt. 137, 00-950 Warszawa, Poland
B.V. Smith Thomas	Department of Mathematical Sciences, Memphis State University, Memphis, Tennessee 38152, USA
G.E. Strecker	Department of Mathematics, Kansas State University, Manhattan, Kansas 66502, USA
W. Tholen	Fachbereich Mathematik, Fernuniversität, Postfach 940, D-5800 Hagen, Fed. Rep. Germany
R. Vázquez	Instituto de Matemáticas U.N.A.M., Ciudad Universitaria, México 20, D.F.
E. Wattel	Subfaculteit Wiskunde, Wiskundig Seminarium, Vrije Universiteit, De Boelelaan 1081, Amsterdam - 1011, The Netherlands
R. Wiegandt	Mathematical Institute of the Hungarian Academy of Sciences, H-1053 Budapest, Reáltanoda u. 13-15, Hungary
M. Wischnewsky	Fachbereich Mathematik, Universität Bremen, Kufsteiner Straße, D-2880 Bremen 33, Fed. Rep. Germany
O. Wyler	Department of Mathematics, Carnegie-Mellon University, Pittsburgh, PA 15213, USA
H. Zidenberg-Spirydonow	Zakład Matematyki, Wyższej Szkoły Pedagogicznej, ul. Wielkopolska 15, 70-387 Szczecin, Poland

CONTENTS

Recovering a space from its Banach sheaves
by B. Banaschewski 1

Completeness is productive
by H.L. Bentley and H. Herrlich 13

Legitimacy of certain topological categories
by R. Börgor 18

On $\underline{E}$-normal spaces
by H. Brandenburg 24

Two procedures in bitopology
by G.C.L. Brümmer 35

Saks spaces and vector valued measures
by J.B. Cooper and W. Schachermayer 44

A question in categorical shape theory: When is a shape-invariant functor a Kan extension?
by A. Froi and H. Kloisli 55

The finest functor preserving the Baire sets
by Z. Frolík 63

Lifting closed and monoidal structures along semitopological functors
by G. Greve 74

On non-simplicity of topological categories
by D.W. Hajek and A. Mysior 84

Kan lift-extensions in C.G. Haus$_*$
by K.A. Hardie 94

Topological functors from factorization
by J.M. Harvey 102

Groupoids and classification sequences
by P.R. Heath and K.H. Kamps 112

Concentrated nearness spaces
by N.C. Heldermann 122

Initial and final completions
by H. Herrlich 137

Algebra ∪ Topology
by H. Herrlich and G.E. Strecker 150

Topological spaces admitting a "dual"
by R.-E. Hoffmann 157

Special classes of compact spaces
by M. Hušek 167

Pairs of topologies with same family of continuous self-maps
by V. Kannan 176

Hereditarily locally compact separable spaces
by V. Kannan and M. Rajagopalan 185

Injectives in topoi, I: Representing coalgebras as algebras
by F.E.J. Linton and R. Paré 196

Injectives in topoi, II: Connections with the axiom of choice
by P.T. Johnstone, F.E.J. Linton, and R. Paré .. 207

Categories of statistic-metric spaces (a co-universal construction)
by R.B. Lüschow 217

A categorical approach to primary and secondary operations in topology
by J.L. MacDonald 225

Limit-metrizability of limit spaces and uniform limit spaces
by T. Marny 234

Banach spaces over a compact space
by C.J. Mulvey 243

A note on (E,M)-functors
by R. Nakagawa 250

Convenient topological algebra and reflexive objects
by L.D. Nel 259

Existence and applications of monoidally closed structures in topological categories
by H.-E. Porst and M.B. Wischnewsky 277

Connection properties in topological categories and related topics
by G. Preuß 293

On projective and injective objects in some topological categories
by T.G. Raghavan and I.L. Reilly 308

An embedding characterization of compact spaces
by S. Salbany 316

Connection and disconnection
by G. Salicrup and R. Vázquez 326

Connections between convergence and nearness
by F. Schwarz 345

Functors on categories of ordered topological spaces
by Z. Semadeni and H. Zidenberg-Spirydonow 358

On the coproduct of the topological groups $\mathbb{Q}$ and $\mathbb{Z}_2$
by B.V. Smith Thomas 371

Lifting semifinal liftings
by W. Tholen 376

Normally supercompact spaces and convexity preserving maps
by E. Wattel 386

Structure functors - Representation and existence theorems -
by M.B. Wischnewsky 395

Function spaces in topological categories
by O. Wyler 411

Recovering a Space from its Banach Sheaves

B. Banaschewski, Hamilton, Ontario, Canada

Motivated by the active interest, shown by various authors during recent years, in Banach sheaves and related topics (Auspitz [1], Banaschewski [2,3], Hofmann [5], Hofmann-Keimel [6], Mulvey [10]) this paper deals with the special case, appropriate to Banach sheaves, of a general natural question which has been considered in many other instances. Examples of the latter which are particularly closely related to the results presented here concern the passage from a ring to its associated module category and from a monoid M to the category of M-sets. Here, we show that a Tychonoff (= completely regular Hausdorff) space X is determined by the category BANShX of Banach sheaves on X (Proposition 2). As a step towards this, we first prove the analogous result for the category EBAN of Banach modules over a commutative Banach algebra E with unit (Proposition 1), and as a further application of our basic step in the proof of Proposition 2 we derive corresponding facts for the category of C-Modules, C the sheaf of continuous functions on a Tychonoff space (Proposition 3) and the category of sheaves of abelian groups on a 0-dimensional Hausdorff space (Proposition 4).

The arguments presented here are based on the endomorphism monoid of the identity functor of the categories considered. This is rather a coarse invariant of a category, and hence it may well be that a different approach will yield a better result concerning BANShX. The cruicial technical steps in our proofs are that the partially defined addition of maps in such categories as EBAN and BANShX can be derived from the category structure (Lemma 2), and that the fixed maximal ideals of C*X are identifiable by means of BANShX (Lemma 4).

1. Background on Banach sheaves.

Recall that a Banach sheaf (also: approximation sheaf in Banaschewski [2], Q-sheaf in Auspitz [1]) on a topological space X is a presheaf S on X with values in the category BAN of Banach spaces (over the scalar field $K = \mathbb{R}$ or $\mathbb{C}$) and linear contractions satisfying the following separation condition (S) and approximation patching condition (AP), where in the latter

S_x is the stalk $\varinjlim SV(x \in V, \text{ open})$ of S at $x \in X$, U any open subset of X, and $s \rightsquigarrow \tilde{s}$ the map $SU \to \Pi S_x(x \in U)$ for which $\tilde{s}$ is the family of images s_x of s under the different colimit maps $SU \to S_x$:

(S) For any open cover $U = \cup U_i$, $\|s\| = \sup_i \|s|U_i\|$ for each $s \in SU$.

(AP) For any $\sigma \in \Pi S_x(x \in U)$, $\sigma = \tilde{s}$ for some $s \in SU$ whenever, for each $\varepsilon > 0$ and each $x \in U$, there exists an open neighbourhood $W \subseteq U$ of x and some $t \in SW$ such that $\|\sigma|W - \tilde{t}\| \leq \varepsilon$.

Note that the map $SU \to SV$ for $V \subseteq U$ is given by $s \rightsquigarrow s|V$, and that the norm in a product of Banach spaces is the supremum norm. Also, the conditions (S) and (AP) together imply the patching condition which says that, for any open cover $U = \cup U_i$, if $s_i \in SU$ are given such that $s_i|U_i \cap U_\kappa = s_\kappa|U_i \cap U_\kappa$, and the set of norms $\|s_i\|$ is bounded, then $s_i = s|U_i$ for some $s \in SU$. In particular, this shows (Banaschewski [2]) that Banach sheaves are indeed sheaves in the general sense of the term (Mitchell [8], Chapter X).

The condition on σ in (AP) singles out those elements of $\Pi S_x(x \in U)$ which are locally approximated by the $\tilde{t}$ for elements t belonging to S. More generally, if S and T are BAN-valued presheaves on X such that $S \subseteq T$ in the sense that each SU is a subspace of TU then $t \in TU$ will be called locally approximated by S iff, for any $x \in U$ and any $\varepsilon > 0$, there exists an open neighbourhood $W \subseteq U$ of x and some $s \in SW$ such that $\|t|W - s\| \leq \varepsilon$. If this holds for all elements of TU, for each open U, then T is called a <u>dense extension</u> of S. With this notion, the Banach sheaves are also characterized as those separated BAN-valued presheaves which have no proper dense extension (Banaschewski [3]).

In the following, BANShX will be the category of Banach sheaves on the space X, the maps being the presheaf maps $\phi: S \to T$, i.e. the natural transformations from the functor S to the functor T. An important aspect of BANShX, first established by Mulvey [9], is that it is equivalent to the category of internally defined Banach spaces in the topos ShX of sheaves of sets on the space X. On the other hand, as originally shown by Auspitz [1], BANShX is equivalent to the category of an appropriate type of Banach fibre spaces over the space X (see also Mulvey [10] and Hofmann [5]).

The second fundamental notion which will concern us here is that of a <u>Banach module</u> over a Banach algebra E; by this is meant a Banach space A together with an action $E \times A \to A$, $(c,x) \rightsquigarrow cx$, making A into a module over

E and such that $\|cx\| \le \|c\|\,\|x\|$. The relevant maps $h: A \to B$ between Banach modules A and B are the E-linear contractions, and EBAN will be the resulting category.

Banach modules naturally appear in the context of Banach sheaves (Banaschewski [2]): for any Banach sheaf S on the space X, each of the Banach spaces SU, U any open set, can be made into a Banach module over the Banach algebra C*U of bounded continuous $\mathbb{K}$-valued functions on U in such a way that $(fs)|V = (f|V)(s|V)$ for all $f \in C^*U$, $s \in SU$, and $V \subseteq U$. Any BAN-valued sheaf on X with this additional property is called a Banach C*-Module. It is shown in [2] that the maps of Banach sheaves preserve the additional module structure, and that on paracompact spaces the Banach sheaves are precisely the Banach C*-Modules; the latter was recently extended to locally paracompact spaces (Hofmann-Keimel [6]).

In particular, for each S ε BANShX, SX belongs to C*XBAN, and the correspondence S ⇝ SX is functorial. The resulting global elements functor G: BANShX → C*XBAN has a left adjoint, and for Tychonoff spaces X it is a full embedding.

It is the latter case which shall specifically interest us here, and for the remainder of this paper X will always be a Tychonoff space. Also, BANX will be the full reflective subcategory of C*XBAN which is the image of the functor G and thus equivalent to BANShX. We note that, in the special case of compact X, BANX is characterized in C*XBAN as follows (Hofmann-Keimel [6]): A ε C*XBAN is isomorphic to its reflection in BANX iff, for any $a,c \in A$ and $u \in C^*X$, $\|a\|, \|c\| \le 1$ and $0 \le u \le 1$ implies $\|ua+(1-u)c\| \le 1$.

Our aim is to show that X is determined by the category BANShX, using the equivalent category BANX.

2. Recovering E from EBAN.

In this section, E will always be a commutative Banach algebra with unit e, and our goal is to carry out what is stated in the heading.

Recall that, for any category $\underset{\sim}{K}$, the natural transformations of the identity functor of $\underset{\sim}{K}$ into itself form a commutative monoid End($\underset{\sim}{K}$), invariant with respect to category equivalence, i.e. any equivalence $\underset{\sim}{K} \to \underset{\sim}{L}$ of categories induces an isomorphism End($\underset{\sim}{K}$) → End($\underset{\sim}{L}$). For EBAN, we now have

Lemma 1. End(EBAN) is isomorphic to the multiplicative submonoid of E given by the unit ball $|E|$ of E.

Proof. For any $c \in |E|$ and $A \in$ EBAN, let $\tilde{c}_A\colon A \to A$ be the map $x \rightsquigarrow cx$. Since E is commutative and $\|c\| \le 1$, $\tilde{c}_A$ is evidently a map in EBAN, depending naturally on A, and hence one has $\tilde{c} \in \mathrm{End}(\mathrm{EBAN})$ with components $\tilde{c}_A$. Moreover, $\tilde{a}\tilde{c} = (ac)^{\sim}$ for any $a,c \in |E|$ and $\tilde{e}$ is the identity transformation; thus $c \rightsquigarrow \tilde{c}$ is a monoid homomorphism, clearly one-one since $\tilde{c}_E(e) = c$. To see that this is onto, let ϕ be any element of End(EBAN) and put $c = \phi_E(e)$. Then, for any Banach module A over E, and any $a \in |A|$, $h\colon E \to A$ defined such that $h(x) = xa$ is a map in EBAN because $\|a\| \le 1$, and since $\phi_A h = h\phi_E$ it follows that $\phi_A(a) = ca$, which shows that $\phi_A = \tilde{c}_A$; in all this proves $\phi = \tilde{c}$.

Our approach now is to reconstruct E by means of properties of the category EBAN as these are reflected in $|E|$. The main step towards this will be to describe the partial operation of addition in $|E|$.

If f and g are any maps $A \to B$ in EBAN such that $\|f(x)+g(x)\| \le \|x\|$ for all $x \in A$ then the map $x \rightsquigarrow f(x)+g(x)$ $(x \in A)$ again belongs to EBAN and will be called $f + g$. The relation between maps $f,g,h\colon A \to B$ given by $h = f+g$ can then be characterized in EBAN by the following considerations: Recall that, in EBAN, $A \times C$ consists of all pairs (x,y), $x \in A$ and $y \in C$, with the usual componentwise operations and the norm $\|(x,y)\| = \max\{\|x\|,\|y\|\}$, whereas $A \oplus C$ differs from $A \times C$ only in its norm which is $\|(x,y)\| = \|x\|+\|y\|$; since $\max\{\|x\|,\|y\|\} \le \|x\|+\|y\|$, the map $j\colon A\oplus C \to A\times C$, with identity effect, belongs to EBAN. Formally, j is defined by the specifications

$$pju = 1_A,\ qju = 0_A,\ pjv = 0_C,\ qjv = 1_C$$

where $u\colon A \to A\oplus C$ and $v\colon C \to A\oplus C$ are the coproduct embeddings and $p\colon A\times C \to A$ and $q\colon A\times C \to C$ the product projections. Now, consider the pullback diagram

$$\begin{array}{ccc} P & \longrightarrow & A\oplus A \\ \downarrow & & \downarrow j_A \\ A & \underset{\Delta_A}{\longrightarrow} & A\times A \end{array}$$

where Δ_A is the diagonal embedding and j_A the map just discussed. Since P may be viewed as the subspace of $A \times (A\oplus A)$ of all $(x,(y,z))$ such that $(x,x) = (y,z)$, it consists of all $(x,(x,x))$, with norm $2\|x\|$. Hence, P can also be represented by $A^{(2)}$ which is obtained from A by doubling its norm. We let $i_A\colon A^{(2)} \to A$ and $\Delta_A^{(2)}\colon A^{(2)} \to A\oplus A$ be the vertical and horizontal map in the corresponding pullback diagram.

With these maps, we obviously have

<u>Lemma 2. For $f,g,h\colon A \to B$, $h = f+g$ iff the following diagram commutes</u>

$$\begin{array}{ccc}
A^{(2)} & \xrightarrow{\Delta_A^{(2)}} & A \oplus A \\
 & & \downarrow{\scriptstyle f\oplus g} \\
{\scriptstyle i_A}\downarrow & & B \oplus B \\
 & & \downarrow{\scriptstyle +} \\
A & \xrightarrow[h]{} & B
\end{array}$$

<u>where $+\colon B\oplus B \to B$ is the addition map.</u>

<u>Corollary. If E and F are any Banach algebras then any functor $\Phi\colon$ EBAN $\to$ FBAN preserving finite limits and coproducts preserves addition of maps.</u>

Now let $\Phi\colon$ EBAN $\to$ FBAN be a category equivalence for commutative Banach algebras E and F with unit. Then, by Lemma 1, one has a map $\kappa\colon |E| \to |F|$ which is a monoid isomorphism relative to multiplication and has the property that $\Phi(\tilde{a}_A) = \widetilde{\kappa(a)}_{\Phi A}$ for all $a \in |E|$. Moreover, since Φ preserves and reflects the ternary relation $h = f+g$ for maps with the same domain and codomain by Lemma 2, and since $c = a+b$ for any $a,b,c \in |E|$ iff $\tilde{c}_A = \tilde{a}_A + \tilde{b}_A$ for all $A \in$ EBAN, it follows that $c = a+b$ iff $\kappa(c) = \kappa(a)+\kappa(b)$ for any $a,b,c \in |E|$. In particular, then $\kappa(0) = 0$, and for any $a,c \in |E|$ and any rational r, $a = rc$ iff $\kappa(a) = r\kappa(c)$. The latter implies that $ra \in |E|$ iff $r\kappa(a) \in |F|$ for each $a \in |E|$ and all rational r, and since $\|a\| = \inf\{\frac{1}{r} \mid ra \in |E|,\ r>0 \text{ rational}\}$ it follows that $\|a\| = \|\kappa(a)\|$ for all $a \in |E|$. Thus, κ is also a homeomorphism and therefore $\kappa(\lambda a) = \lambda\kappa(a)$ for all $a \in |E|$ and all real λ with $|\lambda| \le 1$.

Finally, we can extend κ to a map $E \to F$ by putting $\kappa(a) = \|a\|\kappa(\frac{1}{\|a\|}a)$, which will then also be equal to $\alpha\kappa(\frac{1}{\alpha}a)$ for any $\alpha \ge \|a\|$, and with the aid of this it is easily established that the extended map is an isomorphism. This proves

<u>Proposition 1. For commutative Banach algebras E and F with unit, any category equivalence $\Phi\colon$ EBAN $\to$ FBAN induces an isomorphism $\kappa\colon E \to F$ of Banach algebras over $\mathbb{R}$ such that $\Phi(\tilde{a}_A) = \widetilde{\kappa(a)}_{\Phi A}$ for all $a \in |E|$.</u>

Note that the above argument also shows that E is the essentially unique Banach algebra whose unit ball, with its multiplication and partial addition, is isomorphic to the endomorphism monoid of the identity functor of EBAN with its partial addition.

Remark 1. The above argument extends to suitable subcategories $\underset{\sim}{K} \subseteq$ EBAN and $\underset{\sim}{L} \subseteq$ FBAN. These should be full and reflective, containing E (respectively F), and have the property that the reflection map $A \oplus A \to A \bar{\oplus} A$, the latter the coproduct in the subcategory, is onto for each $A \in \underset{\sim}{K}$. Note that this means $A \bar{\oplus} A$ again consists of the same elements as $A \times A$ but has a norm between the norms of $A \oplus A$ and $A \times A$. Now one can define the counterpart of $A^{(2)}$ by pullback in $\underset{\sim}{K}$, and this again differs from A by a suitable renorming such that the new norm of $x \in A$ is between $\|x\|$ and $2\|x\|$. Evidently, the analogue of Lemma 2 will then hold, as will that of Lemma 1, and hence one arrives at the same conclusion. We shall refer to subcategories of EBAN satisfying these conditions as standard.

Remark 2. In general, the monoid End($\underset{\sim}{K}$) is only a crude invariant of a category $\underset{\sim}{K}$, and our approach clearly cannot work for non-commutative Banach algebras. Thus, if E is as before and F is the Banach algebra of 2-by-2 matrices over E, suitably normed so that E is embedded in F as the multiples of the unit matrix, then End(FBAN) corresponds to the central elements of $|F|$ and hence only lends itself to reconstructing E.

3. Recovering X from BANShX.

Recall that X will always be a Tychonoff space so that BANShX is equivalent to BANX, and hence the result of the previous section becomes applicable provided we can show that BANX is the right kind of subcategory of C*XBAN. Our first step is to do just that.

Lemma 3. BANX is standard in C*XBAN.

Proof. We first argue in BANShX. For any Banach sheaf S, let T be the BAN-valued presheaf such that $TU = SU \oplus SU$ for each open U. Since

$$\|(s,t)_x\| = \inf_{x \in V \subseteq U} (\|s|V\| + \|t|V\|) = \|s_x\| + \|t_x\|$$

by the basic facts about norms in stalks, the image of $TU = SU \oplus SU$ in the stalk T_x embeds into $S_x \oplus S_x$ by the map $T_x \to S_x \oplus S_x$ which results from the maps $SV \oplus SV \to S_x \oplus S_x$ determined by the colimit maps $SV \to S_x$, and since the images of the different $SU \oplus SU$ in T_x have dense union, this shows the map $T_x \to S_x \oplus S_x$ is an isomorphism. Now, let $T^{\#}U \subseteq \Pi(S_x \oplus S_x)(x \in U)$ be the subspace of those σ which are locally approximated by T. For any such σ it then turns out that $\sigma(x) = (\sigma_1(x), \sigma_2(x))$ where $\sigma_i \in \Pi S_x (x \in U)$ are locally approximated by S: Given $\varepsilon > 0$ and any $x \in U$, there exists by hypothesis a pair $(s,t) \in SU \oplus SU$ and a neighbourhood $W \subseteq U$ of x such that

$$\|(\sigma_1(z), \sigma_2(z)) - (s_z, t_z)\| \leq \varepsilon \text{ (all } z \in W)$$

which gives $\|\sigma_1(z)-s_z\| \leq \varepsilon$ and $\|\sigma_2(z)-t_z\| \leq \varepsilon$ for all $z \in W$. Since S is a Banach sheaf, one now has $s_i \in SU$ such that $\sigma_i = \check{s}_i$ $(i = 1,2)$ by (AP), but this says that the map $TU \to T^{\#}U$ is onto. Further, $T^{\#}$ is the Banach sheaf reflection of T (Banaschewski [2]) and therefore the coproduct of S with itself in BANShX.

For BANX this shows that $T^{\#}X$ is the coproduct of SX with itself in BANX, and the ontoness of the map $SX \oplus SX \to T^{\#}X$ considered above therefore establishes the part of Lemma 3 which is not a priori obvious.

By Proposition 1 and the first remark following it we can now also conclude:

<u>Corollary</u>. <u>Any category equivalence Φ: BANX → BANY for Tychonoff spaces X and Y induces an isomorphism κ: $C^*X \to C^*Y$ of Banach algebras over $\mathbb{R}$ such that $\Phi(\tilde{u}_A) = \kappa(u)\tilde{}_{\Phi A}$ for all $u \in |C^*X|$ and all $A \in$ BANX.</u>

Of course, such an isomorphism between C^*X and C^*Y for Tychonoff spaces X and Y does not say much about X and Y: it holds iff $\beta X \cong \beta Y$, and in general this leaves extensive scope for X to differ from Y. Evidently, the problem of finding X via C^*X is to be able to distinguish between fixed and free maximal ideals in C^*X, and this is exactly what can be done by means of BANX. For the following, recall that an episink is any family f_α: $A_\alpha \to A$ $(\alpha \in I)$ of maps such that, for any g,h: $A \to B$, $gf_\alpha = hf_\alpha$ for all α implies that $g = h$ (Herrlich-Strecker [4], p. 127).

<u>Lemma 4</u>. <u>A maximal ideal M of C^*X is free iff $\tilde{u}_A$: $A \to A$ $(u \in |M|)$ is an episink for each $A \in$ BANX.</u>

Proof. (⇒) Let M be free and f,g: $A \to B$ such that $f\tilde{u}_A = g\tilde{u}_A$ for all $u \in |M|$. Consider any $b \in B$ such that $|M|b = 0$; for any $x \in X$, there exists a $u \in |M|$ such that $u^{-1}\{1\}$ is a neighbourhood of x since M is free, and then ub and b have the same germ at x (in the Banach sheaf of which B is the space of global elements) so that $b_x = 0$. Since $\|b\| = \sup\|b_x\|$ $(x \in X)$ (Banaschewski [2]) it follows that $b = 0$. Now, for any $a \in A$, $uf(a) = f(ua) = f\tilde{u}_A(a) = g\tilde{u}_A(a) = ug(a)$ for all $u \in |M|$ shows that $f(a) = g(a)$, i.e. in all $f = g$, proving that $\tilde{u}_A$: $A \to A$ $(u \in |M|)$ is an episink for any A.

(⇐) Let M be the maximal ideal fixed at $x \in X$. Then, define S by taking $SU = \mathbb{K}$ for $x \in U$ and $SU = 0$ for $x \notin U$, and for any $V \subseteq U$ the map $SU \to SV$ as the zero map for $x \notin V$ and the identity map for $x \in V$. Clearly, S is a separated BAN-valued presheaf on X. Also, $S_x = \mathbb{K}$ and $S_y = 0$ for $y \neq x$, and the colimit map $SU \to S_x$ is the identity map whereas all other such maps $SU \to S_y$ $(y \neq x)$ are the zero maps. Now, take any arbitrary $\sigma \in \Pi S_z(z \in U)$; if $x \notin U$ then $\sigma = \tilde{O}$ since $\Pi S_z(z \in U)$ is then just the zero space, and if $x \in U$ then $s = \sigma(x) \in \mathbb{K}$, hence $s \in SU$ and $\sigma = \tilde{s}$. This shows that the map $SU \to \Pi S_z(z \in U)$ by $s \rightsquigarrow \tilde{s}$ is actually onto, and thus S is clearly a Banach sheaf. The Banach module SX over C*X is then isomorphic to $\mathbb{K}_x$, with underlying Banach space equal to $\mathbb{K}$ and C*X acting such that $fa = f(x)a$. Now, consider the diagram

$$C^*X \xrightarrow[(u \in |M|)]{\tilde{u}=\tilde{u}_{C^*X}} C^*X \xrightarrow{h} \mathbb{K}_x$$

where h is given by evaluation at x. One has $h\tilde{u}(f) = h(\tilde{u}f) = u(x)f(x) = 0$, i.e. $h\tilde{u} = 0$ for all $u \in |M|$ but of course $h \neq 0$. This shows that $\tilde{u}_{C^*X}(u \in |M|)$ is not an episink.

Proposition 2. Any category equivalence Φ: BANX $\to$ BANY for Tychonoff spaces X and Y induces a homeomorphism $X \to Y$.

Proof. Let κ: $C^*X \to C^*Y$ be the isomorphism in the corollary of Lemma 3. Now, a maximal ideal M of C*X is free iff all $\tilde{u}_A(u \in |M|)$ are episinks, which holds iff all $\Phi(\tilde{u}_A)(u \in |M|)$ are episinks, and this means that all $\kappa(u)\tilde{}_{\Phi A}(u \in |M|)$ are episinks, which is satisfied iff $\tilde{v}_B(v \in \kappa(M))$ is an episink for each B in BANY, and this holds iff $\kappa(M)$ is free in C*Y. Thus, the isomorphism κ: $C^*X \to C^*Y$ preserves and reflects fixed maximal ideals, and this means it induces a homeomorphism from X to Y.

Remark 1. Any homeomorphism h: $X \to Y$ naturally induces the equivalence Φ: BANX $\to$ BANY resulting from the more directly induced equivalence Ψ: BANShX $\to$ BANShY for which ΨS is the Banach sheaf $U \rightsquigarrow Sh^{-1}(U)$. Explicitly, the effect of Φ is that each ΦA has the same underlying Banach space as A, the action of C*Y given by $(u,a) \rightsquigarrow (uh^{-1})a$. It should be noted, however, that the homeomorphism $X \to Y$ which appears in the above proof need not induce the given equivalence Φ in this sense. For instance, if $\mathbb{K} = \mathbb{C}$ one has an equivalence Φ: BAN $\to$ BAN for which each ΦA has the same additive group and norm as A but the scalar multiplication is twisted by complex conjugation: $(c,a) \rightsquigarrow \bar{c}a$. Viewing BAN as BANX for a singleton space X, the homeomorphism $X \to X$ induced by Φ is necessarily the identity which in turn induces the identity functor on BAN and not the given Φ. Note that the isomorphism κ in this case is complex conjugation. In general, if $\mathbb{K} = \mathbb{R}$ then the above isomorphism κ: $C^*X \to C^*Y$ is given by $u \rightsquigarrow uh^{-1}$, h: $X \to Y$ the homeomorphism induced by κ. and hence one has the additional fact that $\Phi(\tilde{u}_A) = (uh^{-1})^{\sim}_{\Phi A}$ for all $A \in$ BANX and $u \in |C^*X|$, which exactly says that multiplication by u on A corresponds to multiplication by uh^{-1} on ΦA. We do not know whether anything can be said beyond this.

Remark 2. Instead of using the category BAN one could take the category BAN_∞ of Banach spaces with all bounded linear maps between them. In BAN_∞, the product $A \times B$ (as in BAN) is actually a biproduct, i.e. also the coproduct of A and B (with the obvious coproduct embeddings) since any bounded linear maps $f: A \to C$ and $g: B \to C$ determine the bounded linear map $h: A{\times}B \to C$ such that $h(x,y) = f(x)+g(y)$; indeed, $\|h\| \le \|f\|+\|g\|$. This makes BAN_∞ an additive category (though not abelian), and the same holds for the category $BAN_\infty ShX$ of Banach sheaves with all bounded linear maps, i.e. the maps $\phi: S \to T$ where the components $\phi_u: SU \to TU$ are arbitrary bounded linear maps subject to the usual restriction condition. Hence the endomorphisms of the identity functor of $BAN_\infty ShX$ form a ring whose operations result from certain categorical constructions in $BAN_\infty ShX$, and for Tychonoff spaces X this ring is isomorphic to C^*X. Moreover, the free maximal ideals of C^*X are then characterized by the same condition as before, and as a result we have, with a somewhat simpler proof, the counterpart of Proposition 2 for $BAN_\infty ShX$ instead of BANShX.

Remark 3. It might be possible that these methods actually lead to a more general version of Proposition 2, saying that any equivalence between BANShX and BANShY induces a homeomorphism between the Tychonoff reflections of X and Y, but we have not been able to prove this. The essential obstacle is that, for spaces X which are not Tychonoff, it is unclear whether all endomorphisms of the identity functor of BANShS are given by maps $S \to S$ whose components $SU \to SU$ are $s \rightsquigarrow (f|U)s$ for some $f \in C^*X$.

4. Some analogous results.

We now consider some other settings in which arguments of the type used above can be employed to obtain similar results.

For any topological space X, the counterpart in the category ShX of sheaves of sets on X to the category of $\mathbb{K}$-vector spaces in the category of sets is the category VShX of C-Modules for the sheaf C of rings of continuous $\mathbb{K}$-valued functions on X whose objects are the sheaves A of abelian groups on X for which each AU is a CU-module such that $(fa)|V = (f|V)(a|V)$ for all $f \in CU$, $a \in AU$, and $V \subseteq U$, the maps being the sheaf maps $A \to B$ which are CU-linear on each open set U.

In this situation, we have the following counterpart of Proposition 2:

Proposition 3. Any category equivalence Φ: VShX → VShY for Tychonoff spaces X and Y induces a homeomorphism X → Y.

Proof. We first note that the endomorphisms of the identity functor of VShX are given by the maps $\tilde{u}_A$: A → A, with components AU → AU such that a ⇝ (u|U)a, for the u ε CX. That each such $\tilde{u}_A$ is a natural in A is obvious, and that any natural ϕ_A: A → A is of this type follows from the typical feature of Tychonoff spaces that any a ε AU coincides, on a suitable neighbourhood V for any x ε U, with some $\bar{a}$ ε AX: since $\phi_{AX}(\bar{a}) = f\bar{a}$ for $f = \phi_{CX}(1)$, by the fact that there is a map C → A for which CX → AX maps 1 to $\bar{a}$, $\phi_{AU}(a)$ and (f|U)a coincide on an open cover of U and are therefore equal. It follows from this CX is isomorphic to the endomorphism ring of the identity functor of VShX, by the same argument used in the second remark after Proposition 2, and therefore Φ induces an isomorphism κ: CX → CY such that $\Phi(\tilde{u}_A) = \kappa(u)\tilde{}_{\Phi A}$ for each u ε CX.

Now, the freeness of maximal ideals M of CX is charazterized by the condition that the family $\tilde{u}_A$(uεCX) be an episink for each A ∈ VShX, by essentially the same argument as in the proof of Lemma 4; For the first part, one again uses the fact that any free M contains, for each x ε X, a function constantly equal to 1 on some neighbourhood of x, and for the second part one notes that the S defined above (with SX = $\mathbb{K}_X$) is actually a C-Module, the action of CU on SU; for x ε U, being (f,s) ⇝ f(x)s, and that the argument concerning C*X can then be replaced by the analogous one for C. The remainder of the proof is the same as before.

The change from BANShX to VShX corresponds to the change from Banach spaces to vector spaces; we now go one step further: from vector spaces to abelian groups. Thus, we turn to the category AbShX of sheaves of abelian groups on the space X, viewed as Z-Modules for the sheaf Z of rings of continuous integer-valued functions on X.

The spaces for which our arguments work are the 0-dimensional ones, i.e. those X for which the open-closed sets form a basis for the topology. This may be viewed as the discrete counterpart of complete regularity, and it has exactly the effect that, for any A ε AbShX, any open U, and any x ε U, each a ε AU coincides with some $\bar{a}$ ε AX on a suitable neighbourhood of x. By the proof of Proposition 3, this then implies that the endomorphism ring of the identity functor of AbShX is isomorphic to ZX, and that any category equivalence Φ: AbShX → AbShY induces an isomorphism κ: ZX → ZY such that $\Phi(\tilde{u}_A) = \kappa(u)\tilde{}_{\Phi A}$ for each u ε ZX and A ε AbShX, where

$\tilde{u}_A$ is defined in the same way as previously.

We now have to pass to a slightly better ring which is invariantly determined by ZX, namely the ring QX of all locally constant rational-valued functions on X: QX is the classical ring of quotients of ZX obtained by inverting all elements which are not divisors of zero (Samuel-Zariski, [11], §19). Then, for the maximal ideal space MaxQX, whose points are the maximal ideals M of QX and whose basic open sets are $\{M | u \notin M \in MaxQX\}$, one has the embedding $X \to MaxQX$ given by $x \rightsquigarrow M_x = \{u \mid u \in QX, u(x)=0\}$ provided that X is Hausdorff.

Again, the situation here is that AbShX distinguishes the fixed maximal ideals M_x of QX from the free ones: A maximal ideal M of QX is free iff, for each $A \in$ AbShX, the family $\tilde{u}_A(u \in M \cap ZX)$ is an episink. If M is free then, for any $x \in X$, there exists a $v \in M$ such that $v(x) \neq 0$; multiplication by a suitable element of QX then also produces a $w \in M \cap Z(X)$ such that $w(x) = 1$, and $w^{-1}\{1\}$ is then a neighbourhood of x since w is locally constant. By previous arguments, this is enough to show that $\tilde{u}_A(u \in M \cap ZX)$ is an episink. For the converse, one employs the same kind of construction used in the other two cases. Here, one takes S to be defined so that $SU = \mathbb{Z}$ for $x \in U$ and $SU = 0$ for $x \notin U$, and considers the map $\phi: Z \to S$ determined by evaluation at x. As before, all $\tilde{u}_C: C \to C(u \in M \cap ZX)$ have composite zero with the non-zero ϕ, and hence this is not an episink.

The remainder of the argument is now the same as before, and in all this shows:

Proposition 4. Any category equivalence AbShX → AbShY, for zero-dimensional Hausdorff spaces X and Y, induces a homeomorphism X → Y.

Remark. Related, but much more sophisticated results of Howe [7] imply a similar proposition for essentially arbitrary spaces with stronger hypothesis and a stronger conclusion: If Φ: AbShX → AbShY, for sober spaces X and Y, is an equivalence which transforms $Z \in$ AbShX to $Z \in$ AbShY such that products of ideals are preserved then there exists a homeomorphism X → Y which induces Φ.

References.

1. N. Auspitz, Doctoral dissertation, University of Waterloo, Spring 1975.
2. B. Banaschewski, Sheaves of Banach spaces. Quaest. Math. 2 (1977), 1-22.
3. ---------------, Injective Banach sheaves (to appear).

4. H. Herrlich and G.E. Strecker, Category Theory, Allyn and Bacon, Boston 1973.

5. K.H. Hofmann, Sheaves and bundles of Banach spaces are equivalent. Lecture Notes in Mathematics 575 (1977), 53-69.

6. ------------ and K. Keimel, Sheaf Theoretical concepts in analysis: Bundles and sheaves of Banach spaces, Banach C(X) - modules (to appear).

7. D.J. Howe, Abelian group theory in a topos. Doctoral dissertation, University of Pennsylvania, 1977.

8. B. Mitchell, Theory of Categories. Academic Press, New York and London, 1965.

9. C.J. Mulvey, Private communication, November 10, 1976.

10. -----------, Banach sheaves (to appear).

11. O. Zariski and P. Samuel, Commutative Algebra I. Van Nostrand, Princeton, N.J., 1958.

Added in September 1978: The more general version of Proposition 2 mentioned at the end of Section 3 has been proved in the meantime. To see that <u>every</u> endomorphism ϕ of the identity functor of BANShX is obtained from some $f \in C^*X$ (X arbitrary), one first notices, for any given Banach sheaf S, that the component of ϕ for the Banach sheaf $\tilde{S}$, where $\tilde{S}U = \Pi S_x (x \in U)$, is indeed given by $f = \phi_{C^*X}(1)$, and then concludes the same for S since the canonical map $S \to \hat{S}$ is an embedding (Banaschewski [2]).

Department of Mathematics
McMaster University
Hamilton, Ontario L8S 4K1
Canada

COMPLETENESS IS PRODUCTIVE

H. L. Bentley and H. Herrlich

Abstract: It is shown that, for nearness spaces, products of complete spaces are complete.

0. Introduction.

The concept of "nearness" was introduced [7] as a unification of various concepts of "topological structures" in the sense that the category Near of all nearness spaces and uniformly continuous maps contains the categories (a) of all symmetric topological spaces and continuous maps, (b) of all uniform spaces and uniformly continuous maps, (c) of all proximity spaces and proximally continuous maps, and (d) of all contiguity spaces and contiguity maps as nicely embedded (either bireflective or bicoreflective) full subcategories. A concept of completeness is available for nearness spaces which generalizes the concept of completeness in uniform spaces. Moreover, every nearness space has a completion.

Other generalizations of topological, proximity and uniform structure have been offered from time to time, most notably among these are Katětov's [11] merotopic spaces, Császár's [6] syntopogeneous spaces, and Császár's [6] quasi-uniform spaces. While a concept of completeness can be given for some of the above types of spaces, whether or not a completion exists remains unknown.

The concept of completeness and the tool of completion has already been found to be useful in the study of nearness spaces [1], [2], [3], [4], [5], [7], [8], [9], [10]. The purpose of this paper is to show that, for nearness spaces, completeness is productive.

For basics on nearness spaces, the reader is referred to the papers [7] and [8].

1. Completeness is productive.

A uniformly continuous map $f : X \to Y$ between nearness spaces X and Y is called an exclusive map iff whenever $\mathfrak{A}$ is a near grill on X and B is a subset of Y such that $\{f^{-1}B\} \cup \mathfrak{A}$ is far in X, then $\{B\} \cup f\mathfrak{A}$ is far in Y.

1.1 Lemma. If $f : X \to Y$ is an exclusive map and $\mathcal{A}$ is a cluster on X then

$$\{B \subset Y \mid \{B\} \cup f\mathcal{A} \text{ is near in } Y\}$$

is a cluster on Y.

Proof: It suffices to show that

$$\mathcal{B} = \{B \subset Y \mid \{B\} \cup f\mathcal{A} \text{ is near in } Y\}$$

is near in Y, for then maximality of $\mathcal{B}$ is automatic. Suppose $\mathcal{B}$ were far in Y. Then

$$f^{-1}\mathcal{B} = \{f^{-1}B \mid B \in \mathcal{B}\}$$

would be far in X and since $\mathcal{A}$ is near in X,

$$f^{-1}\mathcal{B} \not\subset \mathcal{A}.$$

Therefore, for some $B \in \mathcal{B}$, $f^{-1}B \notin \mathcal{A}$. Since $\mathcal{A}$ is a cluster, $\{f^{-1}B\} \cup \mathcal{A}$ is far in X. Therefore, since f is an exclusive map, $\{B\} \cup f\mathcal{A}$ is far in Y. This statement contradicts that $B \in \mathcal{B}$.

1.2 Lemma. A projection map of a product onto a factor is an exclusive map.

Proof: Let X_1 and X_2 be nearness spaces and let

$$X_1 \xleftarrow{p_1} X_1 \times X_2 \xrightarrow{p_2} X_2$$

be their product in <u>Near</u> with $f = p_1$. We shall show that f is an exclusive map. This situation is general enough since we can write

$$X_2 = \prod_{j \neq j_0} X_j .$$

Let $\mathcal{A}$ be a near grill on $X_1 \times X_2$ and let B be a subset of X_1 such that $\{f^{-1}B\} \cup \mathcal{A}$ is far in $X_1 \times X_2$. If $B = \phi$, then $\{B\} \cup f\mathcal{A}$ if far in X_1 and we are through. So, assume $B \neq \phi$. There exist $\mathcal{H}_1, \mathcal{H}_2$ far in X_1, X_2 respectively for which*

---

*(1) If $\mathcal{A}_1$ and $\mathcal{A}_2$ are collections of subsets of a space, then we write $\mathcal{A}_1 \vee \mathcal{A}_2 = \{A_1 \cup A_2 \mid A_1 \in \mathcal{A}_1 \text{ and } A_2 \in \mathcal{A}_2\}$

(2) We say that $\mathcal{A}_1$ <u>corefines</u> $\mathcal{A}_2$ iff for each $A_1 \in \mathcal{A}_1$ there exists $A_2 \in \mathcal{A}_2$ with $A_2 \subset A_1$.

$$f^{-1}\mathcal{H}_1 \vee p_2^{-1}\mathcal{H}_2 \text{ corefines } \{f^{-1}B\} \cup \mathcal{O}\!\iota.$$

Since $\mathcal{O}\!\iota$ is near in $X_1 \times X_2$, then

$$f^{-1}\mathcal{H}_1 \vee p_2^{-1}\mathcal{H}_2 \not\subset \mathcal{O}\!\iota.$$

Consequently, there exist $H_1 \in \mathcal{H}_1$ and $H_2 \in \mathcal{H}_2$ with

$$f^{-1}H_1 \cup p_2^{-1}H_2 \notin \mathcal{O}\!\iota.$$

Therefore $f^{-1}B \subset f^{-1}H_1 \cup p_2^{-1}H_2$ and, since $f^{-1}B \neq \phi$ and $f^{-1}H_1 \cup p_2^{-1}H_2 \neq X_1 \times X_2$, then we must have $B \subset H_1$. We shall complete the proof by showing that

$$\mathcal{H}_1 \text{ corefines } \{B\} \cup f\mathcal{O}\!\iota.$$

To that end, let $E \in \mathcal{H}_1$ and suppose that E contains neither B nor a member of $f\mathcal{O}\!\iota$. Then $f^{-1}B \subset f^{-1}E \cup p_2^{-1}H_2$ or $f^{-1}E \cup p_2^{-1}H_2 \in \mathcal{O}\!\iota$. Since $B \not\subset E$, it must be that

$$f^{-1}E \cup p_2^{-1}H_2 \in \mathcal{O}\!\iota$$

and, since $\mathcal{O}\!\iota$ is a grill, $f^{-1}E \in \mathcal{O}\!\iota$ or $p_2^{-1}H_2 \in \mathcal{O}\!\iota$. $f_1^{-1}H_1 \cup p_2^{-1}H_2 \notin \mathcal{O}\!\iota$ implies that $p_2^{-1}H_2 \notin \mathcal{O}\!\iota$, so it must be that

$$f^{-1}E \in \mathcal{O}\!\iota.$$

But then $E = ff^{-1}E \in f\mathcal{O}\!\iota$ which is a contradiction.

1.3 Theorem. The product of complete spaces is complete.

Proof: Let $(X_i)_{i\in I}$ be a family of complete nearness spaces, let

$$X = \prod_{i\in I} X_i$$

be their product, and let $(p_i : X \to X_i)_{i\in I}$ be the canonical projections. Let $\mathcal{O}\!\iota$ be a cluster on X and for each $i \in I$, let

$$\mathcal{L}_i = \{B \subset X_i \mid \{B\} \cup p_i\mathcal{O}\!\iota \text{ is near in } X_i\}.$$

By Lemmas 1.1 and 1.2, each $\mathcal{L}_i$ is a cluster on X_i, which is complete. So, there exists $x = (x_i)_{i\in I}$ with

$$\mathcal{L}_i = \{B \subset X_i \mid x_i \in \mathrm{Cl}_{X_i} B\}$$

It is sufficient to show that $\{x\} \in \mathcal{O}\!\iota$ and for this, it is sufficient to show that $\{\{x\}\} \cup \mathcal{A}$ is near in X. Let J be a finite subset of I and let $(\mathcal{G}_j)_{j \in J}$ be a family with each $\mathcal{G}_j$ a uniform cover of X_j. Since $\{\{x_j\}\} \cup \mathcal{L}_j$ is near in X_j, there exists $G_j \in \mathcal{G}_j$ with $x_j \in G_j$ and with $B \cap G_j \neq \phi$ for each $B \in \mathcal{L}_j$. Let

$$G = \bigcap_{j \in J} p_j^{-1} G_j .$$

It must be shown that $x \in G$ and that $G \cap A \neq \phi$ for all $A \in \mathcal{A}$. It is clear that $x \in G$, so let $A \in \mathcal{A}$ and suppose that $G \cap A = \phi$. For each $j \in J$, $X_j - G_j \notin \mathcal{L}_j$ so $\{X_j - G_j\} \cup p_j \mathcal{A}$ is far in X_j and consequently

$$\{X - p_j^{-1} G_j\} \cup \mathcal{A}$$

is far in X which implies that $X - p_j^{-1} G_j \notin \mathcal{A}$. Since $\mathcal{A}$ is a grill on X,

$$\bigcup_{j \in J} (X - p_j^{-1} G_j) \notin \mathcal{A}.$$

Consequently, A cannot be a subset of

$$\bigcup_{j \in J} (X - p_j^{-1} G_j) = X - G$$

so $A \cap G \neq \phi$

1.4 Corollary (Weil [14]). The product of complete uniform spaces is complete.

1.5 Corollary (Herrlich [7]). The product of complete regular nearness spaces is complete.

1.6 Corollary (Steiner and Steiner [13]). The product of complete semi-uniform spaces is complete.

1.7 Corollary (Bentley and Herrlich [3]). The product of complete separated nearness spaces is complete.

Bibliography

1. H.L. Bentley, Normal nearness spaces, Quaest. Math. 2 (1977) 23-43.
2. H.L. Bentley and H. Herrlich, Extensions of topological spaces, in: Topology: Proc. Memphis State Univ. Conf. edited by S.P. Franklin and B.V. Smith Thomas, Marcel Dekker, New York (1976) 129-184.
3. H.L. Bentley and H. Herrlich, Completion as reflection, preprint.

4. H.L. Bentley and H. Herrlich, The reals and the reals, Gen. Top. Appl. (to appear).
5. J.W. Carlson, H-closed and countably compact extensions, preprint.
6. A. Császár, Foundations of General Topology, Macmillan, New York (1963).
7. H. Herrlich, A concept of nearness, Gen. Top. Appl. 4 (1974) 191-212.
8. H. Herrlich, Topological structures, Math. Centre Tracts 52, Amsterdam (1974) 59-122.
9. H. Herrlich, Some topological theorems which fail to be true, Categorical Topology (Proc. Conf., Mannheim, 1975), Lecture Notes in Math. 540, Springer, Berlin (1976) 265-285.
10. H. Herrlich, Products in topology, Quaest. Math. 2 (1977) 191-205.
11. M. Katětov, On continuity structures and spaces of mappings, Comment. Math. Univ. Carolinae 6 (1965) 257-278.
12. K. Morita, On the simple extension of a space with respect to a uniformity, I-IV, Proc. Japan Acad. 27 (1951) 65-72, 130-137, 166-171, 632-636.
13. A.K. Steiner and E.F. Steiner, On semi-uniformities, Fund. Math. 83 (1973) 47-58.
14. A. Weil, Sur les espaces à structure uniforme et sur la topologie générale, Actual. Sci. Industr. fasc. 551, Paris (1937).

H.L. Bentley
University of Toledo
Toledo, Ohio 43606 USA

H. Herrlich
Universität Bremen
2800 Bremen 33
Fed. Rep. Germany

LEGITIMACY OF CERTAIN TOPOLOGICAL COMPLETIONS

Reinhard Börger

AMS (MOS) code: Primary 18B15
Secondary 04A10

Mac Neille completions, universal initial (resp. final) completions and largest initial completions have been characterized by Adámek, Herrlich and Strecker [1], who gave constructions of these completions and showed that the existence of these completions is equivalent to the legitimacy of the constructions, if the base category has small hom-sets.
As Herrlich, Strecker [6] and Tholen [9] showed, semi-topological functors (cf. Hoffmann [7]), and topologically algebraic functors can be characterized by the existence of reflective initial completions.

In this paper all categories are supposed to have small hom-sets. All functors are faithful, amnestic and transportable. For a fixed category $\underline{X}$, an $\underline{X}$-object X and a functor $P : \underline{A} \to \underline{X}$ we define the comma category $\langle X,P\rangle$, whose objects are pairs (l,A) with $A \in Ob(\underline{A})$, $l \in \underline{X}(X,PA)$, and whose morphisms $(l,A) \xrightarrow{f} (k,B)$ are given by an $\underline{A}$-morphism $f : A \to B$ with $P(f)l = k$. We investigate full subcategories $\underline{S} \subset \langle X,P\rangle$ fulfilling some of the following conditions

(a) $(l,A) \in Ob(\underline{S}), f \in \underline{A}(A,B) \Rightarrow (fl,B) \in \underline{S}$.

(b) $\underline{S}$ fulfills (a) and for any P-initial source $(A, A \xrightarrow{f_i} B_i)_{i \in I}$, $(\forall i \in I\ (P(f_i)l, B_i) \in \underline{S}) \Rightarrow (l,A) \in \underline{S}$.

(c) If $(l,A) \in \langle X,P\rangle$, and for all $PC \xrightarrow{r} X$ the condition $(\forall (l',A') \in \underline{S}\ \exists h' \in \underline{A}(C,A')\ P(h') = l'r)$ implies $\exists\, h \in \underline{A}(C,A)\ \ Ph = l$, then $(l,A) \in \underline{S}$.

It is clear that (c) implies (b) and (b) implies (a). Adámek, Herrlich and Strecker [1] proved:

THEOREM: Let $P : \underline{A} \to \underline{X}$ be a functor. Then the following assertions hold:

(i) P admits a MacNeille completion, if for any $X \in Ob(\underline{X})$ the conglomerate of all full subcategories $\underline{S} \subset \langle X,P \rangle$ fulfilling (c) is legitimate.

(ii) P admits a universal initial completion, if the conglomerate of all full subcategories $\underline{S} \subset \langle X,P \rangle$ fulfilling (b) is legitimate.

(iii) P admits a largest initial completion, if the conglomerate of all full subcategories $\underline{S} \subset \langle X,P \rangle$ fulfilling (a) is legitimate.

An analogous theorem holds for reflective completions.

THEOREM: For a functor $P : \underline{A} \to \underline{X}$ the following staments hold:

(i) P admits a reflective MacNeille-completion (i.e. P is semi-topological, see [9]), if for any $X \in Ob(\underline{X})$ any full subcategory $\underline{S} \subset \langle X,P \rangle$ with (c) has an initial object.

(ii) P admits a reflective universal initial completion (i.e. P is topologically algebraic, see [8]), if for any $X \in Ob(\underline{X})$ any full subcategory $\underline{S} \subset \langle X,P \rangle$ with (b) has an initial object.

Remark: If a full subcategory $S \subset \langle X,P \rangle$ with (a) (a fortiori any $\underline{S}$ with (b) or (c)) has an initial object $(1,A)$, then S is the full subcategory of all $(Pf\,1,B)$, where $A \xrightarrow{f} B$ is an $\underline{A}$-morphism.

PROOF of the second theorem: We prove only (i), since (ii) can be proved in an analogous way. Assume P to have a reflective MacNeille completion $T : \underline{C} \to \underline{X}$ and let $E : \underline{A} \to \underline{C}$ be the embedding. Now consider an $\underline{X}$-object and a subcategory $\underline{S} \subset \langle X,P \rangle$ with (c). Then the source $(X, X \xrightarrow{f} PA)_{(f,A) \in Ob(\underline{S})}$ has an initial lift $(C, C \xrightarrow{f^*} E(A))_{(f,A) \in Ob(\underline{S})}$, and by [1, thm 2.4] it follows that an arbitrary $X \xrightarrow{f} PA$ can be lifted to a map from C to A if $(f,A) \in Ob(\underline{S})$. If now $C \xrightarrow{r} EA_o$ is the reflection map, it follows that for $Pr : TC = X \to TEA_o = PA_o$ we have $(r, A_o) \in Ob(\underline{S})$. If $(f,A) \in Ob(\underline{S})$, then $C \xrightarrow{f} EA$ is an $\underline{A}$-morphism, and by the reflection property of (r, A_o) there is a unique $A_o \xrightarrow{f'} A$

in $\underline{A}$ with $Ef'r = f$, i.e. a unique $<X,P>$-morphism from (r, A_o) to (f,A), hence (r,A) is an initial object of $<X,P>$. On the other hand, if for any $X \in Ob(A)$ any $\underline{S} \subset <X,P>$ with (c) has an initial object, consider $C \in Ob(\underline{C})$ and $X := PC$. Then by [1, thm. 2.4] the property (c) holds for the full subcategory $\underline{S} \subset <X,P>$ of all (rf,A), where $A \in Ob(A)$ and $f: C \to EA$ is a $\underline{C}$-morphism. Hence there is an initial (r,A_o) of $\underline{S}$, i.e. a reflection map for $\underline{C}$.

Remark: Adámek's, Herrlich's and Strecker's result [1, thm. 4.2] on largest initial completions implies that the identity on the void category is the only functor which admits a reflective largest initial completion, since for any $P : \underline{A} \to \underline{X}$, $X \in Ob(\underline{X})$ the empty category $\emptyset \subset <X,P>$ has the property (a), and hence for a largest topological completion $T : \underline{C} \to X$, $E : \underline{A} \to \underline{C}$ of P (if it exists) there is no map from the indiscrete object over X to an object EA with $A \in Ob(\underline{A})$ (and thus no reflection map).

Herrlich, Nakagawa, Strecker and Titcomb [5] and the author [2] independently gave countable (and hence small) examples of semi-topological functors not being topologically algebraic. These functors admit reflective MacNeille completions, and they admit (by their smallness) universal topological completions, which of course fail to be reflective. In [2] the author showed that even over the nice category *Ens* of sets there is a strongly fibre-small semi-topological functor such that the universal initial completion fails to exist. In this paper we will point out the main ideas of that counter-example.

Let $\underline{A}$ be the category whose objects are all triples (X,A,φ), where X is a set, A is a subset of X, and φ is a set map from the power set $P(A)$ of A to X, such that $\varphi(\{x\}) = x$ holds for all $x \in A$. An $\underline{A}$-morphism $(X,A,\varphi) \xrightarrow{f} (Y,B,\psi)$ is given by a set map $X \xrightarrow{f} Y$ with $f[A] \subset B$ (i.e. the set-theoretical image of A under f is contained in B) and $f\,\varphi(M) = \psi(f[M])$ for all $M \subset A$. P is the canonical forgetful functor (with $P(X,A,\varphi) = X$).

THEOREM: P has a fibre-small reflective MacNeille completion, but no universal initial completion.

<u>Main ideas of the proof:</u> The construction of a fibre-small reflective MacNeille completion is more or less canonical. For the non-existence of a universal initial completion, we give a class of maps $\{1\} \hookrightarrow P(W_\alpha, W_\alpha, \rho_\alpha)$, where α is in the class $\underline{L}$ of all ordinals greater or equal to 3.

This class of maps has the property that for any subclass $\underline{K} \subset \underline{L}$ the smallest full subcategory $\underline{S}_{\underline{K}}$ of $<\{1\}, P>$ with the property (b) and containing $(\{1\} \hookrightarrow W_\alpha,\ (W_\alpha, W_\alpha, \rho_\alpha))$ for all $\alpha \in \underline{K}$ does not contain any object $(\{1\} \hookrightarrow W_\beta,\ (W_\beta, W_\beta, \rho_\beta))$ for $\beta \in \underline{L} \setminus \underline{K}$. Hence, for $\underline{K}, \underline{K}' \subset \underline{L}$, $\underline{K} \neq \underline{K}'$ we have $\underline{S}_{\underline{K}} \neq \underline{S}_{\underline{K}'}$ and, as $\underline{L}$ is a proper-class, we have a non-legitimate collection of subclasses of $<\{1\}, P>$ with the property (b).

For $\alpha \in \underline{L}$ we define W_α to be the class of all ordinals less than α, and $\rho_\alpha : P(W\ ) \to W_\alpha$ is given by

$$\rho_\alpha(M) := \begin{cases} \xi, & \text{if } M = \{\xi\}, \\ 0, & \text{if } M = W_\alpha, \\ \min(W_\alpha \setminus M) & \text{else.} \end{cases}$$

To show that $1 \hookrightarrow (W_\beta, W_\beta, \rho_\beta)$ is not in $\underline{S}_{\underline{K}}$ for $\beta \in \underline{L} \setminus \underline{K}$, we give a concrete description of $\underline{S}_{\underline{K}}$. It is trivial that for any set-indexed family $(\alpha_i)_{i \in I}$ with $\alpha_i \in \underline{K}$ the induced map $\{1\} \xrightarrow{r} P(\prod_{i \in I}(W_{\alpha_i}, W_{\alpha_i}, \rho_{\alpha_i}))$ is in $\underline{S}_{\underline{K}}$. ($\prod$ denotes the categorical product, which exists and is preserved by P, because P is semi-topological.) Then any composed map $\{1\} \xrightarrow{r} P(\prod_{i \in I}(W_{\alpha_i}, W_{\alpha_i}, \rho_{\alpha_i})) \xrightarrow{Pf} P(X, A, \varphi)$ (with $(W_\alpha, W_\alpha, \rho_\alpha) \xrightarrow{f} (X, A, \varphi)$ in $\underline{A}$) is in $\underline{S}_{\underline{K}}$. Now it remains to show that the class $\underline{S}'_{\underline{K}}$ of all such composed maps fulfills (b) and that $\{1\} \hookrightarrow P(W_\beta, W_\beta, \rho_\beta)$ is not in $\underline{S}'_{\underline{K}}$. The latter statement means:

(*) If $(\alpha_i)_{i \in I}$ is a set-indexed family of elements of $\underline{K}$, and if $\beta \in \underline{L} \setminus \underline{K}$, then there is no map $f : \prod_{i \in I}(W_{\alpha_i}, W_{\alpha_i}, \rho_{\alpha_i}) \to (W_\beta, W_\beta, \rho_\beta)$ with $f((1)_{i \in I}) = 1$.

We have to show that $\underline{S}'_{\underline{K}}$ fulfills (b). The condition (a) is clear. As the full subcategory $\underline{R} \subset <X, P>$ of all maps $\{1\} \xrightarrow{t} P(X, A, \varphi)$ with $(X, A, \varphi) \in Ob(\underline{A})$ and $\varphi[P(A)] \subset A$ has the property (b) and contains all maps $\{1\} \hookrightarrow P(W_\alpha, W_\alpha, \rho_\alpha)$

with $\alpha \in \underline{L}$, we get $\underline{S}'_{\underline{K}} \subset \underline{R}$. If $\{1\} \xrightarrow{t} P(X,A,\varphi)$ is in $<X,P>$, then there is a smallest subset $Z \subset X$ with $t(1) \in Z$ and $\varphi[P(Z \cap A)] \subset Z$. Then one gets the canonical injective $\underline{A}$-morphism $(Z, Z \cap A, \varphi \mid^{Z}_{P(Z \cap A)}) \hookrightarrow (X,A,\varphi)$, which turns out to be P-initial. Then $\{1\} \xrightarrow{t} P(X,A,\varphi)$ is in $\underline{S}_{\underline{K}}$ if $\{1\} \xrightarrow{t \mid^{Z}} (Z, Z \cap A, \varphi \mid^{Z}_{P(Z \cap A)})$ is in $\underline{S}'_{\underline{K}}$. Now the minimality of Z and the inclusion $\underline{S}'_{\underline{K}} \subset \underline{R}$ give $Z \cap A = Z$, i.e. $Z \subset A$. The trivial fact that each injective $\underline{A}$-morphism $(X,X,\varphi) \xrightarrow{m} (Y,B,\psi)$ is P-initial now leads to the proposition, if the following statements are proved:

(**) If $((X,A,\varphi), (X,A,\varphi) \xrightarrow{h_i} (W_{\alpha_i}, W_{\alpha_i}, \rho_{\alpha_i}))_{i \in I}$ is a P-initial source with $h_i \neq h_{i'}$ for all $i, i' \in I$ with $i \neq i'$, then I is a set, $X = A$ holds, and the source is a product source.

(***) If $\prod_{i \in I} (W_{\alpha_i}, W_{\alpha_i}, \rho_{\alpha_i}) \xrightarrow{f} (Y,B,\psi)$ is an $\underline{A}$-morphism, then there is a subset $I' \subset I$ and a commutative diagram

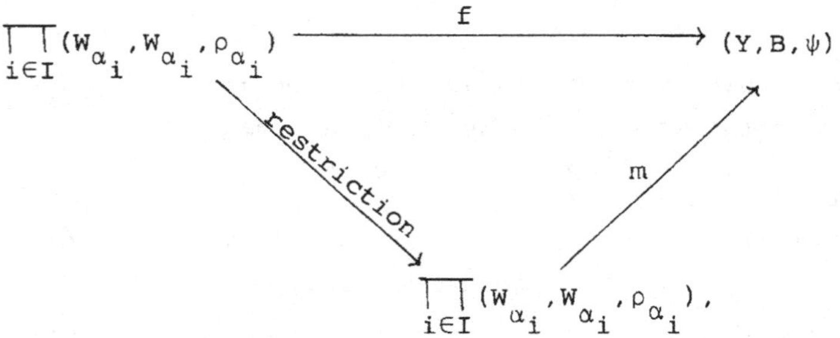

in $\underline{A}$, where m is injective. The smallness of I in (**) follows from the fact that all h_i's are surjective, and hence $|\alpha_i| = |W_{\alpha_i}| \leq |X|$ for all $i \in I$. Hence the source can be factorized into a product source and an initial monomorphism. In (***) we choose for I' the set of all $i \in I$ such that $f(\xi_i)_{i \in I} = f(\xi'_i)_{i \in I}$ always implies $i = i'$. Finally, the statements

(*), (**), (***) follow from the following lemma:

If I is a set, $(\alpha_i)_{i\in I}$ is a family of $\underline{L}$-elements, and $V \subset \prod_{i\in I} W$ is a subset with the following properties:

(i) $(1)_{i\in I} \in V$.

(ii) If $i \neq i'$, $\alpha_i = \alpha_{i'}$, then there are $(\xi_i)_{i\in I}, (\xi'_i)_{i\in I} \in V$ with $\xi_i \neq \xi'_i$.

(iii) If M is a subset of V, then $\tau(M) \in V$, where $(\prod_{i\in I} W_\alpha, \prod_{i\in I} W_\alpha, \tau) = \prod_{i\in I}(W_{\alpha_i}, W_{\alpha_i}, \rho_{\alpha_i})$.

Then it follows that $V = \prod_{i\in I} W_{\alpha_i}$.

References:

[1] Adámek, J., Herrlich, H., Strecker, G.E.: Least and largest initial completions, Preprint, 1978.

[2] Börger, R.: Universal topological completions of semi-topological functors over *Ens* need not exist, Preprint, Fernuniversität Hagen 1978.

[3] Börger, R.: Semitopologisch ≠ topologisch algebraisch, Preprint, Fernuniversität Hagen 1977.

[4] Herrlich, H.: Initial completions. Math. Z. 150 (1976), 101-110.

[5] Herrlich, H., Nakagawa, R., Strecker, G.E., Titcomb, T.: Semi-topological and topologically algebraic functors (are and are not equivalent), Preprint 1977.

[6] Herrlich, H., Strecker, G.E.: Semi-universal maps and universal initial completions, Preprint 1977.

[7] Hoffmann, R.E.: Note an universal completion. Math. Z.

[8] Hong, Y.H.: Studies on categories of universal topological algebras, Thesis, McMaster University, Hamilton (Ontario), 1974.

[9] Tholen, W.: Semi-topological functors I, to appear in J. Pure Appl. Alg.

Reinhard Börger
Fachbereich Mathematik
Fernuniversität
Postfach 940
D-5800 Hagen
West Germany

On E-normal spaces

H.Brandenburg,Berlin

1.Introduction

It is well-known that even for nice topological spaces the two dimension functions ind and Ind do not coincide. In [14] P.ROY constructed a metrizable space X such that ind X = 0 and Ind X = 1. Therefore the following problem naturally arises:

PROBLEM A : Find a necessary and sufficient condition which ensures that a zero-dimensional space (i.e. ind X = 0) is ultranormal (i.e. Ind X = 0).

For paracompact spaces this problem was recently solved by K.A.BROUGHAN , who proved the following theorem.

1.1 THEOREM : ([5] , Theorem 3.22) For a paracompact space X the following statements are equivalent:

(1) ind X = 0 and X is clopen-paracompact.
(2) Ind X = 0.

A space X is clopen-paracompact iff every clopen cover of X has a clopen locally finite refinement. Obviously every zero-dimensional clopen-paracompact space is paracompact. Therefore the condition of clopen-paracompactness is adequate only for paracompact spaces to solve Problem A.

In this paper we are interested in a more general question which includes Problem A as a special case. Consider an arbitrary nonempty class $\underline{E}$ of T_1-spaces. A topological space X is called $\underline{E}$-normal if every pair of disjoint closed subsets of X can be separated by a pair of disjoint $\underline{E}$-closed subsets of X. If $\underline{E}$ has the finite intersection property then every $\underline{E}$-normal R_0-space X belongs to the bireflective hull BH($\underline{E}$) of $\underline{E}$ (see section 2. for definitions) [13].

PROBLEM B : Suppose that $\underline{E}$ has the finite intersection property. Which is a necessary and sufficient condition for a space X in BH($\underline{E}$) to be $\underline{E}$-normal ?

Our main result provides an answer to this question , which holds for all those classes $\underline{E}$ for which interesting characterizations of $\underline{E}$-normal spaces are known. In particular we obtain a complete solution of Problem A.

We are indebted to N.C.HELDERMANN and H.PUST , who carefully read the first draft of this paper.

2. Definitions and preliminary results

Throughout this section $\underline{E}$ denotes an arbitrary nonempty class of topological spaces. A subset B of a topological space X is called $\underline{E}$-closed if there are a continuous mapping $f:X \longrightarrow Y$, $Y \in \underline{E}$, and a closed subset $C \subset Y$ such that $f^{-1}[C] = B$. Complements of $\underline{E}$-closed subsets are called $\underline{E}$-open.

2.1 DEFINITION : A topological space X is called $\underline{E}$-normal if each pair A,B of disjoint closed subsets of X can be separated by a pair F,G of disjoint $\underline{E}$-closed subsets of X (i.e. $A \subset F$, $B \subset G$ and $F \cap G = \emptyset$). The class of all $\underline{E}$-normal spaces is denoted by N($\underline{E}$).

This notion of $\underline{E}$-normality is a natural generalization of the concept of E-normality introduced by S.MRÓWKA in [12]. It should be pointed out that it is different from the concept of $\underline{E}$-normality studied by H.HERRLICH [10].

A subcategory $\underline{C}$ (which we assume to be full and isomorphism-closed) of the category $\underline{Top}$ of all topological spaces is said to be bireflective in $\underline{Top}$ if for every topological space X there is a space $X' \in \underline{C}$ and a bijective continuous mapping $r:X \longrightarrow X'$ such that for every space $Y \in \underline{C}$ and for every continuous mapping $f:X \longrightarrow Y$ there is a unique continuous mapping $f':X' \longrightarrow Y$ such that $f' \circ r = f$. There exists a smallest bireflective subcategory of $\underline{Top}$ containing $\underline{E}$. It is called the bireflective hull of $\underline{E}$ and is denoted by $BH(\underline{E})$. A topological space X belongs to $BH(\underline{E})$ iff the topology of X coincides with the initial topology on X with respect to the class $C(X,\underline{E})$ of all continuous mappings from X into spaces belonging to $\underline{E}$ [11].

The class $\underline{E}$ is said to have the finite intersection property if for every topological space X each intersection of finitely many $\underline{E}$-open subsets of X is $\underline{E}$-open [10],[13]. The following result was essentially proved in [13]. Recall that a space X is called a R_o-space if every open subset of X is a union of closed sets [7].

2.2 PROPOSITION : For a topological space X and a class $\underline{E}$ with finite intersection property the following statements are equivalent , if $\underline{E}$ contains only R_o-spaces :

(1) $X \in BH(\underline{E})$.

(2) The collection of all $\underline{E}$-open subsets of X is a basis for the topology of X.

(3) For each closed subset $A \subset X$ and for each $x \in X \setminus A$ there exists a continuous mapping $f:X \longrightarrow Y$, $Y \in \underline{E}$, such that $cl\{f(x)\} \cap cl\, f[A] = \emptyset$.

We call $\underline{E}$ adequate for functional separation if for every topological space X and for each pair G,H of disjoint $\underline{E}$-closed subsets of X there are a continuous mapping $f:X \longrightarrow Y$, $Y \in \underline{E}$, and two different points $y,z \in Y$ such that $cl\, f[G] \subset \{y\}$ and $cl\, f[H] \subset \{z\}$. Finally a continuous mapping $f:X \longrightarrow Y$ is called $\underline{E}$-closed if f[A] is closed for every $\underline{E}$-closed subset A of X. This notion generalizes the concept of E-closed mappings introduced by J.TSAI [15]. Obviously every closed mapping is $\underline{E}$-closed.

3. Main result

One necessary and sufficient condition which solves Problem B for suitable classes $\underline{E}$ is connected with the collection of all R_o-topologies on a space X. If $\mathfrak{A}$ is a nonempty collection of closed subsets of a topological space $(X,\mathfrak{X})$ we call a topology $\mathfrak{X}' \subset \mathfrak{X}$ T_o-saturated with respect to $\mathfrak{A}$ if $X \setminus \bigcup\{cl_{\mathfrak{X}'}\{a\} \mid a \in A\} \in \mathfrak{X}'$ for every $A \in \mathfrak{A}$. If $\mathfrak{X}' \subset \mathfrak{X}$ is T_o-saturated with respect to the collection of all closed subsets of $(X,\mathfrak{X})$, we call it T_o-saturated with respect to $(X,\mathfrak{X})$. Note that for every topological space $(X,\mathfrak{X})$ the topology $\mathfrak{X}$ and the indiscrete topology on X are T_o-saturated with respect to $(X,\mathfrak{X})$.

<u>3.1 THEOREM</u> : Suppose that $\underline{E}$ is a nonempty class of nonempty T_1-spaces which is adequate for functional separation and has the finite intersection property. For a topological space X the following statements are equivalent:

(1) X is an $\underline{E}$-normal R_o-space.
(2) $X \in BH(\underline{E})$ and every coarser R_o-topology on X which is T_o-saturated with respect to all $\underline{E}$-closed subsets of X is T_o-saturated with respect to X.
(3) $X \in BH(\underline{E})$ and every $\underline{E}$-closed mapping $f:X \longrightarrow Y$ into an arbitrary T_1-space Y is closed.

<u>Proof</u>: (1) implies (2): Consider an arbitrary R_o-topology $\mathfrak{H}$ on X which is coarser than the original topology $\mathfrak{X}$ of X. Let A be a closed subset of $(X,\mathfrak{X})$ and $x \in U := X \setminus \bigcup\{cl_{\mathfrak{H}}\{a\} \mid a \in A\}$. We have to prove the existence of an open neighbourhood V of x with respect to $\mathfrak{H}$ which is contained in U. Since $\mathfrak{H}$ is a R_o-topology , $cl_{\mathfrak{H}}\{x\} \subset U \subset X \setminus A$. The $\underline{E}$-normality of X implies the existence of two disjoint $\underline{E}$-closed subsets G,H of X such that $cl_{\mathfrak{H}}\{x\} \subset G$ and $A \subset H$. Now , if $\mathfrak{H}$ is T_o-saturated with respect to all $\underline{E}$-closed subsets of X , $V := X \setminus \bigcup\{cl_{\mathfrak{H}}\{h\} \mid h \in H\}$

is open with respect to $\mathfrak{W}$. Since $x \in V \subset U$, the implication holds.

(2) implies (3): If $f:X \longrightarrow Y$ is a continuous mapping into a T_1-space Y , the initial topology $\mathfrak{X}'$ on X with respect to f is a R_0-topology. Suppose that f is $\underline{E}$-closed. This implies that $\cup\{cl_{\mathfrak{X}'}\{a\} \mid a \in A\} = f^{-1}[f[A]]$ for every $\underline{E}$-closed subset A of X. Therefore $\mathfrak{X}'$ is T_0-saturated with respect to all $\underline{E}$-closed subsets of X. Hence , by assumption , $\mathfrak{X}'$ is T_0-saturated with respect to X. Consider an arbitrary nonempty closed subset B of X. Then $\cup\{cl_{\mathfrak{X}'}\{b\} \mid b \in B\}$ is closed with respect to $\mathfrak{X}'$, i.e. there is a closed subset F of Y such that $f^{-1}[F] = \cup\{cl_{\mathfrak{X}'}\{b\} \mid b \in B\}$. It is easy to see that $f[B] = F \cap f[X]$. $\underline{E}$ contains at least one nonempty T_1-space. Therefore f[X] (and consequently f[B]) is closed in Y , implying that f is closed.

(3) implies (1): Suppose that $X \in BH(\underline{E})$ is a R_0-space which is not $\underline{E}$-normal. We will show that there exists a T_1-space Y and an $\underline{E}$-closed mapping $f:X \longrightarrow Y$ which is not closed.

The space Y will be obtained in the following way. Denote by $i:X \longrightarrow \omega X$ the T_0-identification (i.e. ωX is the quotient space of X with respect to the equivalence relation "$x \pi y$ iff x has the same neighbourhoods as y"). Then ωX is a T_1-space [7]. Now consider an arbitrary closed subset A of X. Identify $i[A] \subset \omega X$ to a point and denote the corresponding quotient space as usual by $\omega X/_{i[A]}$. Let $j: X \longrightarrow \omega X/_{i[A]}$ be the natural projection. Furthermore let $C(\omega X/_{i[A]},\underline{E})$ be the class of all continuous mappings from $\omega X/_{i[A]}$ into spaces belonging to $\underline{E}$ and consider the set $\omega X/_{i[A]}$ supplied with the initial topology with respect to $C(\omega X/_{i[A]})$. We denote this space by X(A). Note that $X(A) \in BH(\underline{E})$. Put $p = j \circ i$. We claim the following statements to be true:

A. X(A) is a T_1-space.

B. If for every $\underline{E}$-closed set $B \subset X$ which is disjoint from A there exists an $\underline{E}$-closed set $C \subset X$ such that $A \subset C$ and $C \cap B = \emptyset$, then $p:X \longrightarrow X(A)$ is $\underline{E}$-closed.

To prove A. consider two different points $p(x), p(y) \in X(A)$. Without loss of generality assume that $x \notin A$. According to 2.2 there exists a continuous mapping $g: X \longrightarrow W$, $W \in \underline{E}$, such that $cl\{g(x)\} \cap cl\, g[A \cup cl\{y\}] = \emptyset$. Since $\underline{E}$ is adequate for functional separation, one may assume that $cl\{g(x)\}$ respectively $cl\, g[A \cup cl\{y\}]$ consists of exactly one point. Therefore it is possible to define a mapping $g': X(A) \longrightarrow W$ such that $g' \circ p = g$. Note that g' is continuous. Now W is a T_1-space, which implies that $X(A)$ is a T_1-space.

To prove B. we consider an arbitrary nonempty $\underline{E}$-closed subset B of X. We will show that $p[B]$ is closed in $X(A)$. If $p(z) \in X(A) \setminus p[B]$, we have to consider the following cases:

(B.1) $z \in A$;

(B.2) $z \notin A$.

If $z \in A$, then $B \cap A = \emptyset$. By assumption there is an $\underline{E}$-closed subset C of X such that $A \subset C$ and $B \cap C = \emptyset$. Since $\underline{E}$ is adequate for functional separation, there exists a continuous mapping $h: X \longrightarrow Z$, $Z \in \underline{E}$, such that $cl\, h[B] \cap cl\, h[C] = \emptyset$. One may assume that $cl\, h[B]$ respectively $cl\, h[C]$ contains not more than one point.Therefore it is possible to define a mapping $h': X(A) \longrightarrow Z$ such that $h' \circ p = h$. Since h' is continuous, $O = h'^{-1}[Z \setminus cl\, h[B]]$ is an open neighbourhood of $p(z)$ such $O \cap p[B] = \emptyset$, i.e. $p(z) \notin cl\, p[B]$.

If $z \notin A$, there exists a continuous mapping $h: X \longrightarrow Z$, $Z \in \underline{E}$, such that $cl\{h(z)\} \cap cl\, h[A \cup B] = \emptyset$. Arguments similar to those used above can be used to prove that $p(z) \notin cl\, p[B]$. Therefore $p[B] = cl\, p[B]$, which implies that p is $\underline{E}$-closed.

To prove the existence of a T_1-space Y and an $\underline{E}$-closed mapping from X to Y which is not closed we proceed as follows. We distinguish two cases. First we assume that X has the following property:

(*) For each pair G, H of disjoint closed subsets of X such that H is $\underline{E}$-closed there is an $\underline{E}$-closed subset F of X such that $G \subset F$ and $F \cap H = \emptyset$.

Since X is not $\underline{E}$-normal, there are disjoint closed subsets A, B of X which cannot be separated by $\underline{E}$-closed subsets. Define

$Y = X(A)$. Property (*) implies that the conditions of B. are satisfied. Therefore $p:X \longrightarrow Y$ is an $\underline{E}$-closed mapping into a T_1-space. Suppose that p is closed. Thus , since $Y \in BH(\underline{E})$, there exists a continuous mapping $q:Y \longrightarrow E$, $E \in \underline{E}$, such that $cl\ q[p[A]] \cap cl\ q[p[B]] = \emptyset$, which contradicts the fact that A and B cannot be separated by $\underline{E}$-closed sets. Therefore p is not closed.

Now suppose that X does not have property (*). Then there are a closed set B and an $\underline{E}$-closed set A in X such that $A \cap B = \emptyset$ which cannot be separated by $\underline{E}$-closed sets. Define $Y = X(A)$.It follows from B. that $p:X \longrightarrow Y$ is $\underline{E}$-closed , but again it can be shown that p is not closed , which completes the proof.

<u>3.2 COROLLARY</u> : Suppose that in addition to the conditions stated in 3.1 $\underline{E}$ has the property that for every topological space each countable union of $\underline{E}$-open sets is $\underline{E}$-open. Then every Lindelöf space $X \in BH(\underline{E})$ is $\underline{E}$-normal.

<u>Proof</u>: Consider an $\underline{E}$-closed mapping $f:X \longrightarrow Y$ into a T_1-space Y. It is sufficient to show that f is closed. Consider an arbitrary closed set $A \subset X$ and a point $y \in f[X] \setminus f[A]$. For each point $x \in f^{-1}[\{y\}]$ there is an $\underline{E}$-open set $U(x)$ such that $x \in U(x)$ and $U(x) \subset X \setminus A$ (2.2). Since X is a Lindelöf space , there is a sequence $(x_n)_{n \in \mathbb{N}}$ in $f^{-1}[\{y\}]$ such that $f^{-1}[\{y\}]$ is contained in $U := \bigcup\{U(x_n) \mid n \in \mathbb{N}\}$. Since f is $\underline{E}$-closed , $V = Y \setminus f[X \setminus U]$ is an open neighbourhood of y such that $V \cap f[A] = \emptyset$. Therefore $y \notin cl\ f[A]$, which implies that f is closed.

4. Applications

Let us first consider the classes $\underline{E} = \{\mathbb{R}\}$ or $\underline{E}$ = {metrizable spaces } . In these cases the $\underline{E}$-closed mappings are identical with the Z-mappings introduced by Z.FROLÍK [8]. Furthermore in both cases $\underline{E}$ is adequate for functional separation ([9] , 1.15). Therefore from 3.1 we obtain the following result. The equivalence of (1) and (3) is originally due to Ph.ZENOR. In fact , our proof of Theorem 3.1 is modelled after his proof in [16].

4.1 THEOREM : The following statements are equivalent for a topological space X:

(1) X is a normal R_0-space.
(2) X is a completely regular space such that every coarser R_0-topology on X which is T_0-saturated with respect to all zero sets is T_0-saturated with respect to X.
(3) X is a completely regular space such that every Z-mapping $f:X \longrightarrow Y$ into an arbitrary T_1-space Y is closed.

Next we consider the class $\underline{E} = \{D_2\}$, where D_2 is the discrete space with exactly two points. In this case an $\underline{E}$-closed mapping might be called clopen-to-closed mapping. $\{D_2\}$ has the finite intersection property and is adequate for functional separation. It is well-known that a space X belongs to $BH(\{D_2\})$ iff ind X = O and that $X \in N(\{D_2\})$ is equivalent to Ind X = O. Therefore we obtain the following result.

4.2 THEOREM : The following statements are equivalent for a topological space X:

(1) X is an ultranormal (i.e. Ind X = O) R_0-space.
(2) ind X = O and every coarser R_0-topology on X which is T_0-saturated with respect to all clopen sets is T_0-saturated with respect to X.
(3) ind X = O and every clopen-to-closed mapping $f:X \longrightarrow Y$ into an arbitrary T_1-space Y is closed.

Note that 4.2 provides a complete solution of Problem A.

Further applications of Theorem 3.1 are limited by the fact that - to the best knowledge of the author - there are no other essentially different classes $\underline{E}$ in literature for which characterizations of $\underline{E}$-normal spaces are known. Nevertheless in [2] we were able to prove several characterizations of $\underline{D}$-normal spaces , where $\underline{D}$ denotes the class of all developable spaces [1]. They are contained in a paper in preparation. Unfortunately we do not know whether $\underline{D}$ is adequate for functional separation.

To apply Theorem 3.1 in this case we need the following considerations.

4.3 LEMMA : (1) Let $\underline{E}$ be a nonempty class of topological spaces which is closed under the formation of finite products. Then $\underline{E}$ has the finite intersection property.

(2) Suppose in addition that $\underline{E}$ is closed with respect to closed mappings (i.e. $f: X \longrightarrow Y$ closed and $X \in \underline{E}$ implies $Y \in \underline{E}$) and inverse images. Then $\underline{E}$ is adequate for functional separation.

The proof of 4.3 is left to the reader (see [1o] or [13] for the proof of (1)). Since the class $\underline{S}$ of all semistratifiable spaces fullfills all properties stated in 4.3 , it is adequate for functional separation [6] . The following theorem was proved in [3] .

4.4 THEOREM : The following statements are equivalent for a topological space X:

(1) X belongs to the bireflective hull of the class $\underline{D}$ of all developable spaces.
(2) X belongs to the bireflective hull of the class $\underline{S}$ of all semistratifiable spaces.
(3) X has a basis $\mathfrak{B}$ for the closed sets such that for every $B \in \mathfrak{B}$ there is a countable family $(B_n)_{n \in \mathbb{N}}$ in $\mathfrak{B}$ such that $X \setminus B = \bigcup\{B_n \mid n \in \mathbb{N}\}$. 1)

Furthermore it was shown in [2] that a subset A of a topological space is $\underline{D}$-closed iff A is $\underline{S}$-closed. Therefore we obtain from 4.4 , 3.1 and 3.2 the following results:

4.5 THEOREM : The following statements are equivalent for a topological space X :

(1) X is a $\underline{D}$-normal R_o-space.
(2) X has a G_δ- basis for the closed sets and every coarser

---

1) We call a basis $\mathfrak{B}$ for the closed sets with this property a G_δ-basis.

R_o-topology on X which is T_o-separated with respect to all D-closed subsets of X is T_o-separated with respect to X.

(3) X has a G_δ-basis for the closed sets and every D-closed mapping $f: X \longrightarrow Y$ into an arbitrary T_1-space Y is closed.

4.6 THEOREM : Every Lindelöf space which has a G_δ-basis for the closed sets is D-normal.

Note that Theorem 4.6 was announced in [3] without proof. A slightly more general theorem is contained in [4].

References

[1] R.H.Bing , Metrization of topological spaces , Can.J.Math.3 (1951) 175-186.

[2] H.Brandenburg , Hüllenbildungen für die Klasse der entwickelbaren topologischen Räume , Thesis , Free University Berlin (1978).

[3] H.Brandenburg , On a class of nearness spaces and the epireflective hull of developable topological spaces , to appear in the Proc. of Int. Symp. on Top. and its Appl., Beograd (1977).

[4] H.Brandenburg , On spaces with a G_δ-basis , submitted.

[5] K.A.Broughan , Invariants for Real-Generated Uniform Topological and Algebraic Categories , Lect. Notes in Math. 491 , Springer Verlag , Berlin-Heidelberg-New York (1975).

[6] G. Creede , Concerning semi-stratifiable spaces , Pac.J. Math. 32 (1970) 47-54.

[7] A.S.Davis , Indexed systems of neighborhoods for general topological spaces , Amer.Math.Monthly 68 (1961) 886-893.

[8] Z.Frolík , Applications of complete families of continuous functions to the theory of Q-spaces , Czech. Math. J. 11 (86) (1961) 115-132.

[9] L.Gillman , M.Jerison , Rings of continuous functions , Van Nostrand Reinhold Comp. , New York (1960).

[10] H.Herrlich , $\mathfrak{E}$-kompakte Räume , Math.Zeitschrift 96 (1967) 228-255.

[11] Th.Marny , On epireflective subcategories of topological categories , Preprint.

[12] S.Mrówka , Further results on E-compact spaces I , Acta. Math. 120 (1968) 161-185.

[13] G.Preuß , Trennung und Zusammenhang , Monatshefte für Math. 74 (1970) 70-87.

[14] P.Roy , Nonequality of dimensions for metric spaces , Trans.Amer.Math.Soc. 134 (1968) 117-132.

[15] J.Tsai , On E-compact spaces and generalizations of perfect mappings , Pac.J.Math. 46 (1973) 275-282.

[16] P.Zenor , A note on Z-mappings and WZ-mappings , Proc. Amer.Math.Soc. 23 (1969) 273-275.

Harald Brandenburg
Freie Universität Berlin
Institut für Mathematik I
Hüttenweg 9
D 1000 BERLIN 33

TWO PROCEDURES IN BITOPOLOGY

G.C.L. Brümmer
University of Cape Town, 7700 Rondebosch, South Africa

Abstract: Two procedures for extending topological or uniform space concepts to bitopological or quasi-uniform spaces are: (1) spanning subcategories or functors by suitable objects; (2) lifting epireflections. The main theorem relates Cauchy completions of functorial admissible (quasi-) uniformities to generalized compactness reflections. We discuss the non-unique extension of the realcompactness reflection to bitopological spaces and the resulting bitopological version of Shirota's theorem.

AMS(MOS) codes: Primary 54D60, 54E15, 54E55; secondary 18A40.

1. Introduction

A *bitopological space* [15] is an ordered pair $(X_1,X_2) = X$ of topological spaces with the same underlying set. These spaces form a category *Bitop* in which the morphisms are the *bicontinuous* maps $f: X \to X'$, i.e. $f: X_i \to X_i'$ is continuous for $i = 1, 2$. By assigning to X the coarsest topological space finer than X_1 and X_2 we have a faithful functor S: *Bitop* $\to$ *Top*. Assigning to Y in *Top* the pair (Y,Y) is an embedding D: *Top* $\to$ *Bitop*, left adjoint right inverse to S.

A *quasi-uniform space* [18, 10, 16] is a set X with a filter $\mathcal{H}$ of reflexive relations (*entourages*) on X satisfying $(\forall H \in \mathcal{H})(\exists K \in \mathcal{H})(K \circ K \subset H)$. We shall write X for the space and ent X for the filter $\mathcal{H}$. These spaces with the maps $f: X \to X'$ satisfing $(\forall K \in \text{ent } X')((f \times f)^{-1}K \in \text{ent } X)$ form the category *Qun*. Inverting entourages gives an involution c: *Qun* $\to$ *Qun*. The category *Unif* of uniform spaces is the full subcategory of *Qun* which is fixed under c. There is a faithful functor s: *Qun* $\to$ *Unif* assigning to X the coarsest quasi-uniform space finer than X and cX. The inclusion d: *Unif* $\to$ *Qun* is left adjoint right inverse to s.

One procedure will be to replace distinguished objects in *Top* or *Unif* by suitable objects in *Bitop* or *Qun*, then using these to span certain subcategories and functors. In *Top* we have, e.g., R with its usual topology and its subspaces $I = [0,1]$, $D = \{0,1\}$. Carrying their usual uniformity these are written R_u, I_u, D_u, in *Unif*. In *Bitop* we have $R_b = (R_r, R_\ell)$ with subspaces I_b and D_b and generally A_b if $A \subset R$, where R_r carries the right or upper topology with basic open sets $(-\infty,a)$ and R_ℓ the left or lower topology with basic open sets (a,∞). In *Qun* we have R_q with basic entourages $\{(x,y) \mid y < x + \varepsilon\}$, $\varepsilon > 0$, and subspaces I_q, D_q. Clearly $SR_b = R$, $sR_q = R_u$.

The other procedure is the lifting of epireflections against a functor $L: A \to B$. For an epireflective subcategory B' of B we consider the subcategory $L^{-1}[B']$ of A having the objects A for which LA is in B'. (Convention: subcategories are full and isomorphism-closed; reflectors are endofunctors.)

1.1 **Lemma** [2] *Let* $L: A \to B$ *be a functor, and* A *complete, well- and co-wellpowered. Let* B' *be epireflective in* B. *If* L *preserves limits, or preserves products and extremal monomorphisms, then* $L^{-1}[B']$ *is epireflective in* A.

1.2 **Lemma** *Let* $L: A \to B$ *be a functor,* B' *an epireflective subcategory of* B *such that* $L^{-1}[B']$ *is epireflective in* A. *Let* ρ *and* $\hat{\rho}$ *denote the epireflectors corresponding to* B' *and* $L^{-1}[B']$.

(a) [2] *If* L *preserves epis and has a right inverse* F, *then* $L\hat{\rho}F \cong \rho$.

(b) *If* L *reflects epis and has a right inverse* R *with a natural transformation* $\varepsilon: RL \to 1$ *such that* $\varepsilon R = 1$, *then* $\hat{\rho}R \cong R\rho$ *(whence* $L\hat{\rho}R \cong \rho$ *).*

Accordingly we define a bitopological space X to be *S-compact* iff SX is compact, *S-realcompact* iff SX is realcompact; a quasi-uniform space Y to be *s-complete* iff sY is complete; and similarly for other properties.

Given a faithful functor $L: A \to \mathit{Set}$ one may define [2] an object A of A to be *L-separated* iff each L-initial map (or L-initial source) with domain A distinguishes points. Let A_0 denote the subcategory of L-separated objects of A. Then Top_0 consists of the T_0-spaces [2] ; $(\mathit{Top}_0)_0 = \mathit{Top}_0$ [3, 14, 13] ; Unif_0 consists of the separated uniform spaces (i.e. with T_0-topology). Bitop_0 consists of the S-T_0 spaces, and thus has the nice properties given by lemmas 1.1, 1.2. Qun_0 likewise is lifted against s from Unif_0.

The category Crg of completely regular spaces is the initial hull (hence bireflective hull) of I (just as well, R) in Top. We write $T: \mathit{Unif} \to \mathit{Crg}$ for the usual forgetful functor. $\mathit{Crg}_0 = \mathit{Tych}$, the Tychonoff spaces, is the epireflective hull of I (or R) in Top.

imilarly, the category *Pcrg* of *pairwise completely regular* spaces [Lane 16] is the initial (hence bireflective) hull of I_b (or R_b) in *Bitop* [Salbany 21]. To a quasi-uniform space Y one assigns a bitopological space (X_1, X_2) by giving a point x the X_1-neighbourhoods H[x] and X_2-neighbourhoods $H^{-1}[x]$, $H \in$ ent Y. The range of this assignment is ob *Pcrg* [16]; we denote the functor by $\bar{T}$: *Qun* → *Pcrg*. We write *Ptych* (*pairwise Tychonoff*) for $Pcrg_0$, i.e. *Pcrg* ∩ $Bitop_0$, i.e. the epireflective hull in *Bitop* of I_b or R_b.

Salbany [21] showed that a map f: X → Y in *Ptych* is epi in *Ptych* if and only if Sf: SX → SY is epi in *Tych*, i.e. the image is dense in SY (we shall say *S-dense in* Y) (cf. [8]). Likewise the extremal monos in *Ptych* are the S-closed embeddings [2]. Thus both lifting lemmas 1.1, 1.2 are applicable to the restricted S: *Ptych* → *Tych*. Now the compact spaces in *Tych* form the epireflective hull in *Tych* of I, and Salbany's [21] basic discovery was that the S-compact spaces in *Ptych* form the epireflective hull in *Ptych* of I_b. Denoting the corresponding reflector in *Pcrg* by $\bar{\beta}$, we have by 1.2 $\bar{\beta}D \cong D\beta$ (cf. [22]) (from now on we restrict S and D to *Pcrg* ⇄ *Crg*). (S-compactness has also been studied in [10, 28, 9, 22, 23, 24, 26] and elsewhere.)

Let γ denote the completion (always separated) reflector in *Unif*. By 1.1 $\gamma | Unif_0$ lifts against s to an epireflector in Qun_0, which we extend to a reflector $\bar{\gamma}$ in *Qun*. This reflector was described internally by Császár [10] and Salbany [21, 25]. By [10], $\bar{\gamma}$ has the same "rigidity" as γ: any s-completion (always taken s-separated) of X is equivalent to $\bar{\gamma}X$. By 1.2 one has $\bar{\gamma}d \cong d\gamma$, and by the rigidity, $s\bar{\gamma} \cong \gamma s$. By contrast, $S\bar{\beta}$ and βS are not equivalent as Salbany's example $(0,1]_b$ shows [2].

The present paper follows the notations of [5], where some background details omitted here may be found.

2. Generalized compactness and functorial uniformities

The forgetful functor T: *Unif* → *Crg* has a right inverse C*: *Crg* → *Unif*, constructed as follows: for X in *Crg*, C*X is the coarsest uniform space whose *Unif*-maps into I_u coincide with the *Crg*-maps of X into $T(I_u)$. Replacing I_u by R_u, we have a right inverse C of T. (The symbols C*, C are from [12].) More generally, replacing I_u or R_u by any class *U* of uniform spaces, we speak of the functor F: *Crg* → *Unif* *spanned by* *U*. Now F is right inverse to T if and only if $T[\mathcal{U}]$ is initially dense in *Crg* ; and every right inverse of T can be obtained in this way [1, 4, 5]. The *fine functor*, which we denote $R: \mathit{Crg} \to \mathit{Unif}$, left adjoint right inverse to T, is spanned by the class of all uniform spaces, or equally well by the class of all separated complete uniform spaces. R is the finest (and C^* the coarsest) right inverse of T.

We consider the well-known [12] natural equivalences

$$\gamma C^* \cong C^*\beta \ , \quad \gamma C \cong C\upsilon \ , \quad \gamma R \cong R\delta$$

where υ is the Hewitt realcompactification, regarded as a reflector of *Crg* onto the epireflective hull of **R** in *Tych*, and δ is the Dieudonné topological completion, a reflector onto those Tychonoff spaces which admit complete uniformities. These important formulae may be unified as follows.

2.1 Theorem *Let* $\mathcal{U}$ *be a class of separated, complete uniform spaces such that* $T[\mathcal{U}]$ *is initially dense in* *Crg*, *and let* F *be the right inverse of the forgetful functor* $T: \mathit{Unif} \to \mathit{Crg}$ *which is spanned by* $\mathcal{U}$. *Let* α *be the reflector of* *Crg* *onto the epireflective hull of* $T[\mathcal{U}]$ *in* *Tych*. *Then* $\gamma F \cong F\alpha$.

This result, which may well be known, can be proved using the basic idea of [12, Theorem 15.13]. We shall deduce it from the theorem below.

3. Functorial quasi-uniformities

Any class $\mathcal{Q}$ of quasi-uniform spaces spans a functor $\Phi: \mathit{Pcrg} \to \mathit{Qun}$ by the obvious analogue of the above spanning construction; Φ is right inverse to $\overline{T}: \mathit{Qun} \to \mathit{Pcrg}$ if and only if $\overline{T}[\mathcal{Q}]$ is initially dense in *Pcrg* ; each right inverse of $\overline{T}$ can be so obtained [4,5]. The following theorem extends the main result of [7].

3.1 Theorem *Let* $\mathcal{Q}$ *be a class of separated,* s-*complete quasi-uniform spaces such that* $\overline{T}[\mathcal{Q}]$ *is initially dense in* *Pcrg*, *and let* Φ *be the right inverse of the forgetful functor* $\overline{T}: \mathit{Qun} \to \mathit{Pcrg}$ *which is spanned by* $\mathcal{Q}$. *Let* $\overline{\alpha}$ *be the reflector of* *Pcrg* *onto the epireflective hull of* $\overline{T}[\mathcal{Q}]$ *in* *Ptych*. *Then* $\overline{\gamma}\Phi \cong \Phi\overline{\alpha}$.

Proof 1: For clarity we first treat the case that $\mathcal{Q}$ consists of a single space A. 1(a): If Y is in the epireflective hull of $\overline{T}A$ then ΦY is s-complete. Indeed, the map

$$e: \Phi Y \to A^{\mathit{Ptych}(Y,\overline{T}A)}$$

iven by $\pi_f \circ e = f$ is an embedding into an s-complete product, and as $\Phi = 1$, we have an embedding

$$\overline{T}e\colon Y \to (\overline{T}A)^{Ptych(Y,\overline{T}A)}$$

hich, by a construction for the epireflection, is S-closed, so that Y is s-complete. 1(b): Let X be in *Pcrg*. Let η be the unit f $\overline{\alpha}$. Thus for $g\colon X \to \overline{T}A$ there is unique $f\colon \overline{\alpha}X \to \overline{T}A$ with $f \circ \eta_X = g$. ow ΦX is initial for the maps $g\colon X \to \overline{T}A$ to A, whence

(i) ΦX is initial for the maps $f \circ \eta_X$ to A, where f ranges through $Pcrg(\overline{\alpha}X, \overline{T}A)$.

(ii) $\Phi\overline{\alpha}X$ is initial for the maps f to A, where f ranges through $Pcrg(\overline{\alpha}X, \overline{T}A)$.

rom (i) and (ii) one readily sees that $\Phi\eta_X\colon \Phi X \to \Phi\overline{\alpha}X$ is an initial ap. But by 1(a) $\Phi\overline{\alpha}X$ is a separated s-complete space and hence an -completion of ΦX (initial maps rather than embeddings suffice), hence by the rigidity of $\overline{\gamma}$, $\Phi\overline{\alpha}X \cong \overline{\gamma}\Phi X$. 2: We adapt the above roof to the case of general $\mathcal{Q}$. 2(a): If Y is in the epireflec- ive hull of $\overline{T}[\mathcal{Q}]$ then ΦY is s-complete. For, we can reduce the otal *Qun*-source from ΦY to $\mathcal{Q}$ by co-wellpoweredness to a set-indexed ource which still is initial, embed ΦY into a product, and proceed ssentially as in 1(a). 2(b): Adapting 1(b) to $\mathcal{Q}$ is straight- orward. □

We consider three right inverses of $\overline{T}$: $\overline{C}^*$ spanned by I_q (the oarsest right inverse); $\overline{C}$ spanned by R_q; the *fine functor* $\overline{R}$, panned by all (equivalently, all separated s-complete) quasi-uniform paces. These three functors are *odd* in the sense of [5], i.e. commute ith the symmetry involutions in *Qun* and *Pcrg*. Therefore they *re- trict* to C^*, C, R respectively in the sense of [5], i.e.

$$dC^* = \overline{C}^*D \quad , \quad dC = \overline{C}D \quad , \quad dR = \overline{R}D \; .$$

or general right inverses of $\overline{T}$ we have a weaker restriction :

.2 **Proposition** *Let* $\mathcal{Q} \subset$ ob *Qun*, $\Phi\colon Pcrg \to Qun$ *be spanned by* , *and* $F\colon Crg \to Unif$ *be spanned by* $s[\mathcal{Q}]$. *Then* $s\Phi D = F$.

roof This follows since $Crg(X, TsA) = Pcrg(DX, \overline{T}A)$ for $A \in \mathcal{Q}$. □

.3 **Proposition** *Let* A *be a class of Ptych-spaces initially dense n Pcrg. Let* $\overline{\alpha}$ *be the reflector of Pcrg onto the epireflective ull of* A *in Ptych, and* α *that of Crg onto the epireflective hull f* S[A] *in Tych. Then* $\overline{\alpha}D \cong D\alpha$. □

We can now deduce Theorem 2.1 from Theorem 3.1 : If $\mathcal{U}$ satisfies the hypotheses of 2.1, then $\mathcal{Q} = d[\mathcal{U}] \cup \{I_q\}$ satisfies those of 3.1. The same right inverse F of T is spanned by $\mathcal{U}$ and by $s[\mathcal{Q}]$, and thus by 3.2 and 3.3, $\gamma F = \gamma s \Phi D \cong s\overline{\gamma}\Phi D \cong s\Phi\overline{\alpha}D \cong s\Phi D\alpha = F\alpha$. □

In [5] we showed that for any right inverse F of T there exists an odd right inverse Φ of $\overline{T}$ which *extends* F in the sense that $dF = \Phi D$; and that in the case $F = C^*$, there is only one right inverse of $\overline{T}$ which extends F (namely $\overline{C}^*$). In all other cases the uniqueness problem is, to the author's knowledge, still open.

4. Bitopological realcompactness

We consider four reflective properties for bitopological spaces.

(1) X is *birealcompact* if it is in the epireflective hull of R_b in *Ptych* . Let $\overline{\upsilon}$ denote the reflector. By 3.1 and 3.3, $\overline{\gamma}\overline{C} \cong \overline{C}\overline{\upsilon}$ and $\overline{\upsilon}D \cong D\upsilon$.

(2) The S-realcompact spaces in *Ptych* form an epireflective subcategory by 1.1. Let $\hat{\upsilon}$ denote the reflector. By 1.2, $\hat{\upsilon}D \cong D\upsilon$

(3) X is *bitopologically complete* if it admits a separated s-complete quasi-uniformity. Let $\overline{\delta}$ denote the reflector. By 3.1 and 3.3, $\overline{\gamma}\overline{R} \cong \overline{R}\overline{\delta}$ and $\overline{\delta}D \cong D\delta$.

(4) X is *S-topologically complete* if SX is topologically complete i.e. SX admits a separated complete uniformity. By 1.1, the S-topologically complete spaces in *Ptych* form an epireflective subcategory. Let $\hat{\delta}$ denote the reflector. By 1.2, $\hat{\delta}D \cong D\delta$.

4.1 Proposition *For a pairwise Tychonoff space* X *we have the implications*

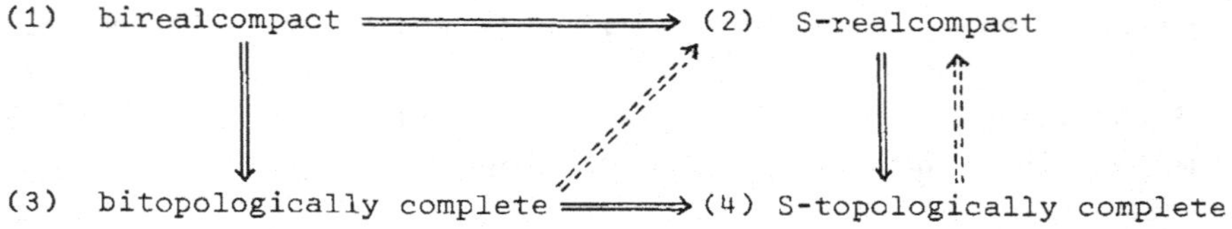

where the broken arrows hold if and only if SX *has no closed discrete subspace of Ulam-measurable cardinality.*

Proof This follows readily from 2.1, 3.1 and Shirota's theorem [27, 12]. □

It was shown in [7] that (2)$\Longrightarrow\!\!\!\!\!\!/\;\;$(1). We shall now show that 3)$\not\Rightarrow$(1). We lack an example for "(2)$\not\Rightarrow$(3)" (which would rule out ny further reversals in the above diagram).

.2 **Example** *Let* X *be the bitopological subspace of* R_b *with point et* [-1,0) ∪ (0,1] . *Then* X *is bitopologically complete but not bi-ealcompact.*

roof It was shown in [7] that X is not birealcompact. The subpaces $X_1 = [-1,0)_b$ and $X_2 = [0,1)_b$ are birealcompact, e.g. by [7: heorem 3.2] or directly by isomorphism to $[0,\infty)_b$ and $(-\infty,0]_b$ respectively. Hence X_i admits an s-complete quasi-uniformity H_i (i = 1,2). et $B = \{U_1 \cup U_2 \cup (X_2 \times X_1) \mid U_1 \in H_1 , U_2 \in H_2\}$. Then B is a base or a filter on X × X and each member contains the diagonal. Given $= U_1 \cup U_2 \cup (X_2 \times X_1)$, $U_i \in H_i$, we may take $V_i \in H_i$ with $V_i \circ V_i \subset U_i$, et $V = V_1 \cup V_2 \cup (X_2 \times X_1)$, and check through nine cases that $V \circ V \subset U$. hus B is a base for a quasi-uniformity H and one easily sees that $(X,H) = X$. Now the uniform space $s(X,H)$ has basic entourages $U \cap U^{-1} =$ $U_1 \cap U_1^{-1}) \cup (U_2 \cap U_2^{-1})$, with U as above. Consequently $s(X,H)$ is the *'nif*-coproduct of the complete uniform spaces $s(X_i,H_i)$ and hence comlete. (This is striking, as X is not the *Bitop*-coproduct of the $_i$.) □

.3 **Remark** The *pair-realcompactness* of Saegrove [19] implies biealcompactness but is strictly stronger : by [24], the two-point space $_b$ is not pair-realcompact. By the methods of [24], the pair-realcomact spaces form a non-reflective subcategory of *Ptych* , closed for the aking of *Bitop*-products.

.4 **Remark** The classical concept of *pseudocompactness* also has nonnique extension to bitopology. Indeed, a Tychonoff-space Y is pseudoompact if and only if C^*Y is T-fine (equivalently : all right inerses of T coincide on Y), whereas the pairwise Tychonoff space X f example 4.2 is such that the maps $X \to R_b$ are bounded (they are preisely the monotone non-decreasing functions which are continuous when iewed as maps $SX \to R$ i.e. in the usual topology) whilst by 4.2 $\overline{C}^*X$ s not $\overline{T}$-fine.

. **The forgetful functor** T_1: *Qun* → *Top*

The functor T_1: *Qun* → *Top* assigns to a quasi-uniform space the irst of the two topologies assigned by $\overline{T}$. The object range of T_1

is ob *Top* [10]. The coarsest right inverse of T_1 , written C_1^* , is spanned by D_q [1]. An open problem is to find an analogue of Theorem 2.1 for the functor T_1 . It seems that, if an analogue exists, the role of the compact spaces in 2.1 will be taken here by the epireflective hull of $T_1(D_q) = D_r$ in Top_0 (i.e. by the *sober* spaces), and that a new completeness concept in *Qun* will be needed. Lack of such a concept led to the study of the non-reflective functor $\beta_1 = T_1\overline{\gamma}C_1^*: Top \to Top$ in [6], to mimick the formula $\overline{\beta} \cong \overline{T}\overline{\gamma}\overline{C}^*$. The functor β_1 is part of a monad in *Top* and assigns T_0-compactifications [6].

More literature has in effect been devoted to the functor T_1 than to $\overline{T}$; cf. the bibliography [17] or, for a sample of such results viewed from our standpoint, see [5]. It seems that the right inverses of T_1 provide a tool for studying topological spaces of very weak separation, i.e. between the axioms T_0 and T_D .

References

[1] Brümmer, G.C.L. Initial quasi-uniformities. *Nederl. Akad. Wetensch Proc. Ser. A* 72 = *Indag.Math.* 31 (1969), 403-409.

[2] ______ A categorial study of initiality in uniform topology. Thesis, Univ. Cape Town, 1971.

[3] ______ Struktuurfunktore en faktorisering. *Proc.S.Afr. Math.Soc.* 4 (1974), 81-83.

[4] ______ Topological functors and structure functors. *Categorical Topology* (Proc.Conf., Mannheim, 1975), pp.109-135. *Lecture Notes in Math.* 540, Springer-Verlag, Berlin, 1976.

[5] ______ On certain factorizations of functors into the category of quasi-uniform spaces. *Quaestiones Math.* 2(1977), 59-84.

[6] ______ On some bitopologically induced monads in *Top*. *Mathematik-Arbeitspapiere Univ.Bremen*, to appear.

[7] ______ and S. Salbany. On the notion of realcompactness for bitopological spaces. *Math.Colloq. Univ.Cape Town* 11(1977),89-99.

[8] Cooke, I.E. Epireflections in the category of bitopological spaces. Thesis, Univ. of London, 1972.

[9] ______ and I.L. Reilly. On bitopological compactness. *J.London Math.Soc.*(2) 9 (1975), 518-522.

[10] Császár, Á. *Fondements de la topologie générale.* Akademiai Kiadó, Budapest, 1960. Revised and extended edition: *Foundations of general topology.* Pergamon Press, Oxford-New York, 1963.

[11] ______ Doppeltkompakte bitopologische Räume. In: G. Asser, J.Flachsmeyer and W. Rinow (ed.): *Theory of Sets and Topology*, pp.59-67. VEB Deutscher Verlag d.Wiss., Berlin, 1972.

12] Gillman, L. and M. Jerison. *Rings of continuous functions.* Van Nostrand, Princeton-New York, 1960.

13] Harvey, J.M. T_0-separation in topological categories. *Quaestiones Math.* 2(1977), 177-190.

14] Hoffmann, R.-E. (E,M)-universally topological functors. Habilitationsschrift, Univ.Düsseldorf, 1974.

15] Kelly, J.C. Bitopological spaces. *Proc.London Math.Soc.*(3) 13 (1963),71-89.

16] Lane, E.P. Bitopological spaces and quasi-uniform spaces. *Proc. London Math.Soc.* (3)17(1967),241-256.

17] Murdeshwar, M.G. Bibliography on quasi-uniform spaces. Preprint, Dept.of Math., Univ. of Alberta, Edmonton, 1974.

18] Nachbin, L. Sur les espaces uniformes ordonnés. *C.R. Acad.Sci. Paris* 226(1948), 774-775.

19] Saegrove, M.J. Pairwise complete regularity and compactification in bitopological spaces. *J.London Math.Soc.*(2)7(1973),286-290.

20] Salbany, S. Quasi-uniformities and quasi-pseudometrics. *Math. Colloq. Univ. Cape Town* 6(1970-71), 88-102.

21] ______ Bitopological spaces, compactifications and completions. Thesis, Univ. Cape Town, 1970. Reprinted as *Math. Monogr. Univ. Cape Town* No. 1, 1974.

22] ______ Compactifications of bitopological spaces. *Math. Colloq. Univ. Cape Town* 7(1971-72), 1-3.

23] ______ On quasi-uniformizability. *Joint Math. Colloq. Univ. of South Africa and Univ. of the Witwatersrand* 1973-74.

24] ______ On compact bitopological spaces. *Proc.S.Afr.Math.Soc.* 4(1974), 219-223.

25] ______ Completions and triples. *Math.Colloq. Univ. Cape Town* 8(1973), 55-61.

26] ______ Reflective subcategories and closure operators. *Categorical Topology* (Proc.Conf.,Mannheim, 1975), pp.549-565. *Lecture Notes in Math.* 540, Springer-Verlag, Berlin, 1976.

27] Shirota, T. A class of topological spaces. *Osaka Math. J.* 4 (1952), 23-40.

28] Swart, J. Total disconnectedness in bitopological spaces and product bitopological spaces. *Nederl.Akad.Wetensch., Proc. Ser.A* 74 = *Indag Math.* 33(1971), 135-145.

Grants to the Topology Research Group from the University of Cape own and from the South African Council for Scientific and Industrial esearch are acknowledged. The author is grateful for the invitation nd financial support received from the Organizers of the present Conerence.

SAKS SPACES AND VECTOR VALUED MEASURES

J.B. Cooper, W. Schachermayer, Linz, Austria

Introduction: The purpose of this note is to give a sample of applications of Saks spaces to the theory of vector measures. For convenience we restrict attention to generalised Riesz representation theorems i.e. we consider representations of operators on spaces of continuous functions by integration with respect to a vector valued measure. A Saks space is a vector space with two structures, a norm and a locally convex topology, which are in some sense compatible. At first sight this may seem a rather strange object and its relevance to measure theory is not at all obvious. However, we hope that this paper demonstrates the thesis that they are a suitable tool for some aspects of the theory. Here we would like to mention one argument which may make this claim more plausible. One of the features of a σ-additive measure with values in a Banach space and defined on a σ-field is that it takes its values in a weakly compact set. This means exactly that it takes its values in a Saks space and indeed of a very special kind - one with compact unit ball. Now such Saks spaces are precisely those which are expressible as projective limits (in a suitable sense) of finite dimensional spaces. This allows us, for example, to reduce the proof of a Riesz representation theorem for operators with values in such Saks spaces to the finite dimensional (i.e. essentially the scalar valued) case by means of a simple formal manipulation with suitable functors. Surprisingly enough, although this result seems very special, it contains as Corollaries three important results and thus we obtain a simple and unified approach to them.

For the convenience of the reader we begin with a brief survey of the results and concepts on Saks spaces which we shall require. A detailed discussion can be found in COOPER [1]. In the second half of the paper we prove the Riesz representation theorem mentioned above and deduce three important representation theorems (cf. DIESTEL and UHL [2]).

This article is an extract from a systematic treatment of vector measures from the point of view of Saks spaces which is now in preparation.

1. <u>Definition</u>: A <u>Saks space</u> is a triple $(E,\|\ \|,\tau)$ where $(E,\|\ \|)$ is a normed space and τ is a locally convex topology on E so that $B_{\|\ \|}$, the unit ball of $(E,\|\ \|)$, is τ-closed and bounded.
We then write $\gamma[\|\ \|,\tau]$ or simply γ for the finest locally convex topology on E which coincides with τ on $B_{\|\ \|}$. We resume the most important elementary properties of γ in the following Proposition (cf. COOPER [1]):

2. <u>Proposition</u>: 1) $\tau \subsetneq \gamma \subsetneq \tau_{\|\ \|}$;
2) the γ-bounded subsets of E coincide with the norm-bounded sets;
3) a sequence (x_n) in E converges to zero with respect to γ if and only if it is norm bounded and τ-convergent to zero;
4) a subset of E is γ-compact if and only if it is norm-bounded and τ-compact;
5) (E,γ) is complete if and only if $B_{\|\ \|}$ is τ-complete;
6) the dual E'_γ of (E,γ) is the norm-closure of $(E,\tau)'$ in the dual of $(E,\|\ \|)$.

3. <u>Examples</u>: I. If E is a Banach space, then the following triples are Saks spaces:

$$(E,\|\ \|,\tau_{\|\ \|}),\quad (E,\|\ \|,\sigma(E,E')),\quad (E',\|\ \|,\sigma(E',E)).$$

II. If X is a completely regular space and $C^b(X)$ denotes the space of bounded, continuous complex-valued functions on X then $(C^b(X),\|\ \|,\tau_K)$ is a Saks space where τ_K is the topology of compact convergence.

III. If H is a Hilbert space and L(H) is the algebra of continuous linear operators on H we denote by τ_w and τ_s the weak resp. strong operator topology on L(H). Hence τ_w is defined by the seminorms $T \to |f(Tx)|$ $(x \in H, f \in H')$ and τ_s is defined by the seminorms $T \to \|Tx\|$ $(x \in H)$. Then

$$(L(H),\|\ \|,\tau_w) \quad\text{and}\quad (L(H),\|\ \|,\tau_s)$$

are Saks spaces.

IV. Let μ be a positive, finite, σ-additive measure on the space (Ω,Σ) and denote by $L^\infty(\mu)$ the corresponding L^∞-space. Then $(L^\infty(\mu),\|\ \|,\tau_1)$ is a Saks space where τ_1 is the topology induced by the L^1-norm and its dual is L^1.

4. <u>Completions</u>: If $(E,\|\ \|,\tau)$ is a Saks space its completion is defined as follows: we let $\hat{B}$ denote the τ-completion of $B_{\|\ \|}$ i.e. the closure of $B_{\|\ \|}$ in the completion $\hat{E}_\tau$ of (E,τ). Then if $\hat{E}$ is the span of $\hat{B}$ and $\|\ \|^\wedge$

denotes the Minkowski functional of $\hat{B}$ $(\hat{E},\| \|^{\wedge},\tau)$ is the required completion. As an example, if E is a normed space then the completion of the Saks space $(E,\| \|,\sigma(E,E'))$ is the Saks space $(E'',\| \|,\sigma(E'',E'))$.

5. Saks space products and projective limits: If $\{(E_\alpha,\| \|_\alpha,\tau_\alpha)\}_{\alpha\varepsilon A}$ is a family of Saks spaces we form their product as follows:
if E denotes the Cartesian product $\prod_{\alpha\varepsilon A} E_\alpha$ we put

$$E_o := \{x = (x_\alpha) \varepsilon E : \|x\| := \sup \|x_\alpha\|_\alpha < \infty\}$$

Then $(E_o,\| \|,\tau)$ is a Saks space where τ is the Cartesian product of the topologies $\{\tau_\alpha\}$. $(E_o,\| \|,\tau)$ is called the Saks space product of $\{E_\alpha\}$ and is denoted by $S \prod E_\alpha$.

Now let $\{\pi_{\beta\alpha} : E_\beta \to E_\alpha, \alpha,\beta \varepsilon A, \alpha \leq \beta\}$ be a projective spectrum of Saks spaces. As usual, we define the projective limit of this spectrum as the subspace of the product formed by the threads i.e. as

$$E_1 := \{(x_\alpha) \varepsilon S \prod E_\alpha : \pi_{\beta\alpha}(x_\beta) = x_\alpha \quad \text{for } \alpha \leq \beta\}$$

E_1 is denoted by $S \varprojlim\{E_\alpha\}$. As an example, if X is a locally compact space and $\mathcal{K}(X)$ denotes the family of compact subsets of X then

$$\{\rho_{K_1,K} : C(K_1) \to C(K), K \subseteq K_1\}$$

forms a projective spectrum of Banach spaces (where C(K) denotes the Banach space of continuous, complex-valued functions on K and $\rho_{K_1,K}$ is the restriction operator) and its Saks space projective limit is naturally identifiable with $(C^b(X),\| \|,\tau_{\mathcal{K}})$.

Now if $(E,\| \|,\tau)$ is a Saks space, we say that a family of seminorms S which generates τ is a suitable family if it satisfies the condition:

1) if $p,q \varepsilon S$ then $\max\{p,q\} \varepsilon S$;
2) $\| \| = \sup S$.

If $p \varepsilon S$, E_p denotes the Banach space generated by p (i.e. the completion of the normed space E/N_p where N_p is the kernel of p) and if $p \leq q$ then ω_{pq} denotes the natural mapping from E_q to E_p. Then $\{\omega_{qp} : E_q \to E_p\}$ forms a projective spectrum of Banach spaces.

6. Proposition: If $(E,\| \|,\tau)$ is a complete Saks space, then E is naturally identifiable with $S \varprojlim_{p\varepsilon S} E_p$.

7. Proposition: Let $(E,\| \|,\tau)$ be a Saks space. Then the following are equivalent:

1) $B_{\| \|}$ is τ-compact;

2) E is a Saks space projective limit of finite dimensional Banach spaces;

3) E has the form $(F',\| \|,\sigma(F',F))$ for some Banach space F.

Then $\gamma = \tau_c(F',F)$, the topology of uniform convergence on the compact sets of E, is the finest topology on E which agrees with τ on $B_{\| \|}$.

In fact, if 1) is fulfilled, then E is naturally identifiable with $S \varprojlim_{F \in \mathcal{F}(E'_\gamma)} \{F'\}$ where $\mathcal{F}(E'_\gamma)$ denotes the family of finite dimensional subspaces of E'_γ.

Further, the following are equivalent:

1) B is τ-compact and metrisable;

2) E is the Saks space projective limit of a sequence of finite dimensional Banach spaces;

3) E has the form $(F',\| \|,\sigma(F',F))$ for a separable Banach space F;

4) $B_{\| \|}$ is τ-compact and normable (i.e. there is a norm $\| \|_1$ on E so that $\tau = \tau_{\| \|_1}$ on $B_{\| \|}$).

8. The Hom functor: If $(E,\| \|,\tau)$, $(F,\| \|_1,\tau_1)$ are Saks spaces, then Hom (E,F) denotes the set of γ-continuous linear operators from E into F. Note that, as a vector space, this coincides with the space of norm-bounded linear operators from E into F if E is a Banach space. We regard Hom (E,F) as a Saks space with the supremum norm and τ_p, the topology of pointwise convergence, with respect to τ.

9. Proposition: 1) If $\{E_\alpha\}$ is an inductive system in BAN_1 and F is a complete Saks space then there is a natural isomorphism between the Saks spaces

$$\mathrm{Hom}\,(B \varinjlim_\alpha E_\alpha,\ F) \quad \text{and} \quad S \varprojlim_\alpha \mathrm{Hom}\,(E_\alpha,\ F).$$

In particular, if E is a Banach space, we have

$$\mathrm{Hom}\,(E,F) = \mathrm{Hom}\,(B \varinjlim_{G \in \mathcal{F}(E)} G,\ F) = S \varprojlim_{G \in \mathcal{F}(E)} \mathrm{Hom}\,(G,F).$$

2) if $\{F_\alpha\}$ is a projective system of Saks spaces, E a Saks space, then there is a natural isomorphism between the Saks spaces

$$\mathrm{Hom}\,(E,\ S \varprojlim_\alpha F_\alpha) \quad \text{and} \quad S \varprojlim_\alpha \mathrm{Hom}\,(E,F_\alpha).$$

In particular, if F is a Saks space with $B_{\| \|}$ τ-compact, we have

$$\mathrm{Hom}\,(E,F) = \mathrm{Hom}\,(E,\, S \varprojlim_{G \in F(F'_\gamma)} G') = S \varprojlim_{G \in F(F'_\gamma)} \mathrm{Hom}\,(E,G').$$

10. Remarks on the Saks space $C^b(X)$:

The space $C^b(X)$ is one of the most important Saks spaces and we shall be interested in representations of operators on it. We note here that the dual of $(C^b(X),\gamma)$ is the space of bounded Radon measures on X, the duality being established by integration. This follows easily from 2. For the dual of $(C^b(X),\tau_K)$ is the space of Radon measures with compact support and the bounded Radon measures are just those which can be approximated by such measures. We remark in passing that there are natural Saks space structures on $C^b(X)$ so that the corresponding dual spaces consist of the bounded σ-additive Borel measures (resp. τ-additive Borel measures).
For the theory of $C^b(X)$ cf. COOPER [1], Ch.II.

11. Radon measures with values in a Saks space:

Let $(E,\| \|,\tau)$ be a Saks space, S a suitable family of τ-seminorms on E and X a completely regular space. A bounded Borel measure on X with values in E is a (finitely additive) norm bounded set function μ from Bo(X), the Borel field of X, into E. Such A measure is a Radon measure if it is inner regular with respect to τ i.e. satisfies the condition that

$$\lim_{\substack{K \in K(X) \\ K \subseteq A}} \omega_p \circ \mu(K) = \omega_p \circ \mu(A) \quad \text{in } E_p$$

for each Borel set A in X and each $p \in S$.

Note that we can then integrate functions in $C^b(X)$ with respect to such a measure. The integral $\int x\, d\mu$ is then in $\hat{E}_\gamma$, the completion of E.

We use the notation $M_R(X;E)$ for the (vector) space of E-valued Radon measures on X.
We regard $M_R(X;E)$ as a Saks space with the following structures. The norm is the semivariation norm and the subsidiary locally convex topology is that defined by the seminorms

$$\mu \mapsto p(\int x\, d\mu) \qquad (p \in S,\ x \in C^b(X))$$

12. Proposition: If $(E,\|\ \|,\tau)$ is a complete Saks space, then there are natural identifications:

$$M_R(X;E) = S \varprojlim_{p\in S} M_R(X;E_p).$$

Proof: If μ is in the left hand side, then the elements $(\omega_p \circ \mu)$ form a thread which defines an element of the right hand side. On the other hand, a thread (ω_p) on the right hand side pieces together to form a bounded E-valued measure which is clearly Radon.

We consider Riesz representation theorems for continuous linear operators from the Saks space $C^b(X)$ (cf. 3.II) into a Saks space. We prove one main theorem for operators with values in a Saks space with compact unit ball. Using the machinery developed above, this can be proved in a few lines. We then show how the classical representation theorems for bounded, weakly compact and compact operators from C(K) into a Banach space follow immediately from this.

First we recall that if F is a Banach space, X a completely regular space, then if $\mu : Bo(X) \to F$ is a Radon measure

$$T_\mu : x \mapsto \int x \, d\mu$$

is a β-continuous linear operator from $C^b(X)$ into F. $\mu \mapsto T_\mu$ is an isometry from $M_R(X;F)$ into $Hom(C^b(X),F)$. In general it is not onto as the example of the identity operator on C[0,1] shows. However, if F is finite dimensional it clearly is (once again, we can easily reduce to the case where F is one-dimensional and then the result can be found, for example, in COOPER [2], II.3.3).
Hence we have, for finite dimensional Banach spaces F,

$$M_R(X;F) \cong Hom(C^b(X),F).$$

Once again, this is a natural isomorphism of functors.
The above isomorphism is a Saks space isomorphism if we give the left hand side the topology of pointwise convergence on $C^b(X)$ as auxiliary topology.

13. Proposition: Let $(E,\|\ \|,\tau)$ be a Saks space with $B_{\|\ \|}$ τ-compact, X a completely regular space. Then if $T : C^b(X) \to E$ is a β-γ-continuous linear operator there exists a Rdon measure $\mu : Bo(X) \to E$ representing

T i.e. T is the operator

$$T_\mu : x \mapsto \int x \, d\mu$$

Conversely, every Radon measure μ defines a β-γ-continuous linear operator T_μ in the above manner.
Hence integration establishes a Saks space isomorphism

$$M_R(X;E) = \mathrm{Hom}(C^b(X),E).$$

<u>Proof</u>: We put $G := E'_\gamma$ and calculate:

$$\begin{aligned} \mathrm{Hom}(C^b(X),E) &= \mathrm{Hom}(C^b(X),\ S \varprojlim_{F \in \mathcal{F}(G)} F') \\ &= S \varprojlim_{F} \mathrm{Hom}(C^b(X),F') \\ &= S \varprojlim_{F} M_R(X,F') \\ &= M_R(X,\ S \varprojlim_{F} F') \\ &= M_R(X,E). \end{aligned}$$

14. <u>Remark</u>: A less formal demonstration of the above result goes as follows: since T maps bounded sets in $C^b(X)$ into relatively compact subsets of (E,γ), then

$$T'' : C^b(X)'' \to E''$$

actually takes its values in E.

Noting that if A is a Borel set in X then its characteristic function χ_A defines, by integration, an element of $C^b(X)'' = (M_R(X),\| \ \|)'$ we may define

$$\mu(A) := T''(\chi_A)$$

which is an E-valued measure.

By the continuity of T'' (with respect to the norm in $C^b(X)''$) we can deduce that

$$T''(x) = \int x \, d\mu$$

for every bounded, Borel-measurable function on S and so in particular, for $x \in C^b(X)$. The converse fact is easy.

Using the above result, we can now easily obtain a result for general operators T with values in a locally convex space.

15. <u>Proposition</u>: Let E be a locally convex space, X a completely regular space. Then any continuous, linear operator $T : C^b(X) \to E$ may be represented by integration with respect to a Radon measure μ from Bo(X) into $(E'',\sigma(E'',E'))$. In fact, μ takes its values in the $\sigma(E'',E')$-closure of $T(B(C^b(X)))$.

If T maps the unit ball of $C^b(X)$ into a relatively weakly compact subset of E, then μ takes its values in E (actually in $T(B(C^b(X)))$ and is a Radon measure with respect to the original topology in E.

<u>Proof</u>: For the first assertion, let B be the $\sigma(E'',E')$-closure of $T(B(C^b(X)))$ in E" and let F be the Saks space spanned by B in E" with $\| \|_B$ as norm and $\sigma(E'',E')$ as auxiliary topology. Then this is a Saks space with compact unit ball and the result follows immediately from 14.

In the second case, take $B := \overline{T(B(C^b(X)))}$, the closure now being taken in E, and define the Saks space F to be $(E_B, \| \|_B, \sigma(E,E'))$. Then T is represented by an F-valued Radon measure μ (i.e. Radon with respect to $\sigma(E,E')$). Now the Orlicz-Pettis theorem for Radon measures implies that μ is a Radon measure with respect to the finest topology on F compatible with the duality between (F,F'_γ) and in particular with the respect to the topology of E.

16. <u>Remark</u>: We note that if $\mu \in M_R(X;E)$ is such that its range is contained in a weakly compact, absolutely convex set, then the associated integration operator $T : C^b(X) \to E$ sends the $C^b(X)$-ball into this same set and T is weakly continuous and so continuous if $(C^b(X),\beta)$ is a Mackey space (e.g. if S is locally compact, paracompact).

17. <u>Proposition</u>: Let (E,τ) be a quasi-complete locally convex space, $T : C^b(X) \to E$ a β-continuous linear operator. If T does not map the unit ball of $C^b(X)$ into a relatively weakly compact subset of E there is a sequence (x_n) of functions in $C^b(X)$ with mutually disjoint supports so that if j is the mapping $(\lambda_n) \to \Sigma \lambda_n . x_n$ from c_o into $C^b(X)$ and A denotes the β-closed span of $\{x_n\}$ in $C^b(X)$ then in the following diagram

$$\begin{array}{ccccc} & & C^b(X) & \xrightarrow{T} & E \\ & & \uparrow & & \uparrow \\ c_o & \xrightarrow{j} & A & \xrightarrow{T|A} & T(A) \end{array}$$

j and $T|_A$ are isomorphisms. More informally, T fixes a subspace of $(C^b(X),\beta)$ which is isomorphic to c_o.

Consequently, if E fails to contain a copy of c_o then every continuous linear operator $T : C^b(X) \to E$ takes the unit ball into a relatively weakly compact subset of E.

Proof: If T fails to satisfy the given condition, then by a standard result, $T' : E' \to M_R(X)$ takes some equicontinuous set H in E' to a subset of $M_R(X)$ which is bounded but not relatively $\sigma(M(X),M(X)')$-compact. Then by GROTHENDIECK's characterisation of weakly compact sets in $M_R(X)$ (cf. BUCHWALTER and BUCCHIONI [3], p.76) there exists a sequence (f_n) in H and a sequence (U_n) of disjoint open sets in S and an $\varepsilon > 0$ so that $|T'(f_n)(U_n)| > \varepsilon$ $(n \in \mathbb{N})$. i.e. $|f_n \circ \mu(U_n)| > \varepsilon$ (where μ represents T). By ROSENTHAL's Lemma (cf. DIESTEL-UHL [2], I.4.1) we may suppose that

$$|f_n \circ \mu| \, (\bigcup_{m \neq n} U_m) < \varepsilon/2 \qquad (n \in \mathbb{N}).$$

Now choose a sequence (x_n) in $C^b(X)$ so that $|x_n| \leq \chi U_n$ and

$$|f_n \circ T(x_n)| = |\int_S x_n d(f_n \circ \mu)| > \varepsilon$$

which is possible since $f_n \circ \mu$ is a Radon measure. Then j, as defined in the statement of the theorem, is clearly a well-defined, continuous injection. We claim that

$$\|T \circ j((\lambda_n))\|_H \geq \varepsilon/2 \, \|(\lambda_n)\|_{c_o}$$

for each $(\lambda_n) \in c_o$, where $\| \; \|_H$ denotes the seminorm of convergence on the equicontinuous set H. Indeed for any $(\lambda_n) \in c_o$ and any $k \in \mathbb{N}$,

$$\begin{aligned}
\|T \circ j((\lambda_n))\|_H &\geq |\langle T \circ j((\lambda_n)), f_k \rangle| = |\int_S \sum_{n \in \mathbb{N}} \lambda_n x_n d(f_k \circ \mu)| \\
&\geq |\int_S \lambda_k x_k d(f_k \circ \mu)| - \|(\lambda_n)\|_{c_o} \, |f_k \circ \mu| (\bigcup_{l \neq k} U_l) \\
&\geq |\lambda_k| \varepsilon - \|(\lambda_n)\|_{c_o} \, . \varepsilon/2
\end{aligned}$$

Taking the supremum over k on the right hand side we get the required estimate.

This shows that $(T \circ j)^{-1}$ is well-defined and continuous on $T \circ j(c_o)$, from which it follows that j, as an operator from c_o to $j(c_o)$ and T, as an operator from $j(c_o)$ to $T \circ j(c_o)$ are isomorphisms (by the following trivial Lemma).

18. <u>Lemma</u>: Let X,Y,Z be topological spaces, $f : X \to Y$ and $g : Y \to Z$ continuous, surjective mappings such that $g \circ f$ is an isomorphism. Then f and g are also isomorphisms.

<u>Proof</u>: The injectivity of $g \circ f$ implies that of g and f so that f^{-1} and g^{-1} are well-defined. But as $g^{-1} = f \circ (g \circ f)^{-1}$ and $f^{-1} = (g \circ f)^{-1} \circ g$ it is clear that f and g are continuous.

With these results it is now easy to prove the following Proposition:

19. <u>Proposition</u>: Let $(E, \| \|, \tau)$ be a complete Saks space, $T : C^b(X) \to E$ a β-γ-continuous linear operator with representing measure $\mu \in M_R(X;E'')$. Then the following are equivalent:

1) T does not fix a copy of c_o;
2) T maps the unit ball of $C^b(X)$ into a relatively weakly compact set;
3) T maps weakly summable sequences to summable sequences;
4) T maps weakly Cauchy sequences into convergent sequences;
5) T maps sequences which tend weakly to zero convergent sequences;
6) if (x_n) is a bounded sequence of functions in $C^b(X)$ with mutually dusjoint supports, then $Tx_n \to 0$ in E;
7) T maps weakly compact sets in $C^b(X)$ into compact sets;
8 μ takes its values in E;
9) μ is a Radon measure with values in E;
10) μ is a strongly additive measure with values in E.

<u>Proof</u>: 1) $\Longleftrightarrow$ 2) is Proposition 17
2) $\Longrightarrow$ 9) is Proposition 15
9) $\Longrightarrow$ 2) - 6) are all simple applications of the Lebesgue-dominated convergence theorem. The reverse implications all follow from the fact that if μ is not Radon then the (x_n) constructed in the proof of Proposition 17 supply counter-examples.
9) $\Longrightarrow$ 8) is clear and 8) $\Longrightarrow$ 10) follows from the weak σ-additivity of μ and ORLICZ-PETTIS. 10) implies 2) is embedded in the proof of Proposition 17.
9) $\Longrightarrow$ 7) Note that $(C^b(X),\beta)$ is not necessarily complete so that we cannot use EBERLEIN-SMULIAN. Firstly, we may assume that E is a Banach space (using the characterisation of compactness in Saks spaces) (cf. COOPER [1], I.1.12). Then $\mu : Bo(X) \to E$ has tight semi-variation

norm so that for $\varepsilon > 0$ we may find $K \in \mathcal{K}(X)$ so that if T_K denotes the operator associated to $\mu|_K$ then $\|T - T_K\| < \varepsilon$. Now if B is weakly compact in $C^b(X)$, then $T_K(B)$ is compact in $(E, \| \|)$ (using factorisation through C(K) and EBERLEIN-SMULIAN) and we may conclude that T(B) is compact in $(E, \| \|)$.

REFERENCES

[1] COOPER, J.B., Saks spaces and applications to func tional analysis (Amsterdam 1978).

[2] DIESTEL, J. and UHL, J.J.Jr., The theory of vector measures (Providence 1977).

[3] BUCHWALTER, H., and BUCCHIONI, D., Intégration vectorielle et théorème de Radon-Nikodym (Département de Mathematiques, Lyon 1975).

Institut für Mathematik
Universität Linz
A-4045 LINZ-Auhof
Austria

A QUESTION IN CATEGORICAL SHAPE THEORY:

WHEN IS A SHAPE-INVARIANT FUNCTOR A KAN EXTENSION?

by

Armin Frei and Heinrich Kleisli *

The notion of shape was first introduced by Borsuk in the study of homotopy properties of compacta, later extended by Mardešić and Segal to compact Hausdorff spaces. Following LeVan, Deleanu and Hilton and the first named author investigated a categorical formulation of shape theory. In particular, they introduced the notion of a shape-invariant functor and showed that a right Kan extension of a functor along K is shape-invariant for the shape of K.

Our principal object is to study the converse question: When is a shape(coshape)-invariant functor a right(left) Kan extension? In order to widen the range of application we do this in the context of enriched categories for an enriched functor K.

As an illustration of the applicability of the general theorems we shall consider the following algebraic situation. Let T be an associative P-algebra with unit element. The natural ring homomorphism $K : P \to T$ can be interpreted as an additive functor of single object additive categories. Likewise a P-module N can be regarded as an additive functor $N : P \to Ab$, and the left Kan extension of N along K is given by the tensorproduct $T \otimes_P N$. A T-module M is coshape-invariant if its T-module structure can be extended to a module structure over the endomorphism ring $S = \mathrm{End}_P\, T$.

In this particular situation our question can be restated: When is a coshape-invariant T-module extended from P? The following answers are obtained. If the P-algebra T is finitely generated and projective over P, then every coshape-invariant T-module is extended from P (Theorem 2.2). If T is the polynomial algebra P [X], a necessary and sufficient condition for a coshape-invariant P [X] -module to be extended from P is given (Theorem 2.4).

---

* Partially supported by the NRC of Canada

1. A sufficient condition for a shape-invariant functor to be a Kan extension

We shall establish some general theorems in the context of enriched categories. We do this in order to widen the range of application, but also to free the proofs of unnecessary restrictions and to take advantage of the duality principle.

Let V be a (fixed) symmetric monoidal closed category, abbreviated closed category (see [D], introduction). All categories, functors, etc., in this section are to be regarded as V-categories, V-functors, etc.. All Kan extensions are supposed to be pointwise Kan extensions, that is, given by their Kan formula (see [D], theorem I.4.3, formula (1)).

Definition 1.1. Let $K : P \to T$ be a functor, P a small category and V a complete closed category. Then the shape of K is the category S_K whose objects are the objects of T and whose V-objects $S_K(X,Y)$ are given by

$$S_K(X,Y) = \int_P V(T(Y,KP),T(X,KP)).$$

It is not difficult to check that the objects of T and the above end between them form a V-category, which is also known by the name clone of operations of K (see [D], chapter II, section 3).

The identity map between the objects of T and S_K can be made into a V-functor $D_K : T \to S_K$ by lifting the family of maps

$$T(-,KP) : T(X,Y) \to V(T(Y,KP),T(X,KP))$$

into the end $S_K(X,Y)$.

Definition 1.2. In the situation of definition 1.1, we say that a functor $F : T \to A$ is shape-invariant (with respect to K) if it factors through D_K.

Theorem 1.3. (Deleanu-Hilton [D-H], Frei [F]). Let $K : P \to T$ be a functor, P a small category and A a complete category. For any functor $H : P \to A$, the right Kan extension $F = \mathrm{Ran}_K H$ is shape-invariant.

Proof Consider the following diagram

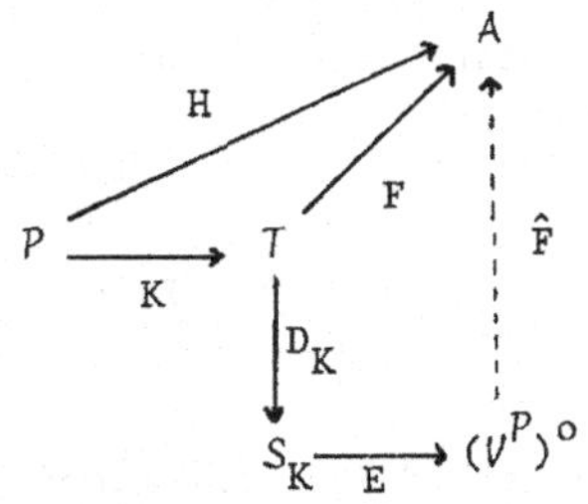

where E is the embedding functor of the shape S_K into the category $(V^P)^o$. We shall construct a functor $\hat{F} : (V^P)^o \to A$ such that $\hat{F}ED_K = F$. By hypothesis, $F = \mathrm{Ran}_K H$, that is,

$$F\cdot = \int_P \overline{A}(T(\cdot,KP),HP).$$

Hence, we define the functor $\hat{F}$ by setting

$$\hat{F}- = \int_P \overline{A}(-P,HP).$$

The existence of this end is guaranteed by the smallness of P and the completeness of A , and we have

$$\hat{F}ED_K\cdot = \int_P \overline{A}(T(\cdot,KP),HP) = F. \ \square$$

Theorem 1.4. Assume the base category V to be complete and cocomplete. Let $K : P \to T$ be a functor, P a small category and A a complete category. Let $\overline{F} : T \to A$ be a shape-invariant functor, that is a functor of the form $F = \overline{F}D_K$ for some functor $\overline{F} : S_K \to A$. If $\overline{F}$ can be extended along the embedding $E : S_K \to (V^P)^o$ to a small continuous functor $F : (V^P)^o \to A$, then F is a right Kan extension along K.

Proof. Since V is cocomplete so is V^P, hence $(V^P)^o$ is complete. Let $L : P \to (V^P)^o$ denote the left Yoneda functor. Then by the V-Yoneda lemma (D, Prop. I.5.2), for each object H in $(V^P)^o$ and G in V^P one has:

$$(V^P)^o(H,\int_P \overline{(V^P)^o}(GP,LP)) \cong \int_P V(GP,(V^P)^o(H,LP)) \cong \int_P V(GP,HP) = (V^P)^o(H,G),$$ whence

$\overline{\int_P (V^P)^o(GP,LP)} \cong G$. Now $\hat{F}$ is continuous, so that $\hat{F}- = \hat{F}\overline{\int_P (V^P)^o(-P,LP)} =$
$= \int_P \overline{A}(-P,\hat{F}LP)$, and therefore

$$F = \hat{F}ED_K\cdot = \int_P \overline{A}(T(\cdot,KP),\hat{F}LP) = \mathrm{Ran}_K\hat{F}L \quad \square$$

Since $E : S_K \to (V^P)^o$ is an embedding, the right Kan extension $\mathrm{Ran}_E\overline{F}$ is an ordinary extension of $\overline{F}$ along E. If T and therefor S_K are small, that Kan extension exists and can be computed by means of the Kan formula.

$$\mathrm{Ran}_E\overline{F}- = \int_S \overline{A}(V^P(ES,-)\overline{F}S).$$

This yields the following corollary which is more manageable than theorem 1.4..

Corollary 1.5. In the situation of theorem 1.4, let T be a small category. If the functor $\mathrm{Ran}_E\overline{F}$ is small continuous, then F is a right Kan extension along K $\square$

Replacing the categories P, T, S_K and A used above by their opposite categories P^o, T^o, S_K^o and A^o, we obtain the following "dual" definitions and theorems.

Definition 1.1^o. Let $K : P \to T$ be a functor, P a small category and V a complete closed category. Then the coshape of K is the category ${}_KS$ whose objects are the objects of T and whose V-objects ${}_KS(X,Y)$ are given by

$${}_KS(X,Y) = \int_P V(T(KP,X),T(KP,Y)).$$

The identity map between the objects of T and ${}_KS$ can be made into a V-functor ${}_KD : T \to {}_KS$ by lifting the family of maps

$$T(KP,-) = T(X,Y) \to V(T(KP,X),T(KP,Y))$$

into the end ${}_KS(X,Y)$.

Definition 1.2^o. In the situation of definition 1.1^o, we say that a functor $F : T \to A$ is coshape-invariant (with respect to K) if it factors through ${}_KD$.

Theorem 1.3^o. Let $K : P \to T$ be a functor, P a small category and A a cocomplete category. For any functor $H : P \to A$, the left Kan extension $F = \mathrm{Lan}_K H$ is coshape-invariant $\square$

Theorem 1.4°. Assume the base category V to be complete and cocomplete. Let $K : P \to T$ be a functor, P a small category and A a cocomplete category. Let $F : T \to A$ be a coshape-invariant functor, that is, a functor of the form $F = \overline{F}{}_K D$ for some functor $\overline{F} : {}_K S \to A$. If $\overline{F}$ can be extended along the embedding $E : {}_K S \to V^{P^o}$ to a small cocontinuous functor $\check{F} = V^{P^o} \to A$, then F is a left Kan extension along K □

Corollary 1.5°. In the situation of theorem 1.4° let T be a small category. If the functor

$$\mathrm{Lan}_E\overline{F}- = \int^S V^{P^o}(ES,-)\otimes_A \overline{F}S$$

is small cocontinuous, then F is a left Kan extension along K □

2. Application to coshape-invariant modules.

We shall apply theorem 1.4° and its corollary 1.5° to the following special situation. As base category V we choose the category Ab of abelian groups, that means, all categories and functors are to be regarded as additive categories and additive functors. Furthermore, let P and T be rings with unit elements, and $K : P \to T$ a (unitary) ring homomorphism. We consider P and T as (additive) single object categories and $K : P \to T$ as an (additive) functor.

Then the coshape of K is again an (additive) single object category, given by the endomorphism ring

$$S = \mathrm{End}_{P^o} T$$

of T considered as a right P-module. The ring homomorphism $D : T \to S$ which associates to each element t of T the left multiplication $x \to tx$ in T yields the corresponding functor ${}_K D$ (see definition 1.2°).

A (left) T-module M can be considered as an (additive) functor $M : T \to Ab$. Hence the following definition suggests itself.

Definition 2.1. We say that a T-module $M : T \to Ab$ is coshape-invariant (with respect to $K : P \to T$) if it factors through $D : T \to S$; in other words, if the T-module M admits an S-module structure extending its T-module structure.

A T-module $M : T \to Ab$ is a left Kan extension along K of a P-module $N : P \to Ab$ if it is isomorphic to the T-module $T\otimes_P N$; in other words, if it is extended from P.

Theorem 1.3^o expresses therefore the obvious fact that a T-module of the form $T\otimes_P N$ is coshape-invariant, that is, admits an S-module structure given by

$$s(t\otimes n) = s(t)\otimes n \quad \text{for all} \quad s \in S, \quad t \in T \quad \text{and} \quad n \in N.$$

Theorem 1.4^o and its corollary 1.5^o give the following answers to the question: When is a coshape-invariant T-module extended from P?

Theorem 2.2. Let $K : P \to T$ be a ring homomorphism. If T considered as a right P-module (via K) is finitely generated and projective, then every coshape-invariant T-module M is extended from P.

Proof. We compute the left Kan extension $\mathrm{Lan}_E M$ of the S-module M along the embedding $E : S \to P^o\text{-Mod}$ and apply corollary 1.5^o.

Clearly E is the functor which associates to the single object of S the right P-module T. The Kan formula for $\mathrm{Lan}_E M$ (see corollary 1.5^o) yields

$$\mathrm{Lan}_E M- = \mathrm{Hom}_{P^o}(T,-)\otimes_S M.$$

By assumption T is finitely generated and projective over P^o so that the functor $\mathrm{Hom}_{P^o}(T,-) : P^o\text{-Mod} \to S\text{-Mod}$ and therefore also the functor $\mathrm{Hom}_{P^o}(T,-)\otimes_S M : P^o\text{-Mod} \to \mathcal{Ab}$ are cocontinuous. By corollary 1.5^o the T-module M is a left Kan extension along K, that is, extended from P $\square$

Remark 2.3. Theorem 2.2. can be obtained in a simpler way by using the Morita equivalence $T\otimes_P - : P\text{-Mod} \to S\text{-Mod}$ (see, for instance, [P], section 4.11). On the other hand, the theorem permits a short proof of that special Morita equivalence, at least under the additional assumption that $K : P \to T$ is split monic. Indeed, it only remains to verify that the functor $T\otimes_P-$ is full and faithful, which is not difficult.

For a ring T which is not finitely generated and projective over P^o, corollary 1.5^o ceases to be of help. However, theorem 1.4^o can be exploited to give an answer to our question in the case where T is the polynomial algebra P[X] over a commutative ring P.

Theorem 2.4. Let P be a commutative ring, T the polynomial algebra P[X] and $K : P \to T$ the natural embedding of P into T. A necessary and sufficient condition for a coshape-invariant T-module M to be extended from P is given by

$$S_o M = M,$$

where S_o is the left ideal of the endomorphism ring $S = \mathrm{End}_P T$ consisting of the endomorphisms s for which $s(X^n) = 0$ for almost all exponents n.

Proof. The necessity of the condition $S_oM = M$ for M to be of the form $T\otimes_P M$ is obvious.

In order to show that the condition is also sufficient, we consider the diagram of P-homomorphisms

(D) $$T_o \leftarrow \dots \leftarrow T_n \xleftarrow{r_n} T_{n+1} \leftarrow \dots ,$$

where

$$T_n = \{t \in T; \quad \deg t \le n\},$$

and

$$r_n\left(\sum_{k=0}^{n+1} \alpha_k X^k\right) = \sum_{k=0}^{n} \alpha_k X^k \quad \text{for every polynomial} \quad \sum_{k=0}^{n+1} \alpha_k X^k \text{ in } T_{n+1}.$$

For a P-module N, consider the induced diagram

(D^*N) $$\mathrm{Hom}_P(T_o,N) \to \dots \to \mathrm{Hom}_P(T_n,N) \xrightarrow{r_n^*} \mathrm{Hom}_P(T_{n+1},N) \to \dots$$

and form the colimit

$$\mathrm{Hom}_P(T,N)_o = \mathrm{colim}\, D^*N,$$

which can be made into a functor $\mathrm{Hom}_P(T,-)_o : P\text{-Mod} \to P\text{-Mod}$. Unfortunately, the P-modules $\mathrm{Hom}_P(T,N)$ do not admit a right S-module structure. However, they have a natural right module structure over the subring S_1 of S whose elements are the endomorphisms s for which there exists a λ in P such that $s(X^n) = \lambda X^n$ for almost all exponents n. Indeed, we may identify the P-modules $\mathrm{Hom}_P(T_n,N)$ with submodules of $\mathrm{Hom}_P(T,N)$ so that the colimit $\mathrm{Hom}_P(T,N)_o$ can be regarded as the union of all those submodules. Now it is easily seen that this union is stable under the right action of S_1 on $\mathrm{Hom}_P(T,N)$ given by

$$(fs)t = f(s(t)) \text{ for all } s \in S_1, \quad f \in \mathrm{Hom}_P(T,N) \quad \text{and} \quad t \in T.$$

We may therefore form the tensorproduct $\mathrm{Hom}_P(T,N)\otimes_S M$ and thus obtain a functor

$$\check{F}- = \mathrm{Hom}_P(T,-)_o \otimes_{S_1} M : P\text{-Mod} \to Ab$$

which by construction is small cocontinuous. Moreover,

$$\check{F}E- = Hom_P(T,T)_o \otimes_{S_1} M = S_o \otimes_{S_1} M \cong S_o M.$$

Indeed, the colimit $Hom_P(T,T)_o$ can be identified with the left ideal S_o of S, and it is not difficult to see that the S-homomorphism $i \otimes id : S_o \otimes_{S_1} M \to S_1 \otimes_{S_1} M$ induced by the embedding i of S_o into S_1 is a monomorphism. In other words, the functor $\check{F}$ is an extension along E of each S-module M satisfying the condition $S_o M = M$. By theorem 1.4^o, such a T-module M is a left Kan extension along K, that is, extended from P ▯

Remark 2.5. Looking at the proof of theorem 1.4^o we can give explicitly a P-module N and a T-isomorphism $R : T \otimes_P N \to S_o M$, namely by setting $N = S_o^o M$, where S_o^o is the P-submodule of S_o consisting of the endomorphisms s in S_o for which $\deg s(X^n) \leq 0$, and by defining h as the P-homomorphism determined by

$$h(t \otimes n) = tn \quad \text{for all} \quad t \in T \quad \text{and} \quad n \in N = S_o^o M.$$

The verification that h is a T-isomorphism can then be carried out without referring to the categorical results.

References

[D-H] A. Deleanu and P. Hilton, On the categorical shape of a functor, Fund. Math. 97 (1977), 157-176.

[D] E. Dubuc, Kan extensions in enriched category theory, Lecture Notes in Mathematics 145, Springer Verlag, Berlin-Heidelberg 1970.

[F] A. Frei, On categorical shape theory, Cahiers Top. et Geom. Diff., XVII-3 (1976), 261-294.

[P] B. Pareigis, Kategorien und Funktoren, Mathematische Leitfäden, BG. Teubner, Stuttgart 1969.

University of British Columbia
Vancouver, B.C.

Université de Fribourg
Fribourg, Switzerland

The finest functor preserving the Baire sets

by Zdeněk Frolík

Let Ba_- be the finest functor of uniform spaces into itself which preserves the Baire sets, and let M be the coreflection of uniform spaces onto measurable uniform spaces (= hereditarily metric-fine spaces, or equivalently: σ-discrete completely Baire-additive covers are uniform). The aim of my talk is to explain why $Ba_- = M$ under a strong set-theoretic assumption (each completely Baire-additive partition of a metric space is σ-discretely decomposable), see Theorem 1 below, and $Ba_- \neq M$ under some set-theoretical assumptions (e.g. MA + $\neg$ CH), see Theorem 2 below.

In § 1 we discuss several general categorial notions to explain the concept of the minus functor. In § 2 we recall the basic properties of cozero sets and Baire sets in uniform spaces. The main results are contained in § 3.

Notation. U is the category of uniform spaces, so U(X,Y) is the set of all uniformly continuous maps from X into Y; we write simply U(X) for U(X,R), R the space of reals, and $U_b(X)$ is the set of all bounded functions in U(X). The cardinal reflections are denoted by p^α : p^α is the reflection on the spaces which admit a basis for uniform covers consisting of covers of cardinal less than $\aleph_\alpha$. We shall need just the cases $\alpha = 0$, i.e. the precompact reflection, and $\alpha = 1$ (J. Isbell writes e for p^1).

Convention. If we speak about a set X (e.g. a cardinal) as a uniform space we have in mind the finest uniformity on X (i.e. $\{(x) \mid x \in X\}$ is a uniform cover). So $p^1\omega_1$ is the uniform space $\langle X, \mathcal{V} \rangle$ where $X = \omega_1$, and all the countable covers of ω_1 form a basis for $\mathcal{V}$.

§ 1. The minus and plus functors.

By a category we mean a concrete category, and by a functor we mean a concrete functor (covaciant), i.e. a functor preserving the underlying sets. An object X is finer than Y in a category $\mathcal{K}$ if the underlying sets coincide, and if the identity mapping from X onto Y is a morphism in $\mathcal{K}$. This order induces an order on functors K of $\mathcal{K}$ into itself; F is finer than G if FX is finer than GX for each object X. A functor is said to be negative or positive if it is finer or coarser, resp., than the identity functor. If $\mathcal{F}$ is a class of functors, we denote by $\mathcal{F}_-$ the class of all negative functors in $\mathcal{F}$, and similarly for $\mathcal{F}_+$.

Now let Φ be a functor of U into a category $\mathcal{L}$ such that the object-map is onto. Let

$$\mathrm{inv}\,(\Phi) = \{F \mid F:U \to \mathcal{L}\,,\quad \Phi \circ F = \Phi\}$$

A cross-section of Φ is a functor $\Psi : \mathcal{L} \to U$ such that $\Phi \circ \Psi$ is an identity. The class of all $\Psi \circ \Phi$, Ψ being a cross-section of Φ, is denoted by cross (Φ). Clearly

$$\mathrm{cross}\,(\Phi) \subset \mathrm{inv}\,(\Phi),$$

and

$$\mathrm{cross}\,(\Phi) = \{F \in \mathrm{inv}\,(\Phi) \mid U(X,Y) = \mathcal{L}(X,Y)\}.$$

It is easy to check that $F = \Psi \circ \Phi \in \mathrm{cross}_-(\Phi)$ iff

$$\forall X\ \forall Y : U(\Psi K,Y) = \mathcal{L}(K,\Phi Y),$$

iff each ΨK is Φ-fine. Recall that X is Φ-fine means that $U(X,Y) = \mathcal{L}(\Phi X,\Phi Y)$ for each Y, and since the class of Φ-fine objects is closed under inductive generation (a formation of final structures), it is necessarily coreflective; the coreflection is denoted by Φ_f. Thus either $\mathrm{cross}_-(\Phi) = \emptyset$ or $\mathrm{cross}_-(\Phi) = (\Phi_f)$, and the latter case appears iff Φ has the left-adjoint Ψ (and then $\Phi_f = \Psi \circ \Phi$). Of course similar results hold for + and Φ_c

where Φ_c is the reflection in Φ-coarse objects.

Definition. If there exists the finest functor in $inv_-(\Phi)$, it is denoted by Φ_- , and it is called the minus-functor of Φ . Similarly we define Φ_+.

Thus Φ_- is the finest negative functor preserving Φ . In general $inv_-(\Phi)$ is fine-directed (and $inv_+(\Phi)$ is coarse-directed).

It is clear that always Φ_f is finer than Φ_-, and $\Phi_f = \Phi_-$ iff $\Phi_f \in inv(\Phi)$, and this is the case just when Φ has a left adjoint.

Warning. If, for some X, $\Phi\Phi_f X = \Phi X$, then $\Phi_f X = \Phi_- X$ does not hold, in general.

Examples. (a) Sets. Let set is the forgetful functor of U onto the category of sets. Then $set_f = set_-$ is the coreflection on uniformly discrete spaces, $set_c = set_+$ is the reflection on uniformly accreet (= indiscrete) spaces.

(b) Topology. Let t be the functor of U onto uniformizable topological spaces. Then $t_f = t_-$, $t_+ = p^o$, $t_c = set_c$.

(c) Proximity. Let p be the functor of U onto proximity spaces. Then $p_f \neq$ identity, p_- is the identity, $p_+ = p_c = p^o$.

There are many interesting examples of Φ "from life" such that $cross_-(\Phi) = o$ and $cross_+(\Phi) = 0$, hence we have no adjoint situation, but cross (Φ), inv (Φ), $inv_-(\Phi)$ and $inv_+(\Phi)$ are rich. This is the case of several functors in descriotive theory of sets on uniform spaces, e.g. coz and Ba which are associated with the cozero sets and the Baire sets on uniform spaces.

For details see the seminar notes "Seminar Uniform Spaces", e.g. [F 4].

§ 2. Cozero sets and Baire sets. The zero sets in a space X are the sets of the form $\{x \mid fx = 0\}$ with f in U(X) (equivalently, in $U_b(x)$). The complements of zero sets are called the cozero sets. We denote by Z(X) or COZ(X) the collection of all zero sets or cozero sets respectively. Denote by Ba(X) the smallest σ-algebra containing coz (X) (equivalently, z(X)).

Like in the topological case (i.e. $t_f X = X$), both coz (X) and z(X) are closed under finite intersections and finite unions, the former under countable unions, and the latter under countable intersections. Also we have the usual Baire classification

$$Ba(X) = \bigcup \{coz_\alpha(X) \mid \alpha < \omega_1\} = \bigcup \{z_\alpha(X) \mid \alpha < \omega_1\}$$

where $coz_0(X) = coz(X)$, $z_0(X) = z(X)$, and $coz_\alpha(X)$ and $z_\alpha(X)$ are defined by induction just by iterating the operations of taking all countable unions or countable intersections.

It should be noted that neither of the considered collections of sets is closed under the formation of discrete unions in general (consider uncountable products like $\omega_1^{\omega_1}$), However, if $X = t_f X$ (i.e., in the topological case) each of the collections $coz_\alpha(X)$ and $z_\alpha(X)$ is closed under the formation of discrete unions. Of course, even in the topological case Ba is not closed under discrete unions in general (consider a subset $\sum B_\alpha$ in $\sum \{I \mid \alpha < \omega_1\}$ where B_α is of class at least α in I). A collection $\mathcal{M}$ of sets is said to be of bounded Baire-class if $\mathcal{M} \subset coz_\alpha(X)$ for some $\alpha < \omega_1$. Now we may say that Ba (X) is closed under taking the unions of discrete collections of bounded Baire class. In § 3 we shall need the following result of Preiss [P]:

Lemma 1. In a metric space X every disjoint completely Ba (X)-additive family is of bounded Baire class.

Of course, completely Ba (X)-additive means that each subunion is a Baire set in X.

For convenience (following P. Meyer) let a paved space be a air $\langle X, \mathcal{Q} \rangle$ such that $\mathcal{Q} \subset \exp X$; the collection $\mathcal{Q}$ is called he pavement and its elements are called the stones. The catego- y of the paved spaces has the paved space for its objects; and he morphisms are the "measurable maps", i.e. the maps of the nderlying sets such that the preimages of the stones are the sto- es. If we assign to each uniform space X the paved space $\langle |X|, \operatorname{coz}(X) \rangle$ we obtain a functor which will be denoted by oz. Similarly, we define the functor Ba. Note that since now on a (X) may mean the collection of all Baire sets in X or the un- erlying set of X endowed with the σ-algebra Ba (X).

It is easy to see that the two functors preserve subspaces since the unit interval is injective) and do not preserve pro- ucts (e.g. $\operatorname{coz}\,\omega_1 \times \operatorname{coz}\,\omega_1 = \operatorname{coz}(p^{\circ}\omega_1 \times p^{\circ}\omega_2) \neq \operatorname{coz}(\omega_1 \times \omega_1)$ nd case of Ba will be considered in § 3). This implies that $\operatorname{oz}_f \neq \operatorname{coz}_-$, $\operatorname{Ba}_f \neq \operatorname{Ba}_-$. On the other hand, the two functors preser- e projective limits (directed!). This follows from the following imple

Lemma 2. Let $\{\mathcal{U}_a\}$ be a fine-directed family of uniformi- ies on a set X, and let $\mathcal{U} = \bigcup\{\mathcal{U}_a\}$. Then $\langle X, \mathcal{U} \rangle = \varprojlim \langle X, \mathcal{U}_a \rangle$ in U, and

(a) $\operatorname{coz}\langle X, \mathcal{U} \rangle = (\bigcup\{\operatorname{coz}\langle X, \mathcal{U}_a \rangle\})_\sigma$

(b) $\operatorname{Ba}\langle X, \mathcal{U} \rangle$ is generated in σ-algebras by $\bigcup\{\operatorname{coz}\langle X, \mathcal{U}_a \rangle\}$

Corollary. Let $\{f_a : X_a \to Y_a \mid a \in A\}$, and for $B \subset A$ let

$$f_B : \prod\{X_a \mid a \in B\} \to \prod\{Y_a \mid a \in B\}$$

here $f_B\{x_a \mid a \in B\} = \{f_a x_a \mid a \in B\}$. Then f is Baire measurable f for each finite $B \subset A$ the map f_B has the property that the preimages of the cozero sets are the Baire sets.

Since $\mathrm{inv}_{-}(\Phi)$ is always fine-directed we obtain from Lemma 2 that both coz_{-} and Ba_{-} exist. The properties of coz are quite clear ([F5],[F6]) :

coz_{-} = metric-fine $\ast$ coz_{f} , and $\mathrm{coz}_{-}X$

is projectively generated by all $f:X \to Y$ such that $f \times f \in \mathrm{coz}\,(\mathrm{coz}\,(X \times X),\ \mathrm{coz}\,(Y \times Y))$, and σ-discrete completely coz(X)-additive covers of X form a basis for the uniform covers of $\mathrm{coz}_{-}\,X$.

Remark. It is proved in [F5]:

$\mathrm{coz}_{+} = \mathrm{Ba}_{+} = p^{o}$, $\mathrm{coz}_{c} = \mathrm{Ba}_{c} = \mathrm{set}_{c}$.

In the concluding section we consider Ba_{-}.

§ 3. Evaluation of Ba_{-}.

Several functors in $\mathrm{inv}_{-}(\mathrm{Ba})$ have been studied.

For example the coarsest one among all $F \in \mathrm{inv}_{-}\,\mathrm{Ba}$ such that coz (FX) = Ba (X). This is a coreflection, and its value at X is obtained by adding all finite partitions of X by Baire sets to the uniform covers. Recently J. Vilímovský and G. Tashjian described the corresponding coreflective class of spaces as the largest one among the coreflective classes which do not contain the unit interval of reals.

We obtain a finer coreflection in inv_{-} Ba if we add all countable partitions by Baire sets (this class was studied, e.g., by A. Hager for the case of separable X).

We obtain yet finer coreflection in $\mathrm{inv}_{-}(\mathrm{Ba})$ if we add all σ-discrete completely Ba-additive partitions. Since this is a possible candidate for Ba_{-} (compare with the description of coz_{-}

n § 2), we should describe it more explicitly. In [F 1], details n [F 3], a (uniform!) space is called measurable if U(X,M), M etric, is closed under taking the pointwise limits of sequences, nd it is shown that X is measuable iff the σ-discrete complete-y Ba (X)-additive covers are uniform: Moreover, these covers form basis for the uniform covers of the coreflection M of X in mea-urable spaces.

Remark. Measurable spaces are just the hereditarily metric-ine (=coz_-) spaces, and in this connection were also studied by . Rice [R].

Since M $\in$ inv_- (Ba), M $\circ$ $Ba_- \in inv_-$ (Ba), and hence the values a_- X are measurable spaces. So Ba_- X has a basis consisting of artitions.

To show that Ba_- X = MX, it is enough to show that every uni-orm partition of Ba_- X has a refinement which is a σ-discrete ompletely Ba (X)-additive cover of X. So it would be useful to ave a description of partitions which are uniform covers of MX they need not be σ-discrete).

Definition. A family $\{X_a \mid a \in A\}$ in a space X is called σ-discretely decomposable (abb. σ-d.d.) if there exists a fami-y $\{X_{an} \mid a \in A,\ n \in \omega\}$ such that

$$X_a = \bigcup \{X_{an} \mid n \in \omega\},$$

nd each family $\{X_{an} \mid a \in A\}$ is discrete. (This notion is due to . Hansell [H1] for the case of metric spaces.)

Lemma 3. If X is metric them every σ-d.d. completely Ba (X)-dditive partition of X is a uniform cover of MX.

Proof. Use Lemma 1 and invariance of Baire classes under iscrete unions to show that X_{an} can be chosen so that $\{X_{an}\}$ is

σ-d.d. and completely Ba (X)-additive, i.e. a uniform cover of MX.

Consider the following statement which will be denoted by (P):

If X is metric then

(*) each completely Ba (X)-additive family is σ-d.d. (and hence uniform on MX).

Theorem 1. Under (P), $Ba_- = M$.

Remark. (*) is true for each complete metric space X by a non-trivial result of Hansell [H1]. On the other hand, Q-set $X \subset R$ does not satisfy (*). Recall that a Q-set is an uncountable subspace X of R (the reals) such that each subset of X is a G_σ in X (i.e. coz_1 (X)), and hence $\{(x) \mid x \in X\}$ is completely Ba (X)-additive. Since X is separable, every σ-discrete family is countable, and so $\{(x) \mid x \in X\}$ is not σ-d.d.

It should be noted that Fleissner [F1] showed that (P) holds in the model constructed by collapsing a supercompact cardinal to ω_2 using Levy forcing.

Proof of Theorem 1. Fix a space X, and let $\mathcal{P}$ be a partition pf X which is a uniform cover of Ba_-X. We will show that $\mathcal{P}$ is a uniform cover of MX, and to do that, it is enough to show that $\mathcal{P}$ is σ-d.d. in X, because obviously $\mathcal{P}$ is completely Ba (X)-additive. Let Y be exp X endowed with the discrete uniformity (i.e. setfine), and consider the following set

$$B = \bigcup \{(y) \times \bigcup y \mid y \in Y\}$$

in $X \times Y$. Since B is a Baire set in $Ba_- X \times Y$, it is necessarily a Baire set in $X \times Y$. It follows (Lemma 2) that there exists a uniformly continuous pseudometric d on X such that B is a Baire set n $\langle X,d\rangle \times Y$. Hence $\mathcal{P}$ is completely $Ba(\langle X,d\rangle)$-additive, and ence by (P) the partition $\mathcal{P}$ is σ-d.d. in $\langle X,d\rangle$, and so n X. By Lemma 3 $\mathcal{P}$ is a uniform cover of MX.

Now we shall show that it is consistent that $Ba\_ \neq M$.

The range of Ba consists just from the so called "measurab-e spaces" or "Borel spaces" or "Baire spaces" in various termi-ologies, i.e. from paved spaces such that the paving is a σ-algebra. We denote by x_{Ba} the product in that category. Hence $\langle X, \mathcal{A}\rangle \times_{Ba} \langle X, B\rangle$ is $\langle X\times Y, \mathcal{C}\rangle$ such that $\mathcal{C}$ is the smallest σ-algebra containing all rectangles $A\times B$, $A\in\mathcal{A}$, $B\in\mathcal{B}$. If $\varkappa$, λ are cardinals, denote by K ($\varkappa$, λ) the statement:

$$\langle \varkappa, \exp \varkappa\rangle \times_{Ba} \langle \lambda, \exp \lambda\rangle = \langle \varkappa\times\lambda, \exp(\varkappa\times\lambda)\rangle$$

It is easy to check that K ($\varkappa$, λ) is false if $\varkappa, \lambda > 2^{\omega}$, t is easy to prove that $K(\omega_1, \omega_2)$ holds, and it was shown by K. Kunnen in his dissertation that K(c,c) holds under MA, and ence under Ma + $\neg$ CH we have that $K(\omega_1, 2^{\omega_1})$ holds.

Theorem 2. Assume $K(\varkappa, 2^{\varkappa})$. Then each completely Ba (X)-additive partition of X of cardinal $\varkappa$ is a uniform cover of Ba_X.

Corollary. If $K(\omega_1, 2^{\omega_1})$ holds, then $Ba\_ \neq M$, e.g. $Mp^1\omega_1 = p^1\omega_1$, but $Ba\_p^1\omega_1 = \omega_1$.

Proof of Theorem 2. It is enough to show that there exists a functor F in inv_ (Ba) such that

$$Fp^o\varkappa = \varkappa .$$

For each X define

$$FX \hookrightarrow X \times \varkappa^{U(X,p^1\varkappa)}$$

Clearly F is a negative functor, and we must show that

$$Ba\ (FX) = Ba\ (X)$$

for each X. Since $M \in inv\_(Ba)$, it is enough to check the relation for measurable X, i.e. for X with MX = X. Assume that X is measurable; hence X can be embedded into a product λ^{τ} for some cardinals λ and τ. So it is enough to show (Ba preserves subspaces!) that the identity mapping

$$J:\ \lambda^{\tau} \times \varkappa^{U(X,p^1\varkappa)} \longrightarrow \lambda^{\tau} \times (p^1\varkappa)^{U(X,p^1\varkappa)}$$

is a Baire isomorphism (i.e. an isomorphism in Ba), and since J is uniformly continuous, it is enough to show that J^{-1} is Baire-measurable (i.e. a morphism in Ba). In view of Lemma 2 it is enough to show that for each $n \in \omega$ the identity map

$$\lambda^n \times (p^1\varkappa)^n \longrightarrow \lambda^n \times \varkappa^n$$

is Baire measurable, and for this, it is enough to show that

$$Ba\ (p^1\lambda^n \times p^1\varkappa^n) = \exp\ (\lambda^n \times \varkappa^n)$$

and this follows from $K(\varkappa, 2^{\varkappa})$ if $\lambda \leq 2^{\varkappa}$, and the rest follows from the following simple

Lemma 4. $K(\varkappa, 2^{\varkappa})$ implies $K(\varkappa, \lambda)$ for each λ.

Concluding remarks. (a) Using the proof of Theorem 2 one can characterize U(ba_X,Y) in the spirit of one characterization of U (coz X,Y) recalled in § 2.

(b) I think that the functors like Ba or coz may be of purely categorial interest even if there are no adjoints, because there are various relevant coreflections, reflections and idempotent functors (e.g. cross) around.

References:

W.G. Fleissner:

[FL] An axiom for nonseparable Borel theory, Preprint 1978.

Z. Frolík:

[F1] Interplay of measurable and uniform spaces, in: Topology and its applications (Proc. 2nd Yugoslavia Symposium, Budva 1972), Beograd 1973, pp. 96-99.

[F2] A note on metric-fine spaces, PAMS 49(1974), 111-119.

[F3] Measurable uniform spaces, Pacif. J. Math. 55(1974), 93-105.

[F4] Three technical tools in uniform spaces, in: Seminar Uniform Spaces 1973-4, MÚ ČSAV Praha 1975, pp. 3-26.

[F5] Four functors into paved spaces, in: SUS 1973-4, pp. 27-72.

[F6] Cozero sets in uniform spaces, Soviet Math. Dokl., Vol. 17(1976), 1444-1448.

A.W. Hager:

[Hg] Measurable uniform spaces, Fund. Math. 77(1972), 51-73.

R. Hansell:

[H1] Borel measurable mappings for nonseparable metric spaces, TAMS 161(1971), 145-169.

[H2] On Borel mappings and Baire functions, TAMS 194(1974), 195-211.

D. Preiss:

[P] Completely additive disjoint system of Baire sets is of bounded class, Comment. Math. Univ. Carolinae 1 (197), 341-344.

M. Rice:

[R] Covering and function-theoretic properties of uniform spaces, Bull. AMS 80(1974), 159-163.

LIFTING CLOSED AND MONOIDAL STRUCTURES ALONG SEMITOPOLOGICAL FUNCTORS

Georg Greve

In the present paper some of the new results concerning semitopological functors are applied to get characterizations of inner homfunctors in categories. The methods described in the following sections yield all relevant closed structures in algebra and topology like *simple convergence*, *compact open topology* or *uniform convergence*. Before coming to the point we repeat shortly some properties of semitopological functors. Detailed descriptions of this matter can be found for example in [10],[11].

1. SEMITOPOLOGICAL FUNCTORS

1.1 A functor $T:A \to V$ is called *semitopological* if one of the following two equivalent internal characterizations is satisfied:

(ISF) Every T-cocone $(D, \xi: T\circ D \to \Delta Y)$ has a semifinal extension, i.e. there is a T-epimorphism $(B, r: Y \to TB)$ and a cocone $\beta: D \to \Delta B$ in A with $(\Delta r)\xi = T\circ\beta$, such that the universal semifinal property described in the diagram on the right holds:

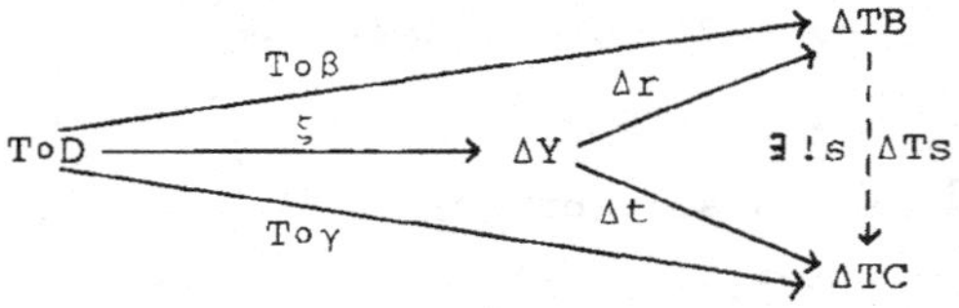

(ISI) Every T-cone $(D, \phi: \Delta X \to T\circ D)$ has a semiinitial extension, i.e. there is a T-quotient $(A, e: X \to TA)$ (a semifinal prolongation of some T-cocone like the r in (ISF)), and a cone $\alpha: \Delta A \to D$ with $(T\circ\alpha)\Delta e = \phi$, such that the semiinitial property described on the right is fulfilled:

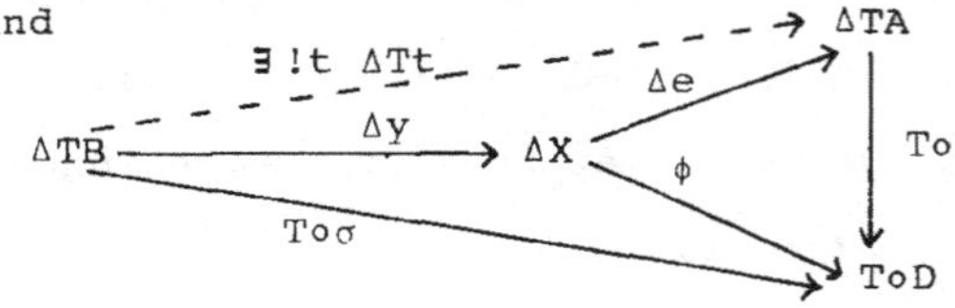

Assuming e in (ISI), resp. r in (ISF) to be identities, one gets the definitions of initial, resp. final liftings and thus the notion of a topological functor.

The following two external characterizations of a semitopological functor $T:A \to V$ are equivalent to (ISF) and (ISI):

(ESF) For all functors K, L, S and any natural transformation $\psi: T\circ L \to S\circ K$, there exists a functor F and natural transformations $\pi: L \to F\circ K$, $\kappa: S \to T\circ F$ with

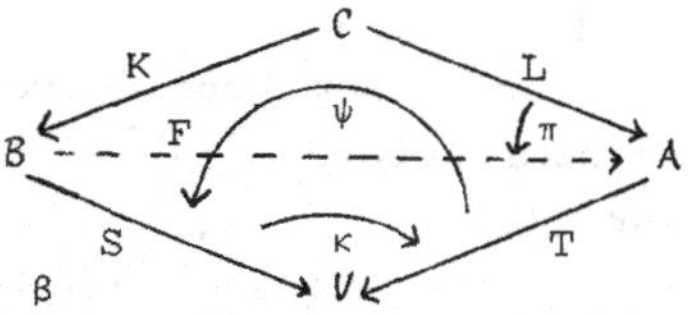

. $T\circ\pi = (\kappa\circ K)\psi$ and

I. For all functors G and natural ransformations $\alpha: L \to G\circ K$, $\beta: S \to T\circ G$ ith $(\beta\circ K)\psi = T\circ\alpha$ there exists a nique $\delta: F \to G$ with $(\delta\circ K)\pi = \alpha$ and $(T\circ\delta)\kappa = \beta$

ESI) For all functors K, L, S and any natural transformation $:S\circ K \to T\circ L$ there exists a functor F and natural transformations $\rho: F\circ K \to L$ nd $\sigma: S \to T\circ F$, where σ is pointwise a T-quotient, with

. $(T\circ\rho)(\sigma\circ K) = \psi$ and

I. For all functors G and natural transformations $\alpha: G\circ K \to L$, $\beta: T\circ G \to S$ ith $\psi(\beta\circ K) = T\circ\alpha$ there exists a unique $\delta: G \to F$ with $\rho(\delta\circ K) = \alpha$ and $\circ\delta = \sigma\beta$.

gain assuming σ in (ESI) resp. κ in (ESF) to be identities one gets xternal characterizations of topological functors (cp. also [2] and [12]).

he dual notion of semitopological is cosemitopological. Of course there re dual characterizations $(ISI)^{op}$, $(ISF)^{op}$, $(ESI)^{op}$ and $(ESF)^{op}$.

rom now on we assume $V = (V_o, V, H, \otimes, I, i, r, l)$ to be a symmetric monoidal losed category in the sense of Eilenberg and Kelly [3], A to be a -category with underlying category A_o and external homfunctor $(\square,\square): A_o^{op} \times A_o \to A_o$ and $T := (T_o: A_o \to V_o, \hat{T}_{A,B}: A(A,B) \to H(T_oA, T_oB))$ to be a -functor. The reader who is not familiar with this terminology may lways take A to be an ordinary category, T an ordinary functor and = SET. Furthermore, for technical reasons, all topological functors are ssumed to be amnestic (cp. [2]).

. (CO)-SEMITOPOLOGICAL LIFTED HOMFUNCTORS (external characterization)

he aim of the following section is to provide A_o with a closed strucure, i.e. to get a functor $h: A_o^{op} \times A_o \to A_o$, such that $T_o \circ h$ is a good pproximation of $A(\square,\square)$. Now in our basic situation one has the ollowing diagram:

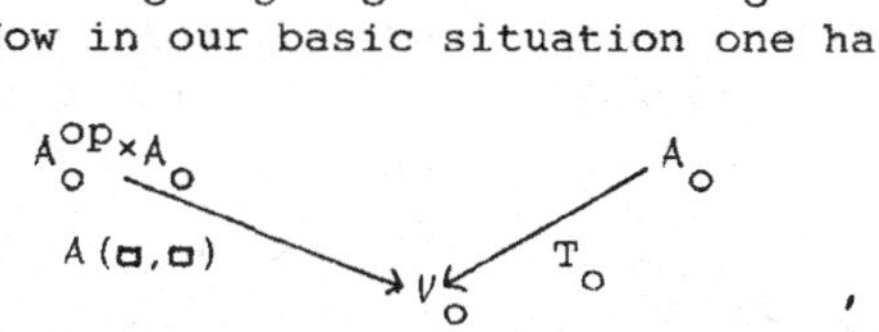

,

nd in order to get a homfunctor $h: A_o^{op} \times A_o \to A_o$ from the external characerizations of semitopological resp. cosemitopological functors one has to put a roof on this diagram". Thus, considering the pullback (P, P_1, P_2) f $(A(\square,\square), T_o)$, we get the following

2.1 <u>THEOREM</u>

(a) Take T_o to be semitopological. Then for every subcategory $S < P$ there are homfunctors $h_S : A_o^{op} \times A_o \to A_o$ and $q_S : A_o^{op} \times A_o \to A_o$ determined by (ESI) resp. (ESF).

(b) If T_o is cosemitopological, then for every $S < P$ there are homfunctors h_S^* resp. q_S^* determined by $(ESI)^{op}$ resp. $(ESF)^{op}$.

Proof: (a) Put $S_1 := P_1/S$ and $S_2 := P_2/S$ and apply (ESI) to the diagram

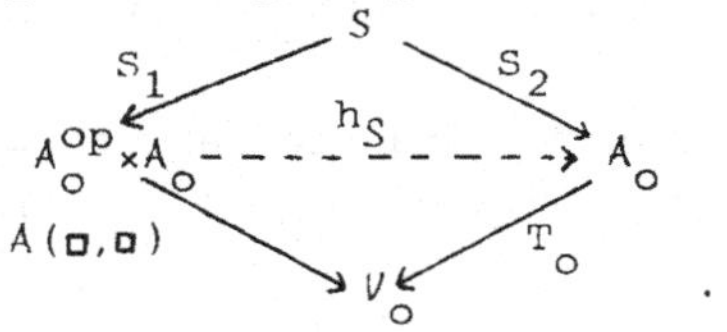

Analogously applying (ESF) one gets q_S.
h_S^* and q_S^* are obtained in a dual way.

We call h_S a *semiinitially lifted homfunctor*, q_S a *semifinally lifted homfunctor*, h_S^* a *cosemiinitially lifted homfunctor* and q_S^* a *cosemifinally lifted homfunctor*.

2.2 <u>THEOREM</u>

Take T_o to be topological.

(a) The assignment S to h_S establishes a surjective contravariant functor from the lattice of all subcategories of P to the lattice of all inner homfunctors of A_o.

(b) The assignment S to q_S gives a surjective covariant functor from the lattice of all subcategories of P to the lattice of all inner homfunctors of A_o.

Proof: If $S' < S < P$, then (ESI) yields a natural transformation from h_S to $h_{S'}$, therefore the assignment S to h_S is contravariant. Furthermore following Brümmer and Hoffman [2], any inner homfunctor $h : A_o^{op} \times A_o \to A_o$ is the coarsest diagonal of the diagram

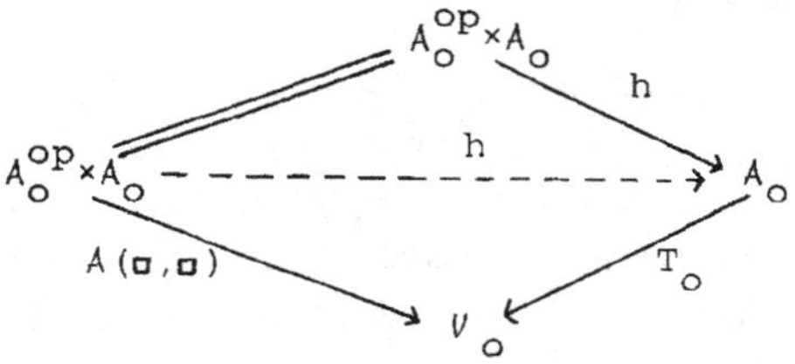

Thus $h = h_S$ with $S := \{(f,g,h(f,g)\} < P$, whence the assignment is surjective. The proof of (b) works analogously.

ote, that the pullback P of $(A(\square,\square),T_o)$ is a well known category, so
.2 admits a nice classification of function spaces in topology.

. INTERNAL CHARACTERIZATION

or every subcategory $S < P$ there are homfunctors h_S, h_S^*, q_S and q_S^* for
uitable T_o, but one does not know , how they look like. Therefore in
his section the results of Tholen and Wischnewsky [11] are applied to
et an internal description of these functors. As T_o is faithful, we
an restrict our investigations to discrete subcategories $S < P$.

.1 For $A,B \in \text{Ob } A_o$ consider the discrete cone of all morphisms
$f,g) \in A_o^{OP} \times A_o$ with $(f,g):(A,B) \to S_1(C,D,E)$ (cp.2.1). By applying $A(\square,\square)$
ne gets the discrete T_o-cone consisting of all T_o-morphisms
$E,A(f,g):A(A,B) \to A(C,D)=T_oE)$, and provided that T_o is semitopological,
semiinitial extension of this cone yields $h_S(A,B) \in \text{Ob } A_o$:

$$
\begin{array}{ccc}
 & \xrightarrow{\quad T_o h_S(A,B)\quad} & \\
\sigma_{A,B}\nearrow & & \searrow T_o\mu_{f,g} \\
A(A,B) & \xrightarrow[\quad A(f,g)\quad]{} & A(C,D) = T_oE
\end{array}
$$

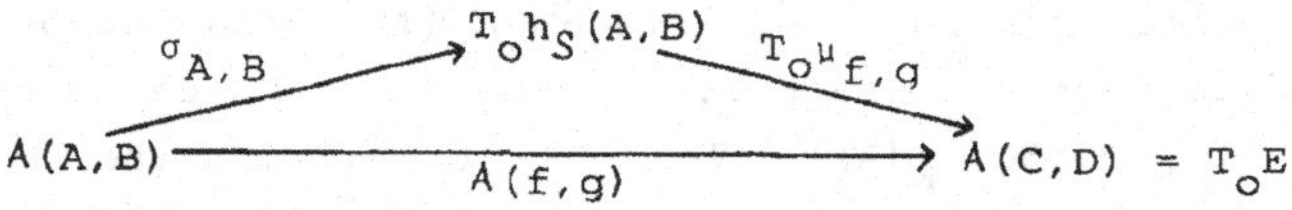

ow applying (ISI) h_S is extended to a functor $h_S:A_o^{OP} \times A_o \to A_o$.

.2 Analogously one obtains the functor q_S by considering the cocone
f all morphisms $(f,g):S_1(C,D,E) \to (A,B)$ and taking a semifinal exten-
ion of the T_o-cocone consisting of all T_o-comorphisms
$E,A(f,g):T_oE=A(C,D) \to A(A,B))$:

$$
\begin{array}{ccc}
 & \xrightarrow{\quad T_o q_S(A,B)\quad} & \\
T_o\nu_{f,g}\nearrow & & \nwarrow \kappa_{A,B} \\
T_oE = A(C,D) & \xrightarrow[\quad A(f,g)\quad]{} & A(A,B)
\end{array}
$$

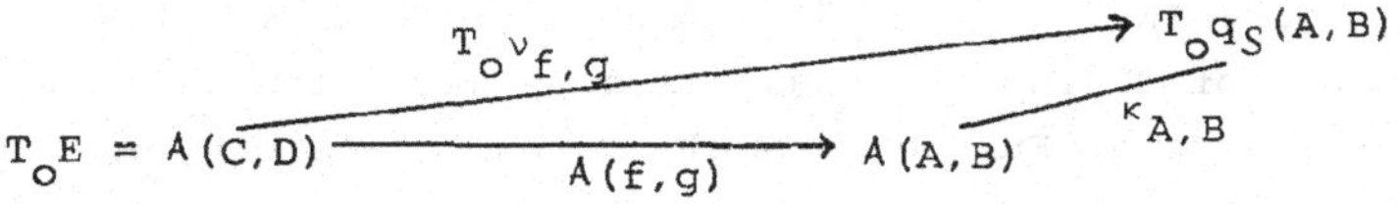

. APPLICATIONS AND EXAMPLES

he internal characterizations, which at first sight do not look very
opeful, turn out to be very appropriate and useful in practice. We
ive a short selection of interesting examples:

.1 Consider the forgetful functor $T:TOP \to SET$.
.1.1 For $S = P$, $h_S(A,B) = q_S^*(A,B)$ is the set of all continuous
unctions from A to B with the *discrete topology*,because following 3.1
$_S(A,B)$ is obtained by providing $TOP(A,B)$ with the initial topology
ith respect to the cone of all maps $TOP(f,g)$, where the codomain of f
s A and the domain of g is B.
$_S^*(A,B) = q_S(A,B)$ is the same set with the *indiscrete topoloy*.

4.1.2 For $S = \{(I,B,B') \mid I = \{\emptyset\},\ B \cong B' \text{ and } (I,B,B') \in \mathrm{Ob}\,P\} < P$, $h_S(A,B) = q_S^*(A,B)$ is the set of all continuous functions from A to B with the *simple topology*, because $h_S(A,B)$ is obtained by providing $TOP(A,B)$ with the initial topology with respect to the cone of all maps $TOP(a,g):TOP(A,B) \to TOP(\{\emptyset\},D)$, where $TOP(\{\emptyset\},D)$ carries the topology of D. The initial topology with respect to the cone of all $TOP(a,g)$ coincides with the initial topology with respect to the cone of all $TOP(a,B)$. This cone however is nothing else than the cone of all evaluation maps $\omega_a^{AB}:TOP(A,B) \to TB$, $\omega_a^{AB}(f) = Tf(a)$, $a \in TA$.
$h_S^*(A,B) = q_S(A,B)$ has a very fine topology: Open sets are all sets M of continuous functions, such that the set of values of the constant functions in M is open in B.

4.2 Consider the forgetful functor $T:TYCH \to SET$, $TYCH$ being the category of TYCHONOFF spaces. For a topological space X define $C_K(X)$ to be the set of all continuous functions $f:X \to \mathbb{R}$ ($\mathbb{R}$ = real numbers), with the topology of compact convergence. For the discrete category $S := \{(A,\mathbb{R},C_K(A)) \mid (A,\mathbb{R},C_K(A)) \in \mathrm{Ob}\,P\} < P$, $h_S(A,B)$ is the set of all continuous functions from A to B with *compact open topology*, because $h_S(A,B)$ is the set $TOP(A,B)$ provided with the initial topology with respect to the cone of all maps $TOP(f,g):TOP(A,B) \to TC_K(D)$. A subbasis of the topology of $h_S(A,B)$ therefore consists of all sets of continuous functions M with $M = \{x \mid g \circ x \circ f \in N\}$, where N is open in $C_K(D)$. One can put $N = U(K,O) = \{y \mid y(K) \subset O\}$ with K compact in D and O open in $\mathbb{R}$. That means $M = \{x \mid g \circ x \circ f(K) \subset O\} = \{x \mid x \circ f(K) \subset g^{-1}(O)\}$. Now $TYCH$ is the epireflective hull of $\mathbb{R}$ in TOP, thus all open sets of B are unions of sets $g^{-1}(O)$ for some g. Furthermore all compact sets can be written as f(K) and this completes the proof.

4.3 Consider the forgetful functor $T:UNIF \to SET$, where $UNIF$ is the category of uniform spaces. For $X \in \mathrm{Ob}\,UNIF$ define C(X) to be the set of all uniformly continuous functions $f:X \to \mathbb{R}$ with the uniformity of uniform convergence. Then a calculation similar to 4.2 shows, that for the discrete category $S = \{(A,\mathbb{R},C(A)) \mid (A,\mathbb{R},C(A)) \in \mathrm{Ob}P\}$, $h_S(A,B)$ is the set of all uniformly continuous functions $f:A \to B$ with the uniformity of *uniform convergence*, provided that B is in the epireflective hull of $\mathbb{R}$.

4.4 Take GR to be the category of groups with the forgetful functor $T:GR \to SET$. For the discrete subcategory $S = \{(Z,B,B') \mid (Z,B,B') \in \mathrm{Ob}P, B \cong B'\}$ (Z = integers), $h_S(A,B)$ is the set of all maps $f:A \to B$ with $f(a) = \prod_{i=1}^{n} g_i(a)$ for all $a \in A$, $g_i:A \to B$ being homomorphisms or antihomomorphisms.

he multiplication in $h_S(A,B)$ is pointwise.

here are many more interesting examples and there are a lot of possible eneralizations, but we have to confine us to the following case, which eneralizes 4.1.2.

. FUNCTIONAL HOMFUNCTORS

he notion of a functional inner homfunctor was introduced 1976 by anaschewski and Nelson in [1] and 1975 by Greve in [4]. In this ection it is shown, that functional inner homfunctors are cosemifinally ifted homfunctors.

n our basic situation one has for all objects $A,B \in \mathrm{Ob}\,A_o$ and each $:I \to T_oA$ in A_o a morphism

$$\omega_a^{AB}: A(A,B) \xrightarrow{\hat{T}_{A,B}} H(T_oA,T_oB) \xrightarrow{H(a,1)} H(I,T_oB) \xrightarrow{i^{-1}_{T_oB}} T_oB .$$

n the set based case ω_a^{AB} is the *evaluation map*: $\omega_a^{AB}(f) = T_of(a)$. The amily of all morphisms ω_a^{AB} yields a discrete T_o-cone $(\Delta B, \omega^{AB})$. Now aking T_o to be cosemitopological, a cosemifinal extension of $(\Delta B, \omega^{AB})$ ives an object $q^*(A,B)$, a T_o-comorphism $(q^*(A,B), \sigma_{A,B})$ and a discrete one $(\alpha_a^{AB})_{a \in V_o(I,T_oA)}$ making the following diagram commutative:

$$\begin{array}{ccccc} T_oq^*(A,B) & & \xrightarrow{\quad T_o\alpha_a^{AB} \quad} & & T_oB \\ & {\scriptstyle \sigma_{A,B}} \searrow & & \nearrow {\scriptstyle \omega_a^{AB}} & \\ & & A(A,B) & & \end{array} .$$

q^* can be canonically extended to a functor $q^*: A_o^{op} \times A_o \to A_o$.

e call q^* a *functional homfunctor* of A_o. It is easy to see that func-ional homfunctors are unique. Moreover in the set based case this efinition is consistent with the corresponding notion in [1] , provided hat $\sigma_{A,B}$ is the identity.

urthermore assuming T to be V-representable (there is an object $J \in \mathrm{Ob}\,A_o$ nd a natural V-isomorphism $j_A: T_oA \to A(J,A)$), we define

$$\lambda_e^{AB}: T_oB \cong A(J,B) \xrightarrow{A(e,B)} A(A,B)$$

or every $e:A \to J$ in A_o. Thus one gets a discrete T_o-cocone $(\Delta B, \lambda^{AB})$ nd assuming T_o to be semitopological a semifinal extension of $(\Delta B, \lambda^{AB})$ yields a functor $q: A_o^{op} \times A_o \to A_o$ described by the following iagram:

$$\begin{array}{ccccc} & & & & T_oq(A,B) \\ & & {\scriptstyle T_o\beta_e^{AB}} \nearrow & & \nearrow {\scriptstyle \kappa_{A,B}} \\ T_oB & & \xrightarrow{\quad \lambda_e^{AB} \quad} & & A(A,B) \end{array} .$$

q is something like a "cofunctional" homfunctor.

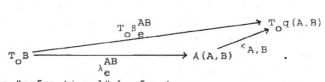

n particular every topological functor $T: A \to SET$ is representable by a natural isomorphism $j_A : TA \to A(J,A)$. Thus taking again P to be the pullback of $(A(\square,\square),T)$ and $S := \{ (J,A,A') \mid (J,A,A') \in \mathrm{Ob}\, P,\ A \triangleq A' \} < P$ one gets the following

5.1 THEOREM

Take $T: A \to SET$ to be a topological functor, which allows only one structure on the one element set $\{\emptyset\}$.

(a) q^* coincides with q_S^* and q with q_S (terminology as above).

(b) The functor q^* provides A with a symmetric closed structure in the sense of Eilenberg-Kelly [3] and Schipper [9], which is the coarsest closed structure on A.

(c) The functor q yields the finest closed structure on A.

Proof: (a) is a generalization of 4.1.2.

(b): Composition law and exponential law are obtained by initial liftings of the corresponding morphisms in SET. Moreover it is easy to see that also j_A can be lifted. Given another closed structure on A with inner homfunctor r, the evaluation maps $\omega_a^{AB} : Tr(A,B) \to TB$ are A-morphisms, thus one gets a morphism $f_{A,B} : r(A,B) \to q^*(A,B)$ with $Tf_{A,B} = A(A,B)$. A similar argument yields (c).

In general q is not symmetric and does not fulfill an exponential law.

The rest of this paper deals with the lifting of monoidal structures along T_o. In the preceding sections we have lifted external homfunctors of an enriched category A. So it is not appropriate to lift the monoidal structure of V along T_o, but one has to lift an external tensorproduct of A, i.e. a functor $\oplus : V_o \times A_o \to A_o$. Therefore we assume from now on the V-category A to be tensored (every V-funtor $A(A,\square)$ has a V-left adjoint) and following Kelly [6] we get a functor $\oplus : V_o \times A_o \to A_o$, which even is the underlying functor of a V-functor $\oplus : V \otimes A \to A$. But first we have to make some remarks about a class of functors including the semitopological functors:

6. PRESEMITOPOLOGICAL FUNCTORS

There are many categories looking very algebraically, which are not semitopological, i.e. finite groups, - rings etc., simple groups, - rings etc., semi-simple rings, - modules etc., cyclic groups, Noetherian rings and many more. All these categories are not described by semitopological functors but subsumed under the following notion:

.1 DEFINITION (internal charaterization)

functor $T:K \to B$ is called *presemitopological*, if every T-cone $(D, \alpha: T\circ D \to \Delta TA)$ has a presemifinal extension with respect to A, i. e. here exists a K-morphism $t:A \to B$ and a cocone $\beta: D \to \Delta B$ in K with $(\Delta Tt)\alpha = T\circ\beta$, such that the universal property described on the right holds:

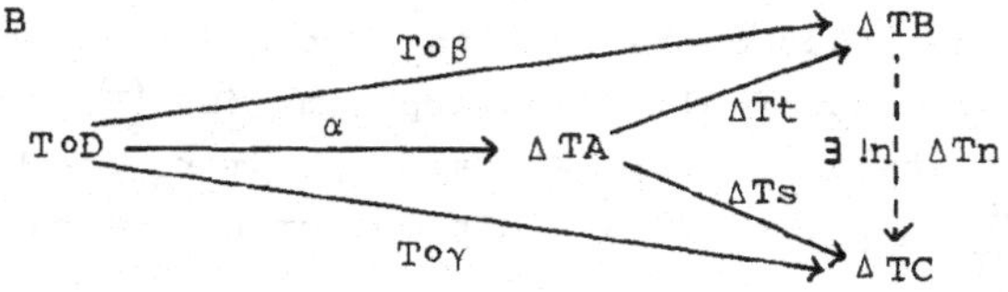

he following external characterization is equivalent to 6.1:

EPS) For all functors S, L, K and any natural transformation $\phi: T\circ L \to T\circ S\circ K$, there exist a functor F and natural transformations $\rho: L \to F\circ K$, $\sigma: S \to F$ with $T\circ\rho = (T\circ\sigma\circ K)\phi$, such that the following universal property holds:

or all functors G and natural transformations $\alpha: L \to G\circ K$, $\beta: S \to G$ with $T\circ\alpha = (T\circ\beta\circ K)\phi$, there exists a unique $\delta: F \to G$ with $\alpha = (\delta\circ K)\rho$ and $\beta = \delta\sigma$.

he proof of (EPS) works analogously to the corresponding proof in [11]. functor T is semitopological, iff T is presemitopological and has a eft adjoint.

. PRESEMITOPOLOGICALLY LIFTED TENSORPRODUCTS

n our basic situation with A tensored, we are given functors $T_o: A_o \to V_o$ nd $\oplus: V_o \times A_o \to A_o$. Thus one has a diagram

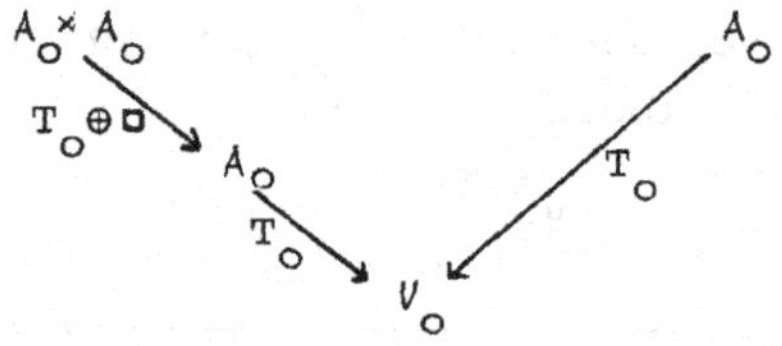

,

nd in order to apply (EPS) to get tensorproducts in A_o, one has to ut a roof on it. There are definitions, methods and results corres- onding to the statements about homfunctors. We shall describe here nly one presemitopologically lifted tensorproduct by giving its nternal characterization:

s A is tensored, we have adjuctions $\square\oplus A \xrightarrow[\varepsilon^A]{\eta^A} A(A,\square)$, which are under- ying adjunctions of V-adjunctions, and as V is monoidal closed,there is canonical isomorphism $\pi_{X,A}: V_o(X, H(T_oA, T_o(X\oplus A))) \to V_o(X\otimes T_oA, T_o(X\oplus A))$ or all $X \in \mathrm{Ob}\, V_o$, $A \in \mathrm{Ob}\, A_o$. Define $\tilde{T}_{X,A} := \pi_{X,A}(\hat{T}_{A,X\oplus A}\eta^A_X)$, then $\tilde{T}: \square\otimes T_o \to T_o(\square\oplus\square)$ is a natural transformation. Now for $A, B \in \mathrm{Ob}\, A_o$ and

$a: I \to T_o A$ one has a T_o-comorphism

$$\zeta_a^{AB}: T_o B \xrightarrow{r^{-1}} T_o B \otimes I \xrightarrow{T_o B \otimes a} T_o B \otimes T_o A \xrightarrow{\tilde{T}_{T_o B, A}} T_o(T_o B \oplus A) .$$

In the set based case ζ_a^{AB} is a *section* into the copower $\coprod_{b \in T_o B} A$. Now consider a presemifinal extension of the discrete T_o-cocone consisting of all T_o-comorphisms (B, ζ_a^{AB}) with respect to $T_o B \oplus A$. Thus one gets an object $B \boxtimes A$, a morphism $\kappa_{A,B}: T_o B \oplus A \to B \boxtimes A$ and a discrete cocone $(\gamma_a^{AB})_{a \in V_o(I, T_o A)}$ as in the commutative diagram below:

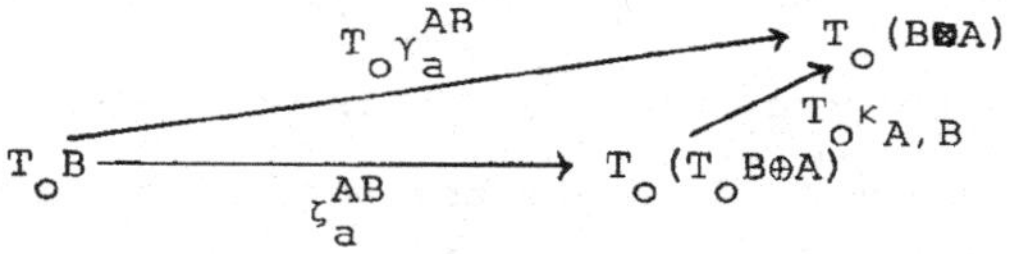

$\boxtimes$ can be canonically extended to a tensorproduct $\boxtimes: A_o \times A_o \to A_o$.

8. FUNCTIONAL ADJUNCTION

Finally a theorem of Pumplün [8] and Banaschewski-Nelson [1] is generalized. They found that tensorproducts constructed with universal bimorphisms are left adjoint to inner homfunctors in "structured categories" resp. to functional inner homfunctors (in the set based case). For the presemitopologically lifted tensorproduct $\boxtimes$ of 7. and the functional homfunctor q^* of 5. one gets the following

8.1 THEOREM

$\boxtimes$ is left adjoint to q^*.

Proof: One has to show that for each $A \in \mathrm{Ob}\, A_o$ $q^*(A, \square)$ is right adjoint to $\square \boxtimes A$. Consider the following diagram:

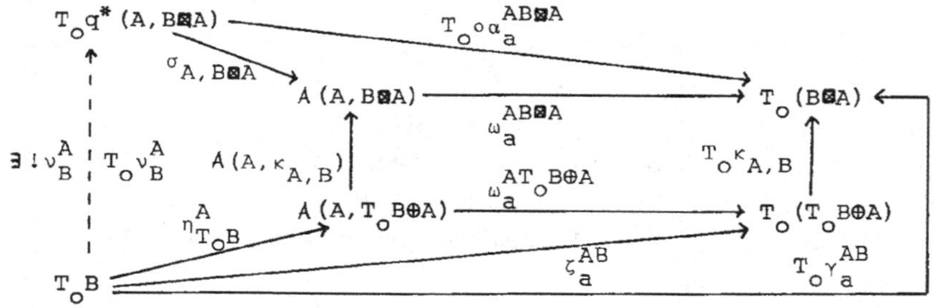

A straightforward diagram chasing shows that in the diagram above all inner diagrams are commutative. Thus there is a unique morphism $\nu_B^A: B \to q^*(A, B \boxtimes A)$, which turns out to be a $q^*(A, \square)$-universal arrow, moreover ν_B^A is natural in B and extraordinary natural in A, thus $\boxtimes$ is left adjoint to q^*.

In the set based topological case 8.1 yields the well known symmetric monoidal closed structure of pointwise convergence, and in the set based algebraic case one gets the symmetric monoidal closed structure of "pointwise multiplication", provided that the evaluation maps can be lifted. Also if $V \neq SET$, 8.1 defines under weak additional assumptions a symmetric monoidal closed structure in the sense of Eilenberg and Kelly [3]. Of course one does not always get a monoidal closed structure, for example take TAB_c to be the category of connected abelian topological groups considered as a V-category over TOP, then 8.1 provides TAB_c with a structure, which is only closed up to a natural transformation.

LITERATURE

[1] Banaschewski-Nelson Tensorproducts and bimorphisms, Canad.Math. Bull.19 (4),385-401,1976

[2] Brümmer-Hoffmann An external characterization of topological functors, Proc.Conf.Categ.Top. (Mannheim 1975), LN 540,Springer,Berlin,Heidelberg,New York, 136-151, 1976

[3] Eilenberg-Kelly Closed categories, Proc.Conf.Categ.Alg. (La Jolla 1965),Springer,Berlin,Heidelberg, New York, 421-562, 1966

[4] Greve Algebraische Konstruktionen von Linksadjungierten, Diplomarbeit,Münster 1975

[5] Greve M-geschlossene Kategorien, Thesis,Hagen 1978

[6] Kelly Adjunction for enriched categories, LN 106, Springer,Berlin,Heidelberg,New York,1969

[7] Porst-Wischnewsky Every topological category is convenient for Gelfand-Naimark-Duality, Manuscripta Mathematica 25, 1978

[8] Pumplün Das Tensorprodukt als universelles Problem, Math.Ann. 171,247-262,1967

[9] Schipper Symmetric closed categories, Mathem.Centre Tracts 64, 1975

[10] Tholen Semitopological functors I, to appear in J.Pure Appl.Alg.

[11] Tholen-Wischnewsky Semitopological functors II, to appear in J.Pure Appl.Alg.

[12] Wolff On the external characterization of topological functors,Manuscripta Mathematica 22,1977

Georg Greve, Fernuniversität, Lützowstr.125, 58 Hagen, West Germany

ON NON-SIMPLICITY OF TOPOLOGICAL CATEGORIES

by

Darrell W. HAJEK and Adam MYSIOR*

The following symbols will be used to denote categories of topological spaces (and continuous functions) :

T	-	all topological spaces
T_0	-	T_0-spaces
T_1	-	T_1-spaces
T_2	-	Hausdorff spaces
$T_{2\frac{1}{2}}$	-	completely Hausdorff spaces
T_3	-	regular spaces
$T_{3\frac{1}{2}}$	-	completely regular spaces
D_1	-	hereditarily disconnected spaces
D_2	-	totally disconnected spaces
D_3	-	zero-dimensional spaces

0. INTRODUCTION

H.Herrlich has introduced in [6] the concept of $\mathfrak{S}$-regular and $\mathfrak{S}$-compact space where $\mathfrak{S}$ is an arbitrary class of topological spaces. Namely, a space X is said to be <u>$\mathfrak{S}$-regular</u> (<u>$\mathfrak{S}$-compact</u>) provided that it is homeomorphic to a subspace (a closed subspace) of a product of spaces from $\mathfrak{S}$.

The category of all $\mathfrak{S}$-regular and $\mathfrak{S}$-compact spaces will be denoted by Reg$\mathfrak{S}$ and Comp$\mathfrak{S}$ respectively.

---

* Based on a talk given at the conference by the second named author.

There are two natural problems connected with every epi-reflective (= productive and hereditary) subcategory $\mathfrak{A}$ of T :

1. Is there a single space E such that $\mathfrak{A} = \mathrm{Reg}\{E\}$?
   In this case the category $\mathfrak{A}$ is said to be <u>T-simple.</u>

2. (Raised by S.Mrówka in [13] and [14]) Is there a <u>minimal</u> class $\mathfrak{E}$ with $\mathfrak{A} = \mathrm{Reg}\,\mathfrak{E}$? The minimality of $\mathfrak{E}$ means here that for every space $E \in \mathfrak{E}$ we have $\mathrm{Reg}\,\mathfrak{E} - \{E\} \subsetneq \mathrm{Reg}\,\mathfrak{E}$ or, equivalently, $E \notin \mathrm{Reg}\,\mathfrak{E} - \{E\}$.
   In this case the category $\mathfrak{A}$ will be called <u>T-semisimple</u>.

Similar problems can be considered for every epi-reflective (= productive and closed hereditary) subcategory $\mathfrak{A}$ of T_2 :

1. Is it <u>T_2-simple</u> ? - i.e. is there a space E such that $\mathfrak{A} = \mathrm{Comp}\{E\}$?

2. Is there a minimal class $\mathfrak{E}$ with $\mathfrak{A} = \mathrm{Comp}\,\mathfrak{E}$?
   In other words : is $\mathfrak{A}$ <u>T_2-semisimple</u> ?

In the first section of this paper we review some examples and known results relative to the above problems. Section II introduces a straightforward and general method of verifying that various categories are neither simple nor semisimple.

All undefined notions are from [1] and [7] .

We assume throughout this paper that all considered categories are full homeomorphism-closed subcategories of T and each of them contains at least one space with more than one point.

For each ordinal α by W_α we will denote the set of all ordinals less than α supplied with the order topology.

I. EXAMPLES AND SOME KNOWN RESULTS

1. T-simple are : T, T_0, $T_{3\frac{1}{2}}$, D_3 .

2. T_2-simple are the categories of : compact Hausdorff spaces, compact zero-dimensional spaces, realcompact spaces, k-compact spaces [9], zero-dimensionally k-compact spaces [8,9] .

3. Several "every-day life" categories are known not to be T-simple.

The first result of this sort comes from S.Mrówka who showed in [12] non-T-simplicity of T_1. The same for T_2 has been proved by A.Birula-Białynicki in 1958 but never published.

H.Herrlich showed in [4] that for every T_1-space E exists a regular space X (with more than one point) such that all continuous functions from X into E are constant. Obviously such an X cannot be E-regular. It follows that neither T_1 , nor T_3 , nor any category lying between them can be T-simple. The same was obtained independently by A.Ramer [18] .

D.Hajek and R.Wilson [3] strengthened this result by showing that in fact, no category between T_1 and $T_3 \cap D_2$ is T-simple.

4. If a productive and hereditary subcategory of T_2 is not T-simple then, clearly, it is not T_2-simple.
The converse is false - T-simple categories $T_{3\frac{1}{2}}$ and D_3 are not T_2-simple by H.Herrlich's observation in [5] .

Determining whether or not a given category is T_2-simple is often related to the measure problem. Under assumption that all cardinals are non-measurable the following categories are not T_2-simple :

(i) [10] any epi-reflective subcategory $\mathcal{U}$ of T_2 that contains the countable discrete space and is preserved by perfect mappings onto $\mathcal{U}$-regular spaces (for example the categories of k-ultracompact spaces where $k > \aleph_0$ and all productive and hereditary subcategories of T_2),

(ii) [17] the category of zero-dimensional realcompact spaces and, likevise, all categories of zero-dimensional k-compact spaces where $k > \aleph_0$.

5. In reply to the Mrówka's question the second author proved in [15] that the categories T_1, T_2, T_3 are not T-semisimple. D.Hajek [2] showed the non-T-semisimplicity of $T_{2\frac{1}{2}}$, D_1 and D_2.

In Section II we introduce a straightforward method of obtaining theorems of a more general nature. As a corollary we obtain some new results : non-T-simplicity and, moreover, non-T-semisimplicity of $T_{3\frac{1}{2}} \cap D_1$ and $T_{3\frac{1}{2}} \cap D_2$; non-T-semisimplicity of $T_3 \cap T_{2\frac{1}{2}}$, $T_3 \cap D_1$, $T_3 \cap D_2$ and non-T_2-semisimplicity of $T_{3\frac{1}{2}}$ and D_3.

6. Nothing in the above prevents one from supposing that notions of simplicity and semisimplicity coincide. However, this is not the case, as indicated by the following two examples :

(i) Denote by $\mathfrak{S}$ a proper class of Hausdorff spaces in which the only continuous functions between any two distinct spaces are constants - as constructed in [11] . It is easy to see that Reg$\mathfrak{S}$ is T-semisimple but is not T-simple.

(ii) Category Comp$\{W_\alpha : \alpha$ is regular$\}$ is not T_2-simple and is T_2-semisimple as observed by S.Mrówka in [13] .

II. THE METHOD AND THEOREMS

Let E be a Hausdorff space.

Denote by E_D the set E with the discrete topology. Let α be the first ordinal with cardinality greater than the cardinality of E . Denote by E^* the set $W_\alpha \times \beta E_D \cup \{\alpha\} \times E$ with topology generated by the open subsets of $W_\alpha \times \beta E_D$ and all sets of the form $U_\gamma = \{\xi : \gamma < \xi < \alpha\} \times cl_{\beta E_D} U \cup \{\alpha\} \times U$ where U is open in E and $\gamma \in W_\alpha$.

LEMMA. For every Hausdorff space E and continuous function $f : E^* \longrightarrow E$ the set $f(E^*)$ is compact.

P r o o f . For any point $e \in E$ there exists $\gamma_e \in W_\alpha$ such that the function f is constant on the segment $\{\xi : \gamma_e < \xi < \alpha\} \times \{e\}$. Let $\gamma = \sup\{\gamma_e : e \in E\}$. The cardinality of $W_\gamma = \bigcup_{e \in E} W_{\gamma_e}$ is less than α . Hence $\gamma < \alpha$ and $f(E^*) \subset cl_E f(W_\alpha \times E_D) \subset cl_E f(W_{\gamma+1} \times E_D) \subset f(W_{\gamma+1} \times \beta E_D)$.

A category $\mathfrak{U}$ will be called <u>$*$-closed</u> if $E^* \in \mathfrak{U}$ for every $E \in \mathfrak{U}$.

Using elementary properties of the Čech-Stone extension it is easy to prove that T_2, T_3, $T_{2\frac{1}{2}}$, $T_{3\frac{1}{2}}$, D_1, D_2, D_3 and all their intersections are $*$-closed.

THEOREM 1. $T_{3\frac{1}{2}}$ is the largest T-simple and $*$-closed subcategory of T_2.

P r o o f . Let $\mathfrak{U} = \mathrm{Reg}\{E\}$ be an arbitrary $*$-closed subcategory of T_2 . We show that the space E is completely regular.

By the assumption there is an embedding $h : E^* \longrightarrow E^{m}$ for some cardinal m. It follows from LEMMA that all projections of the space $h(E^*)$ are compact subsets of E . Hence the space E^* and its subspace $E \simeq \{\alpha\} \times E$ can be embedded into a product of compact Hausdorff spaces. This ends the proof.

THEOREM 1′. $T_{3\frac{1}{2}}$ is the only T-semisimple subcategory of T_2 which is $*$-closed and contains the interval $[0,1]$.

P r o o f . Let $\mathfrak{U}$ be an arbitrary subcategory of T_2 which is T-semisimple, $*$-closed and contains $[0,1]$. We prove that $\mathfrak{U} = T_{3\frac{1}{2}}$. By THEOREM 1 it suffices to show that $\mathcal{U}$ is T-simple.

Let $\mathfrak{E}$ be a minimal class with $\mathcal{U} = \operatorname{Reg}\mathfrak{E}$. At least one space from $\mathfrak{E}$ contains a non-degenerate continuous image of $[0,1]$. Hence there is a space $E \in \mathfrak{E}$ which contains the whole interval itself. We show that $\mathfrak{E} = \{E\}$.

Assume that there is a space $E_1 \in \mathfrak{E} - \{E\}$. For some cardinal m the space E_1 can be embedded into $E_1^{m} \times Y$ where Y is a suitable product of spaces from $\mathfrak{E} - \{E_1\}$. Similarly as in the proof of THEOREM 1, one can show that E_1^* is homeomorphic to a product of Y and a completely regular space. Hence the space E_1^* and its subspace E_1 are $\mathfrak{E} - \{E_1\}$ -regular and this contradicts the minimality of $\mathfrak{E}$.

Compact hereditarily disconnected spaces are zero-dimensional. So the same reasonong yields following

THEOREM 2. D_3 is the only T-semisimple and $*$-closed subcategory of D_1.

By means of the above theorems one can state non-T-simplicity and non-T-semisimplicity of various subcategories of T_2 (cf. examples I.3 and I.5).

In particular, the categories of m-regular and m-completely regular spaces defined in [18] are not T-simple for any $m \geq \aleph_0$. This gives an answer to questions raised by A.Ramer in [18].

THEOREM 3. No $*$-closed subcategory of T_2 is T_2-simple.

P r o o f . Assume that there is a Hausdorff space E such that the category $\operatorname{Comp}\{E\}$ is $*$-closed. Since $\operatorname{Comp}\{E\} = \operatorname{Comp}\{E^{\aleph_0}\}$ we can assume that the space E has infinite cardinality. By analogy to the proof of THEOREM 1 one can show that E^* can be embedded as a closed subspace into a product of compact spaces. But this is impossible since the space E^* is not compact.

Using the same argument as in the proof of THEOREM 1′ one can prove following

THEOREM 3′. (i) No $*$-closed subcategory of T_2 which contains $[0,1]$ is T_2-semisimple.
(ii) No $*$-closed subcategory of D_1 is T_2-semisimple.

COROLLARY. $T_{3\frac{1}{2}}$ and D_3 are not T_2-semisimple.

III. FINAL REMARKS

1. The method presented in the previous section was applied only to subcategories of T_2. However a little strengthening is possible.

Let us say that a space E has property HI if for any compact Hausdorff space X and continuous function $f : X \to E$ the space $f(X)$ is Hausdorff. This property looks a little odd, but HI-spaces form an epi-reflective subcategory of T_1 which includes all KC spaces. The same reasoning as in the proof of THEOREM 1′ yields that $T_{3\frac{1}{2}}$ is the only T-semisimple subcategory of HI which is $*$-closed and contains the interval $[0,1]$. As new corollaries we obtain the non-T-semisimplicity of RegKC and HI .

2. For the sake of completness we give here a short proof that T_1 is not T-semisimple.

Assume that $T_1 = \mathrm{Reg}\,\mathfrak{E}$. Let E be an arbitrary space from $\mathfrak{E}$. We show that $T_1 = \mathrm{Reg}\,\mathfrak{E} - \{E\}$. Let Z be a set of cardinality greater than the cardinality of E. Denote by $\tilde{E}$ the set $E \cup Z$ supplied with topology consisting of sets $U \cup C$ where U is open in E and $C \subset Z$ is co-finite. It is easy to see that every continuous function $f : \tilde{E} \to E$ is constant. Hence for any topological space X and any cardinal $\mathfrak{m}$ $\tilde{E} \subset E^{\mathfrak{m}} \times X$ implies $\tilde{E} \subset X$. It follows that $\tilde{E} \subset \mathrm{Reg}\,\mathfrak{E} - \{E\}$. Since $E \subset \tilde{E}$ so $E \in \mathrm{Reg}\,\mathfrak{E} - \{E\}$ and $\mathrm{Reg}\,\mathfrak{E} - \{E\} = T_1$.

3. The applicability of THEOREMS 3 and 3′ is, in fact, very narrow. The reason is that only a few categories are simultaneously epi-reflective in T_2 and $*$-closed.

In particular, we do not know whether or not the category of zero-dimensional realcompact spaces is T_2-semisimple, however our bets are against.

4. Sometimes, the categories which are neither simple nor semisimple can have a nice generating classes. Let us bring the following examples ($\mathfrak{M}$ stands for an arbitrary proper class of cardinals) :

(i) [12] $T_1 = \mathrm{Reg}\{L_{\mathfrak{m}} : \mathfrak{m} \in \mathfrak{M}\}$ where $L_{\mathfrak{m}}$ is a set of cardinality $\mathfrak{m}$ with co-finite topology.

(ii) [16] $T_{2\frac{1}{2}} = \mathrm{Reg}\{I_{\mathfrak{m}} : \mathfrak{m} \in \mathfrak{M}\}$ and $D_2 = \mathrm{Reg}\{D_{\mathfrak{m}} : \mathfrak{m} \in \mathfrak{M}\}$ where $I_{\mathfrak{m}}$ and $D_{\mathfrak{m}}$ are the products $[0,1]^{\mathfrak{m} \times \mathfrak{m}}$ and $\{0,1\}^{\mathfrak{m} \times \mathfrak{m}}$ respectively with $\bigcap_{s \in \mathfrak{m}} \bigcup_{t \in \mathfrak{m}} p_{st}^{-1}(0)$ as a new closed set.

(iii) [9] $T_{3\frac{1}{2}} = \mathrm{Comp}\{P_{\mathfrak{m}} : \mathfrak{m} \in \mathfrak{M}\}$ and $D_3 = \mathrm{Comp}\{Q_{\mathfrak{m}} : \mathfrak{m} \in \mathfrak{M}\}$ where $P_{\mathfrak{m}}$ and $Q_{\mathfrak{m}}$ are the cubes $[0,1]^{\mathfrak{m}}$ and $\{0,1\}^{\mathfrak{m}}$ each with one point deleted.

(iv) Comp{metrizable spaces} = the category of spaces with complete uniformities.

It seems to be an interesting problem to find nice generating classes for such categories as T_2, T_3, D_1 or $T_{3\frac{1}{2}} \cap D_2$. The analogous problem for the category of zero-dimensional realcompact spaces is probably difficult.

REFERENCES

[1] R.Engelking, General Topology. Warsaw 1977.

[2] D.W.Hajek, Some categories which have no minimal generating classes, preprint.

[3] D.W.Hajek and R.G.Wilson, The non-simplicity of certain categories of topological spaces, Math.Z. 131 (1973) 357-359.

[4] H.Herrlich, Wann sind alle stetigen Abbildungen in Y konstant?, Math.Z. 90 (1965) 152-154.

[5] H.Herrlich, Fortsetzgarkeit stetiger Abbildungen und Kompaktheitsgrad topologischer Räume, Math.Z. 96 (1967) 64-72.

[6] H.Herrlich, $\mathfrak{E}$-kompakte Räume, Math.Z. 96 (1967) 228-255.

[7] H.Herrlich, Categorical topology, Gen.Top.Appl. 1 (1971) 1-15.

[8] S.S.Hong, On k-compactlike spaces and reflective subcategories, Gen.Top.Appl. 3 (1973) 319-330.

[9] M.Hušek, The class of k-compact spaces is simple, Math.Z. 110 (1969) 123-126.

[10] M.Hušek, Perfect images of E-compact spaces, Bull. Acad.Pol.Sci. 20 (1972) 41-45.

[11] V.Kannan and M.Rajagopalan, Constructins and applications of rigid spaces I, preprint.

[12] S.Mrówka, On universal spaces, Bull.Acad.Pol.Sci. 4 (1956) 479-481.

[13] S.Mrówka, Further results on E-compact spaces I, Acta Math. 120 (1968) 161-185.

[14] S.Mrówka, E-complete regularity and E-compactness, Proceedings of the Kanpur Topological Conference 1968 Prague 1971, 207-214.

[15] A.Mysior, On generalized classes of complete regularity, Bull.Acad.Pol.Sci. 24 (1976) 341-342.

[16] A.Mysior, Some remarks on embedding properties of completely Hausdorff and totally disconnected spaces, Bull.Acad.Pol.Sci. 25 (1977) 555-558.

[17] A.Mysior, The category of all zero-dimensional realcompact spaces is not simple, Gen.Top.Appl. 8 (1978) 259-264.

[18] A.Ramer, Some problems on universal spaces, Bull.Acad. Pol.Sci. 13 (1965) 291-294.

Darrell W.Hajek
Department of Mathematics
University of Puerto Rico
Mayaguez, Puerto Rico 00708

Adam Mysior
Institute of Mathematics
University of Gdańsk
80952 Gdańsk, Poland

KAN LIFT-EXTENSIONS IN C.G. HAUS$_*$

K.A. Hardie, Cape Town.

Much progress in algebraic topology during the last twenty-five years has resulted from the discovery of homotopy equivalences (sometimes weak) of endofunctors of Top and Top$_*$. Perhaps the first result of this kind was the suspension splitting of a product

(0.1) $$\Sigma(X \times Y) \simeq \Sigma X \vee \Sigma Y \vee \Sigma(X \# Y)$$

attributed to Peter Hilton [14]. This was followed by I.M. James's approximation $X_\infty \simeq \Omega\Sigma X$ [9], the suspension splitting of $\Omega\Sigma X$ [7] and more recently by the $\Omega^n\Sigma^n$ approximations due to Milgram [12] and May [13] and the splittings due to Snaith and others.

In an attempt to understand and hopefully to discover such equivalences the author introduced the concept of an X-functor (for a particular space X in Top$_*$) motivated by the view that a natural transformation between X-functors ought to be a homotopy equivalence if it was so at X itself. Such considerations in the case $X = P$, the discrete two-point space, did indeed suffice to explain (0.1) and to prove others [6]. Although not recognized as such initially the X-functors were enriched left Kan extensions along the functor including the full subcategory of Top$_*$ with the single object X. At least this is the case if one replaces Top$_*$ by the convenient category C.G. Haus$_*$. Now there are indications that one will have to consider Kan extensions along the inclusion of non-full subcategories. For example the configuration spaces that feature in May's approximations [13] are functorial only if one restricts to monomorphisms. Moreover they take values in C.G. Haus, <u>not</u> C.G. Haus$_*$. Thus it seems that certain significant endofunctors can only be interpreted as relative Kan extensions and as Kan lift-extensions. In this paper the necessary concepts are formulated and a recognition principle given. There are no immediate applications, but the partial successes achieved in the simplest case [6], [5] lead one to expect analogous results in due course.

1. Relative Kan extension and lift-extension.

Let $\Phi : \underset{\sim}{V} \to \underset{\sim}{V}'$ be a normal closed functor between symmetric monoidal closed categories. As pointed out [10;5], Φ induces a 2-functor $\Phi_* : \underset{\sim}{V}\text{-Cat} \to \underset{\sim}{V}'\text{-Cat}$. Let $S : \underset{\sim}{A} \to \underset{\sim}{B}$ be a $\underset{\sim}{V}$-functor, let $J : \underset{\sim}{X} \to \Phi_*\underset{\sim}{A}$ and $\bar{S} : \underset{\sim}{X} \to \Phi_*\underset{\sim}{B}$ be $\underset{\sim}{V}'$-functors and let $i : \bar{S} \to (\Phi_*S)J$ be a $\underset{\sim}{V}'$-natural transformation. The pair (S, i) is a <u>(left) relative Kan extension of</u> $\bar{S}$ <u>along</u> J if given any $\underset{\sim}{V}$-functor $T : \underset{\sim}{A} \to \underset{\sim}{B}$ and any $\underset{\sim}{V}'$-natural transformation $\bar{u} : \bar{S} \to (\Phi_*T)J$ there exists a unique $\underset{\sim}{V}$-natural transformation $u : S \to T$ such that $(\Phi_*u)J.i = \bar{u}$. In this case we denote S as usual by $\mathrm{Lan}_J(\bar{S})$.

Now let $\bar{\bar{S}} : \underset{\sim}{X} \to \underset{\sim}{D}$ and $G : \Phi_*\underset{\sim}{B} \to \underset{\sim}{D}$ be $\underset{\sim}{V}'$-functors and let $j : \bar{\bar{S}} \to G(\Phi_*S)J$ be a $\underset{\sim}{V}'$-natural transformation. The pair (S,j) is a <u>(left) relative Kan lift-extension of</u> $\bar{\bar{S}}$ <u>along</u> J <u>and up</u> G if given any $\underset{\sim}{V}$-functor $T : \underset{\sim}{A} \to \underset{\sim}{B}$ and any $\underset{\sim}{V}'$-natural transformation $\bar{\bar{u}} : \bar{\bar{S}} \to G(\Phi_*T)J$ there exists a unique $\underset{\sim}{V}$-natural transformation $u : S \to T$ such that $G(\Phi_*u)J\cdot j = \bar{\bar{u}}$.

In the case that G has a $\underset{\sim}{V}'$-left adjoint $F : \underset{\sim}{D} \to \Phi_*\underset{\sim}{B}$ the existence of a lift-extension is assured if a relative Kan extension of $F\bar{\bar{S}}$ along J exists.

<u>Proposition 1</u>. If $(\varepsilon,\eta) : F \dashv_{\underset{\sim}{V}'} G$ and if $(\mathrm{Lan}_J(F\bar{\bar{S}}), i)$ exists then $(\mathrm{Lan}_J(F\bar{\bar{S}}), j)$ is a (left) relative Kan lift-extension of $\bar{\bar{S}}$ along J and up G, where $j : \bar{\bar{S}} \to G(\Phi_*S)J$ the unique arrow such that $i = \varepsilon(\Phi_*S)J.Fj$.

The proof of proposition 1 requires only elementary manipulation of the definitions.

2. Free objects.

A relative Kan extension can be interpreted as a free object in the metacategory $\underset{\sim}{B}^{\underset{\sim}{A}}$ of $\underset{\sim}{V}$-functors and $\underset{\sim}{V}$-natural transformations. Let $R : \underset{\sim}{U} \to \underset{\sim}{W}$ be a functor between metacategories and let $j : W \to RU$ be an arrow in $\underset{\sim}{W}$. We recall that (U,j) is <u>free over</u> W <u>with respect to</u> R (in the sense of A. Frei [3]) if given any object U' of $\underset{\sim}{U}$ and any $g : W \to RU'$ there exists a unique $g' : U \to U'$ such that $g = Rg'.j$. If we choose $R : \underset{\sim}{B}^{\underset{\sim}{A}} \to (\Phi_*\underset{\sim}{B})^{\underset{\sim}{X}}$ so that $RS = (\Phi_*S)J$ then, clearly, a relative Kan extension (S,i) is free over $\overline{S}$ with respect to R.

The following coequalizer test for free objects is related to the coequalizer construction for left adjoint [4] and to adjoint interpolation [8], [2]. The proof is omitted. Let $L : \underset{\sim}{W} \to \underset{\sim}{U}$, $N : \underset{\sim}{W} \to \underset{\sim}{W}$ be functors and let $e : N \to 1$, $\alpha : LR \to 1$, $\gamma : N \to RL$ be natural transformations such that

$$eR = R\alpha . \gamma R : NR \to R.$$

<u>Proposition 2</u> (a) If eW is epic and $\alpha U.Lj = \text{coeq}\ (\alpha LW.L\gamma W,\ LeW)$ then (U,j) is free over W with respect to R.

(b) If (U,j) is free over W with respect to R, if $eW = \text{coeq}\ (eNW, NeW)$ and if LeW is epic then $\alpha U.Lj = \text{coeq}\ (\alpha LW.L\gamma W,\ LeW)$.

3. <u>In the metacategory</u>.

Suppose that the underlying category of $\underset{\sim}{B}$ is cocomplete, that $\underset{\sim}{B}$ and $\Phi_*\underset{\sim}{B}$ are tensored and cotensored. Then $\underset{\sim}{B}$ is $\underset{\sim}{V}$-cocomplete and $\Phi_*\underset{\sim}{B}$ is $\underset{\sim}{V}'$-cocomplete. Let $[A, -]$, $[X, -]'$ denote representable $\underset{\sim}{V}$-functors, $\underset{\sim}{V}'$-functors respectively. Since $\underset{\sim}{B}$ is tensored there is an isomorphism $p : [V \otimes B, B'] \to [V,[B, B']]$, $\underset{\sim}{V}$-natural in V, B and B'. Since $\Phi_*\underset{\sim}{B}$ is tensored there is an isomorphism $p' : \Phi[V' \otimes B, B'] \to [V', \Phi[B,B']]'$, $\underset{\sim}{V}'$-natural in V', B and B'. Recall that part of the closed functor Φ is a natural transformation $\hat{\Phi}$ $\Phi[V_1, V_2] \to [\Phi V_1, \Phi V_2]'$. Φ <u>respects tensors</u> if there exists a transformation $\Theta_{VB} : \Phi V \otimes B \to V \otimes B$ $\underset{\sim}{V}$-natural in V and B such that the following diagram is commutative.

$$\begin{array}{ccc}
\Phi[V \otimes B,\ B'] & \xrightarrow{\Phi p} & \Phi[V,\ [B,\ B']] \\
\downarrow{\scriptstyle \Phi[0_{VB},1]} & & \downarrow{\scriptstyle \hat{\Phi}} \\
\Phi[\Phi V \otimes B,\ B'] & \xrightarrow{p'} & [\Phi V,\ \Phi[B,\ B']]'
\end{array}$$

If Φ respects tensors and $\underset{\sim}{X}$ is small then we may set

$$L\bar{S} = \bigvee_{X \in \underset{\sim}{X}} [JX,\ -] \otimes \bar{S}X$$

$$N\bar{S} = \bigvee_{X \in \underset{\sim}{X}} [X,\ -]' \otimes \bar{S}X$$

$$\alpha S = \sum_{X \in \underset{\sim}{X}} \tau_X : LRS \to S$$

$$e\bar{S} = \sum_{X \in \underset{\sim}{X}} \tau'_X : N\bar{S} \to \bar{S}$$

$$\gamma\bar{S} = \bigvee_{X \in \underset{\sim}{X}} \theta_{[JX,\ J-]\bar{S}X} \cdot (J \otimes 1) : N\bar{S} \to RL\bar{S},$$

where $\tau_X A : [JX,A] \otimes SJX \to SA$ corresponds by adjunction to $S : [JX,\ A] \to [SJX,\ SA]$ and $\tau_X{}'Y : [X,Y]' \otimes \bar{S}X \to \bar{S}Y$ to $\bar{S} : [X,\ Y]' \to \Phi[\bar{S}X,\ \bar{S}Y]$.

<u>Proposition 3.</u> If Φ respects tensors and $\underset{\sim}{X}$ is small then $L : (\Phi_*\underset{\sim}{B})^{\underset{\sim}{X}} \to \underset{\sim}{B}^{\underset{\sim}{A}}$ and $N : (\Phi_*\underset{\sim}{B})^{\underset{\sim}{X}} \to (\Phi_*\underset{\sim}{B})^{\underset{\sim}{X}}$ are functors and $\alpha : LR \to 1$, $e : N \to 1$, $\gamma : N \to RL$ are natural transformations such that $eR = R\alpha . \gamma R$.

<u>Proof</u>. That $L\bar{S}$ is a $\underset{\sim}{V}$-functor and $N\bar{S}$ a $\underset{\sim}{V}'$-functor is a consequence of the dual of [1; Proposition I.1.3]. In the obvious sense L and N are functorial. For each fixed $S \in \underset{\sim}{B}^{\underset{\sim}{A}}$, τ_X and hence αS is $\underset{\sim}{V}$- natural. Similarly for each fixed $\bar{S}$, $e\bar{S}$ and $\gamma\bar{S}$ are $\underset{\sim}{V}'$-natural. The naturality of α, e and γ is easily checked. By applying the base functor $\underset{\sim}{V}' \to$ Set to the commutative diagram

$$\begin{array}{ccc}
\Phi[[JX,\ JY] \otimes SJX,\ SJY] & \xrightarrow{\Phi p} & \Phi[[JX,\ JY],\ [SJX,\ SJY]] \\
\Big\downarrow{\scriptstyle \Phi[\Theta_{[JX,\ JY],\ SJX},1]} & & \Big\downarrow{\scriptstyle \hat{\Phi}} \\
\Phi[\Phi[JX,\ JY] \otimes SJX,\ SJY] & \longrightarrow & [\Phi[JX,\ JY],\ \Phi[SJX,\ SJY]] \\
\Big\downarrow{\scriptstyle \Phi[J \otimes 1,\ 1]} & & \Big\downarrow{\scriptstyle [J,\ 1]} \\
\Phi[[X,\ Y]' \otimes SJX,\ SJY] & \xrightarrow{p'} & [[X,\ Y]',\ \Phi[SJX,\ SJY]]
\end{array}$$

one obtains that $\tau_X{}^{JY} \cdot \Theta_{[JX,\ JY],\ SJX} \cdot J \otimes 1 = \tau'_X{}^Y$ from which the desired equality follows.

4. Application to C.G. Haus$_*$

The category C.G. Haus$_*$(=$\underset{\sim}{V}$) of pointed compactly-generated Hausdoff spaces, together with the smash product as tensor product and two-point space P as unit is symmetric monoidal closed [II; VI; 9]. The category C.G. Haus (=$\underset{\sim}{V}'$) together with the Cartesian product and the one-point space Q as unit is symmetric monoidal closed and the forgetful functor $\Phi : \underset{\sim}{V} \to \underset{\sim}{V}'$ is a closed functor. The natural transformation $\Phi : [V,\ V'] \to [\Phi V, \Phi V']'$ is the inclusion of the space of pointed maps into the space of unpointed maps. Choose also $\underset{\sim}{B}$ = C.G. Haus$_*$. Then $\underset{\sim}{B}$ is automatically tensored and cotensored. $\Phi_*\underset{\sim}{B}$ too is tensored, for if we take

$$V' \otimes B = V' \times B \,/\!\!/\, V' \times (*) \qquad (B \in \text{C.G. Haus}_*,\ V' \in \text{C.G. Haus}),$$

the <u>half-smash</u> of V' with B then we have a natural isomorphism $\Phi[V' \otimes B,\ B'] \to [V',\ \Phi[B, B']]'$ in C.G. Haus. Moreover if we choose $\Theta_{VB} : \Phi V \otimes B \to V \# B$ to be the arrow collapsing the other half of the wedge it can be verified that Φ respects tensors. To verify that $\Phi_*\underset{\sim}{B}$ is cotensored, note that the space of unpointed maps of V' into ΦB, endowed with the constant map at $* \in B$ as base-point, yields a cotensor of V' with B.

Now suppose that in this situation we are given a $\underset{\sim}{V}$-functor $S : \underset{\sim}{A} \to \underset{\sim}{B}$, a $\underset{\sim}{V}'$-functor $\bar{S} : \underset{\sim}{X} \to \Phi_* \underset{\sim}{B}$ and a natural transformation $i : \bar{S} \to (\Phi_* S)J$. ($i$ is necessarily $\underset{\sim}{V}'$-natural because the base functor $\underset{\sim}{V}' \to$ Set is faithful.) For each $Y \in \underset{\sim}{X}$ we have $L\bar{S}Y = \bigvee_{X \in \underset{\sim}{X}} [JX, Y] \# \bar{S}X$ and $\alpha SY . LiY : L\bar{S}Y \to SY$ is given by

$$\alpha SY . LiY . \lambda_X (f \# a) = (Sf)(iX(a)) \qquad (f \in [JX, Y],\ a \in \bar{S}X),$$

where λ_X refers to the injection arrow of the X'th summand. Since smash distributes over wedges we have

$$LN\bar{S}Y = \bigvee_{X, X' \in \underset{\sim}{X}} [JX, Y] \# ([X', X]' \otimes SX')$$

and the two arrows $Le\bar{S}Y$, $\alpha L\bar{S}Y . L\gamma\bar{S}Y : LN\bar{S}Y \to L\bar{S}Y$ are given by

$$\left.\begin{aligned} LeSY . \lambda_{X,X'}(f \# (g \otimes a)) &= \lambda_X(f \# (\bar{S}g(a))) \\ \alpha L\bar{S}Y . L\gamma\bar{S}Y . \lambda_{X,X'}(f \# (g \otimes a)) &= \lambda_{X'}(f.Jg \# a) \end{aligned}\right\} \begin{aligned} &(f \in [JX, Y], \\ &g \in [X', X]', \\ &a \in \bar{S}X'). \end{aligned}$$

<u>Theorem 4</u>. The pair (S, i) is a (left) relative Kan extension of $\bar{S}$ along J if and only if, for each $Y \in \underset{\sim}{X}$, $\alpha SY . LiY$ = coequalizer $(LeSY, \alpha L\bar{S}Y . L\gamma\bar{S}Y)$.

<u>Proof</u>. The result will follow from Propositions 2 and 3 once it has been shown that, for each $Y \in \underset{\sim}{X}$, $e\bar{S}Y$ is a split coequalizer of $eN\bar{S}Y$ and $Ne\bar{S}Y$, for then it will follow that $e\bar{S}$ is pointwise an absolute coequalizer of $eN\bar{S}$ and $Ne\bar{S}$. We have $N\bar{S}Y = \bigvee_{X \in X} [X, Y]' \otimes \bar{S}X$ and $N^2\bar{S}Y = \bigvee_{X \in \underset{\sim}{X}} [X, Y]' \otimes (\bigvee_{X' \in \underset{\sim}{X}} [X', X]' \otimes \bar{S}X')$. Suitable splitting arrows

$$\bar{S}Y \xrightarrow{\ \nu\ } N\bar{S}Y \xrightarrow{\ \sigma\ } N^2\bar{S}Y$$

are given by

$$\nu(a) = \lambda_Y (1_Y \otimes a) \qquad (a \in \bar{S}Y)$$

$$\sigma(b) = \lambda_Y (1_Y \otimes b) \qquad (b \in N\bar{S}Y).$$

Recall from [10; §5] that the closed functor Φ induces a $\underset{\sim}{V}'$-functor $\hat{\Phi} : \Phi_*\underset{\sim}{V} \to \underset{\sim}{V}'$ whose value $\hat{\Phi}V$ on objects is ΦV. In our present situation it is easy to check that $\hat{\Phi}$ has a $\underset{\sim}{V}'$-left adjoint F : C.G. Haus $\to \Phi_*$(C.G. Haus$_*$) that supplies each space with a disjoint base-point. One may therefore apply Proposition 1.

It is not hard to verify the following examples. (Q_n denotes the discrete n-point space in C.G. Haus.)

1) S = pointed square, $\bar{\bar{S}}$ = deleted square, $\underset{\sim}{X}$ the isomorphism subcategory of Q_2, $J = F|\underset{\sim}{X}$.

2) S = symmetric pointed square, $\bar{\bar{S}}$ = deleted symmetric square, $\underset{\sim}{X}$ and J as in (1).

3) $S = \Omega$, $\bar{\bar{S}}$ = identity, $\underset{\sim}{X}$ subcategory containing only Q_1, $J = \Sigma F|\underset{\sim}{X}$.

4) P- functors : $\underset{\sim}{X}$ the full subcategory containing P, J = inclusion, $\bar{S} = S|\underset{\sim}{X}$.

5) The functor $C(M, X)$ recently considered by F.R. Cohen : take $\underset{\sim}{X}$ the category containing Q_n, $n \geq 1$ whose arrows are isomorphisms and inclusions $Q_n \to Q_{n+1}$; $\bar{S}\, Q_k = F(M, k)$.

References

[1] E.J. Dubuc. Kan extensions in enriched category theory. Lecture Notes in Math. 145, Springer-Verlag 1970.

[2] E.J. Dubuc. Adjoint triangles. Reports of the Midwest Category Seminar II. Lecture Notes in Math. 61, Springer-Verlag 1968.

[3] A. Frei. Freie Objekte und multiplikative Strukturen. Math. Zeitschr. 93 (1966), 109-141.

[4] K.A. Hardie. A coequalizer construction for left adjoint. Math. Colloq. Univ. Cape Town, 6 (1970-1971) 111 - 114.

[5] K.A. Hardie. Homotopy of natural transformations. Canadian J. Math. 22 (1970), 332 - 341.

[6] K.A. Hardie. Weak homotopy equivalence of P-functors. Quart. J. Math. Oxford (2) 19 (1968), 17 - 31.

[7] P.J. Hilton. Generalizations of the Hopf invariant, Colloque de topologie algébrique, Louvain 1956.

[8] S.A. Huq. An interpolation theorem for adjoint functors. Proc. Amer. Math. Soc. 25 (1970), 880 - 883.

[9] I.M. James. Reduced product spaces. Annals of Math. (2) 62 (1955), 170 - 197.

[10] G.M. Kelly. Adjunction for enriched categories. Reports of the Midwest Category Seminar III. Lecture Notes in Math. 106. Springer-Verlag 1969.

[11] S. MacLane. Categories for the working mathematician. Springer-Verlag 1971.

[12] R.J. Milgram. Iterated loop spaces. Annals of Math. 84 (1966), 386 - 403.

[13] J.P. May. The geometry of iterated loop spaces. Lecture Notes in Mathematics 271. Springer-Verlag 1972.

[14] D. Puppe. Homotopiemengen und ihre induzierten Abbildungen I, Math. Zeitschr. 69 (1958), 299 - 344.

Grants to the Topology Research Group by the University of Cape Town and the South African Council for Scientific and Industrial Research are acknowledged.

Department of Mathematics,
University of Cape Town,
Rondebosch, 7700.
Republic of South Africa.

TOPOLOGICAL FUNCTORS FROM FACTORIZATION

J. Martin Harvey

Department of Mathematics, University of Rhodesia,
Salisbury, Rhodesia.

ABSTRACT: A "topological" functor $T : \underline{A} \to \underline{X}$ gives rise to factorizations of sources in $\underline{A}$ and $\underline{A}^{op}$ by means of initiality (and, resp. coinitiality). Conversely, a topological functor $T : \underline{A} \to \underline{X}$ may be reconstructed, up to a natural degree of uniqueness, given suitable factorizations on $\underline{A}$ and $\underline{A}^{op}$.

AMS Subj. Class: 18A20, 18A40, 18B99, 18D30

topological functor factorization

INTRODUCTION:

Various, more or less, equivalent concepts of what might be generically termed "topological functors" have been introduced and studied by several authors (cf. references in the papers cited here). It is well known that a topological functor induces factorizations of sources in its domain category (and the opposite category) through initial (and, respectively, coinitial) sources. The problem of reconstructing a topological functor $T : \underline{A} \to \underline{X}$ from suitable factorizations of sources in $\underline{A}$ and $\underline{A}^{op}$ has been solved for (fibre-small) absolutely topological functors (in the sense of [8]) by Hoffmann [10].

In this paper, we solve the more general problem, posed in [10; 1.7], of reconstructing $(\underline{E}, \underline{M})$-topological functors (cf. [8]) from suitable factorizations on the domain category and its opposite, and obtain a similar reconstruction of $(\underline{E}, \underline{M})$-universally topological functors (which have been introduced to axiomatize the T_0-axiom of topology. (cf. [5, 9, 11]). These reconstructions are shown to have certain natural uniqueness properties, and Hoffmann's results are recovered by specialization.

No claim of exhaustiveness attaches to our references; a fuller bibliography on topological functors may be obtained from these.

1. Preliminaries

In this section we establish some basic definitions and facts concerning topological functors and factorizations of sources. We assume familiarity with [8] and adopt the notations and conventions thereof, unless the contrary is explicitly stated. All subcategories are assumed full and isomorphism closed.

1.1. Definitions. Let $\underline{X}$ be a category, $\underline{E}$ a class of $\underline{X}$-morphisms and $T : \underline{A} \to \underline{X}$ a functor.

(1) $\underline{E}$ is said to induce a factorization in $\underline{X}$ if $\underline{X}$ is an $(\underline{E}, \underline{M})$-category for some class $\underline{M}$ of sources in $\underline{X}$ (in the sense of [8]).

(2) $\underline{E}$ is called germinal in $\underline{X}$ if $\underline{X}$ has an $\underline{E}$-reflective subcategory $\underline{B}$, with inclusion functor $J : \underline{B} \to \underline{X}$, such that each morphism $e : X \to B$ in $\underline{E}$, with B a $\underline{B}$-object, is J-universal. $\underline{B}$ is called the $\underline{E}$-germ of $\underline{X}$.

(3) An $\underline{X}$-object X is called $\underline{E}$-injective if for each pair of $\underline{X}$-morphisms $f : Y \to X$ and $e : Y \to Z$ with e in $\underline{E}$, there exists $g : Z \to X$ in $\underline{X}$ with $g \cdot e = f$; and $\underline{X}$ is said to have enough E-injectives if for each $\underline{X}$-object X there exists an $\underline{X}$-morphism $e : X \to Y$ in $\underline{E}$ with $\underline{Y}$ $\underline{E}$-injective (cf. [11]).

(4) T is called fibre small if each T-fibre has a small skeleton.

(5) Given that $\underline{X}$ is an $(\underline{E}, M)$-category, T is called an (E, M)-topological functor if each source $(X, f_i : X \to TA_i)_I$ in $\underline{X}$, with $(A_i)_I$ a family of $\underline{A}$-objects, has an $(\underline{E}, \underline{M})$-factorization $f_i = Tm_i \cdot e$, for some T-initial source $(A, m_i ; A \to A_i)_I$ in $\underline{A}$ (cf. [8]).

(6) T is called an absolutely topological functor if it is an

($\underline{E}$, $\underline{M}$)-topological functor for all factorizations of sources ($\underline{E}$, $\underline{M}$) in $\underline{X}$. (cf. [8]).

(7) Given that $\underline{X}$ is an ($\underline{E}$, $\underline{M}$)-universally topological functor iff it is an absolutely topological functor and the class $\underline{E}^T$ of all T-initial morphisms (= $\underline{1}$-indexed sources) $e : A \to B$ with Te in $\underline{E}$ is germinal in $\underline{A}$. (cf. [5, 9, 11]).

1.2 Remark. Every ($\underline{E}$, $\underline{M}$)-topological functor $T : \underline{A} \to \underline{X}$ is faithful and $\underline{A}$ is an ($\underline{E}_T$, $\underline{M}_T$)-category, where $\underline{E}_T$ is the class of all $\underline{A}$-morphisms sent by T to $\underline{E}$ and $\underline{M}_T$ is the class of all T-initial sources in $\underline{A}$ sent by T to sources in $\underline{M}$. (cf. [8]). A faithful functor is absolutely topological if (iso, source)-topological. The following result is an extension of Hoffmann's [12: 1.1, 1.2 and 9: 1, 5] (cf. [6] for a more general result).

1.3 Theorem. Let $\underline{E}$ be a class of epimorphisms in category $\underline{X}$, containing all isomorphisms and closed under composition.
Then the following three conditions are equivalent:

(1) $\underline{E}$ induces a factorization of sources in $\underline{X}$.

(2) (a) Sources consisting of $\underline{E}$-morphisms have (possibly large) multiple pushouts in $\underline{X}$ whose arrows (each) belong to $\underline{E}$.

(b) Each 2-indexed source $(f : X \to Y,\ e : X \to Z)$ in $\underline{X}$ with e in $\underline{E}$ has a pushout $(u : Y \to W,\ g : Z \to W)$ with u in $\underline{E}$.

(3) The functor $\Delta_o : \underline{X}_{\underline{E}} \to \underline{X}$, constructed as follows, is absolutely topological : $\underline{X}_{\underline{E}}$ is the category whose objects are $\underline{E}$-morphisms $e : X \to Y$ and whose morphisms are pairs $(f,g) : e \to e'$ with $f : X \to Y'$, $g : Y \to Y'$ and $e' \cdot f = g \cdot e$, with composition defined component-wise. The functor Δ_o sends $e : X \to Y$ to X and (f,g) to f. Also, if $pq \in \underline{E}$ and q is epic, then $p \in \underline{E}$.

1.4 Proposition. Let $\underline{E}$ be a class of epimorphisms in a category $\underline{X}$. Then the following are equivalent :

(1) $\underline{E}$ is germinal in $\underline{X}$.

(2) $\underline{X}$ has enough $\underline{E}$-injectives.

Proof: (1) implies (2) : Let $\underline{B}$ be the $\underline{E}$-germ of $\underline{X}$, with inclusion functor $J : \underline{B} \to \underline{X}$. We show that all $\underline{B}$-objects are $\underline{E}$-injective, and note, in passing, that the converse also holds. To this end, suppose $(f : X \to Y,\ e : X \to Z)$ is a $\underline{2}$-indexed source with Y in $\underline{B}$ and e in

$\underline{E}$, and let $u : Z \to W$ be a J-universal morphism. Then $u \cdot e : X \to W$ is J-universal and, hence, there exists $h : W \to Y$ in $\underline{X}$ with $h \cdot (u \cdot e) = f$. Thus, there exists $g : h \cdot u : Z \to Y$ in $\underline{X}$ with $g \cdot e = f$: and hence, Y is $\underline{E}$-injective.

(2) implies (1) : It is clear that the full subcategory $\underline{B}$ whose object class consists of the $\underline{E}$-injective objects of $\underline{X}$ is the $\underline{E}$-germ of $\underline{X}$.

1.5 Corollary. [11 : 2.7] Let $\underline{X}$ be an $(\underline{E}, \underline{M})$-category and $T : \underline{A} \to \underline{X}$ be an absolutely topological functor. Then T is an $(\underline{E}, \underline{M})$-universally topological functor if $\underline{A}$ has enough $\underline{E}^T$-injectives.

1.6 Remark. Every class $\underline{E}$ of epimorphisms inducing a factorization $(\underline{E}, \underline{M})$ of sources in some category $\underline{X}$ is germinal, where an $\underline{X}$-object X belongs to the $\underline{E}$-germ of $\underline{X}$ if the empty source $(X, \emptyset)$ is in $\underline{M}$ (cf. [9, 11]). The converse is not true, as seen from considering $\underline{X} = \underline{Set}$ and $\underline{E}$ = {constant epimorphisms}.

1.7 Theorem. [8] Let $\underline{X}$ be an $(\underline{E}, \underline{M})$-category and $T : \underline{A} \to \underline{X}$ a functor. Then the following are equivalent :

(1) T is $(\underline{E}, \underline{M})$-topological.

(2) $\underline{A}$ is an $(\underline{E}_T, \underline{M}_T)$-category and T has a left adjoint F with front-adjunctions $\eta_X : X \to TFX$ belonging to $\underline{E}$.

Further, the following is true :

(3) T has a left adjoint F with back-adjunctions $\varepsilon_A : FTA \to A$ such that $T\varepsilon_A$ is an isomorphism.

(4) $\underline{A}$ is $\underline{E}_T$-cowell-powered if T is fibre-small and $\underline{X}$ is $\underline{E}$-cowell-powered, the converse holding if T is dense.

(5) $\underline{A}^{op}$ is an (iso_T, T-coinitial sink)-category. (cf. [13 : 1.2]).

2. Main Results

2.1 Definition. Let $\underline{A}$ be a category and $(\mathcal{E}, \mathcal{E}^*)$ be a pair of morphism classes in $\underline{A}$. Then $(\mathcal{E}, \mathcal{E}^*)$ is called an intrinsic topological structure on $\underline{A}$ if the following conditions hold :

(1) $\mathcal{E}$ induces a factorization on $\underline{A}$ and $\mathcal{E}^*$ induces a factorization on $\underline{A}^{op}$.

(2) $\mathcal{E}$ and $\mathcal{E}^*$ are compatible in the sense that

(i) $\mathcal{E}^* \subset \mathcal{E}$;

(ii) $p \cdot q \in \mathcal{E} \wedge p \in \mathcal{E}^* \Rightarrow q \in \mathcal{E}$;

(iii) $p \cdot q \in \mathcal{E}^* \wedge q \in \mathcal{E}^* \Rightarrow p \in \mathcal{E}^*$.

2.2 Theorem. Let $\underline{X}$ be an $(\underline{E}, \underline{M})$-category and $T : \underline{A} \to \underline{X}$ be an $(\underline{E}, \underline{M})$-topological functor. Then $(\underline{E}_T, \mathrm{Iso}_T)$ is an intrinsic topological structure on $\underline{A}$.

Proof: Straightforward, noting Theorem 1.7 (4)

2.3 Theorem. Let $\underline{A}$ be an $(\underline{\mathcal{E}}, \underline{\mathfrak{M}})$-category and $(\mathcal{E}, \underline{\mathcal{E}}^*)$ be an intrinsic topological structure on $\underline{A}$. Then, there exists an $(\underline{E}, \underline{M})$-category $\underline{X}$ and a dense $(\underline{E}, \underline{M})$-topological functor $T : \underline{A} \to \underline{X}$ with $\underline{\mathcal{E}} = \underline{E}_T$, $\underline{\mathfrak{M}} = \underline{M}_T$ and $\underline{\mathcal{E}}^* = \mathrm{Iso}_T$.

Proof: Let $\underline{X}$ be the subcategory of $\underline{A}$ such that $\underline{X}^{op}$ is the $\underline{\mathcal{E}}$*-germ of $\underline{A}^{op}$, $J : \underline{X} \to \underline{A}$ be the inclusion functor, with right adjoint $T : \underline{A} \to \underline{X}$ and back adjunctions $\varepsilon_A : JTA \to A$, $\underline{E}$ the class of all $\underline{X}$-morphisms $e : X \to Y$ in $\underline{\mathcal{E}}$ and $\underline{M}$ the class of all sources $(X, m_i : X \to X_i)$ in $\underline{X}$ such that $m_i = \mu_i \cdot \varepsilon$ $(i \in I)$ for some morphism $\varepsilon : X \to Y$ in $\underline{\mathcal{E}}^*$ and source $(Y, \mu_i : Y \to X_i)_I$ in $\underline{\mathfrak{M}}$.

We verify the details in the following series of lemmas.

2.3.1 $\underline{X}$ is an $(\underline{E}, \underline{M})$-category.

Proof: $(\underline{E}, \underline{M})$-factorizability: Let $(X, f_i : X \to X_i)_I$ be a source in $\underline{X}$ with an $(\underline{\mathcal{E}}, \underline{\mathfrak{M}})$-factorization $f_i = \mu_i \cdot \varepsilon$ $(i \in I)$, where $\varepsilon : X \to Y$. Then, there exists $e : X \to TY$ with $\varepsilon_Y \cdot e = \varepsilon$. By choice of $\underline{\mathcal{E}}^*$, $e \in \underline{E}$ and, further, $(TY, \mu_i \cdot \varepsilon_Y)_I$ is in $\underline{M}$, so that $f_i = (\mu_i \cdot \varepsilon_Y)$ $(i \in I)$ is the required $(\underline{E}, \underline{M})$-factorization. $(\underline{E}, \underline{M})$-diagonalization: Suppose f and e are morphisms in $\underline{X}$ and $(Y, \mu_i\ \varepsilon)_I$ and (Z, f_i) are sources in $\underline{X}$, with e in $\underline{E}$, $\varepsilon : Y \to W$ in $\underline{\mathcal{E}}^*$, $(W, \mu_i)_I$ in $\underline{\mathfrak{M}}$ and $f_i \cdot e = (\mu_i \cdot \varepsilon) \cdot f$ $(i \in I)$. Then, by the $(\underline{\mathcal{E}}, \underline{\mathfrak{M}})$-diagonalization property and the J-couniversality of ε, there exist (successively) $g : Z \to W$ and $h : Z \to Y$ making the following diagram commute :

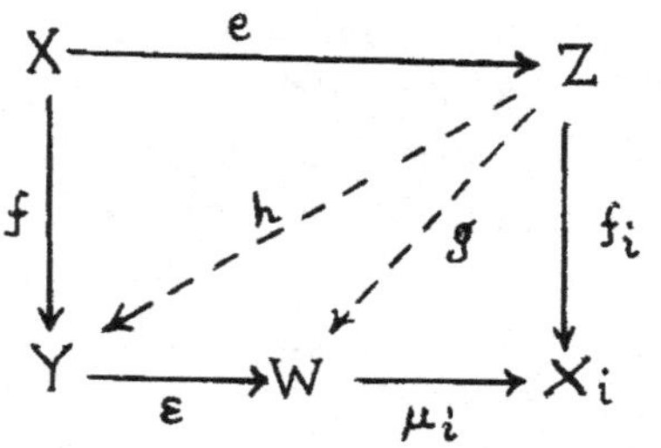

2.3.2 $T : \underline{A} \to \underline{X}$ is a faithful functor and every source $(A, \mu_i : A \to A_i)_I$ in $\mathfrak{M}$ is T-initial.

Proof: That T is faithful is evident from the fact that the back-adjunctions $\varepsilon_A : TA \to A$ are epimorphisms. Now, suppose that $(B, f_i : B \to A_i)_I$ is a source in $\underline{A}$ and $g : TB \to TA$ and $\underline{X}$-morphism with $Tf_i = T\mu_i \cdot g$ $(i \in I)$. Then, by the $(\mathcal{E}, \mathfrak{M})$-diagonalization property, there exists an $\underline{A}$-morphism $f : B \to A$ making the following diagram (in $\underline{A}$) commute, where $\varepsilon_i = \varepsilon_{A_i}$ $(i \in I)$. This makes it clear that $(A, \mu_i)_I$ is T-initial:

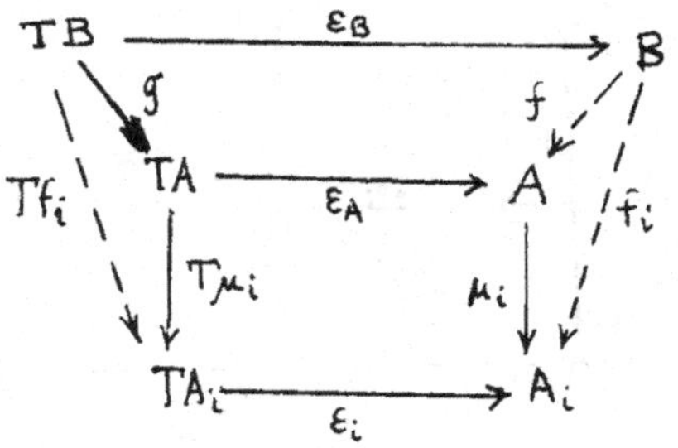

2.3.3 $T : \underline{A} \to \underline{X}$ is an $(\underline{E}, \underline{M})$-topological functor.

Proof: Let $(A_i)_I$ be a family of $\underline{A}$-objects and $(X, m'_i : X \to TA_i)_I$ a source in $\underline{M}$, where $m'_i = \mu_i \cdot \varepsilon (i \in I)$, for some $\varepsilon : X \to Y$ in $\mathcal{E}^*$ and $(Y, \mu_i)_I$ in $\mathfrak{M}$; and let $\mu_i := \varepsilon A_i \cdot m'_i = m_i \cdot e$ $(i \in I)$ be an $(\mathcal{E}, \mathfrak{M})$-factorization of $(X, \mu_i : X \to A_i)_I$, with $e : X \to A$. Then, $(A, m_i : A \to A_i)_I$ is the required T-initial lift of $(X, m_i)_I$. To show this, it clearly suffices to show that $e \in \mathcal{E}^*$, in view of 2.3.2. To this end, consider the (commutative) diagram below, constructed as indicated in the following argument:

(1) $\overline{e} : X \to TA$ is the unique morphism with $\varepsilon_A \overline{e} = e$; we observe that by the compatibility condition for $\underline{\mathcal{E}}$ and $\underline{\mathcal{E}}^*$ above, $\overline{e} \in \underline{\mathcal{E}}$.

(2) $g : Y \to A$ is determined by the $(\underline{\mathcal{E}}, \underline{\mathfrak{M}})$-diagonalization property, noting that $\varepsilon \in \underline{\mathcal{E}}$.

(3) For each $i \in I$, $\varepsilon_i \cdot \overline{m}_i = m_i \cdot \overline{\varepsilon}_i$ in a pull back diagram,

where $\varepsilon_i = \varepsilon A_i$, and $\delta_i : Y \to P_i$ is determined by the pullback property. Here, we note the use of 1.3(2)(b), from which follows that $\overline{\varepsilon}_i \in \underline{\mathcal{E}}^*$, for each $i \in I$.

(4) Hence, for each $i \in I$, $\xi_i = \varepsilon_i \cdot p_i \in \underline{\mathcal{E}}^*$, and so there exists an isomorphism $\eta_i : A \to P_i$ with $\xi_i \cdot \eta_i = \varepsilon_A$.

(5) The fact that, for each $i \in I$, $\overline{\varepsilon}_i$ is a monomorphism implies $\delta_i \cdot \varepsilon = (\varepsilon_{P_i} \cdot \eta_i) \cdot \overline{e}$, for each $i \in I$. Hence, $\overline{e}$ is an isomorphism and (therefore) $e \in \underline{\mathcal{E}}^*$, since $(Y, \delta_i)_I$ is a source in $\underline{\mathcal{M}}$.

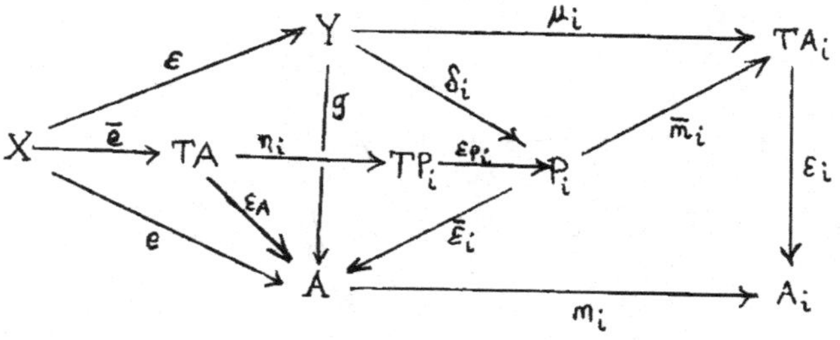

2.3.4 $\quad \underline{\mathcal{E}}^* = \mathrm{iso}_T$ and $(\underline{\mathcal{E}}, \underline{\mathcal{M}}) = (\underline{E}_T, \underline{M}_T)$.

Proof: That $\underline{\mathcal{E}}^* = \mathrm{iso}_T$ and $\underline{\mathcal{E}} = \underline{E}_T$ is evident from the above. In view of 2.3.2, it clearly suffices to show that if $(A, m_i : A \to A_i)_I$ is a source in $\underline{\mathcal{M}}$, then $(TA, Tm_i)_I$ is in $\underline{M}$. Let $Tm_i = \mu_i \cdot e$ $(i \in I)$ be an $(\underline{\mathcal{E}}, \underline{\mathcal{M}})$-factorization of $(TA, Tm_i)_I$. We show that $e \in \underline{\mathcal{E}}^*$. By the $(\underline{\mathcal{E}}, \underline{\mathcal{M}})$-diagonalization property, there exists $d : B \to A$ making the following diagram, in which $\varepsilon_i = \varepsilon_{A_i}$, commute (for each $i \in I$); and hence $e \in \underline{\mathcal{E}}^*$, since $d \cdot e = \varepsilon_A \in \underline{\mathcal{E}}^*$ and $\underline{\mathcal{E}}^*$ induces a factorization of sources in $\underline{A}^{op}$:

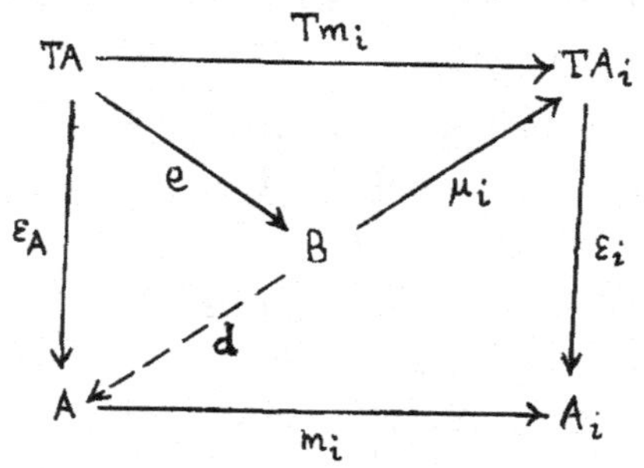

2.4 Remarks. In view of Theorem 1.7, in the above results, T is fibre-small and $\underline{X}$ is $\underline{E}$-cowell-powered iff $\underline{A}$ is $\underline{\mathcal{E}}$-cowell-powered. Further, $\underline{\mathcal{E}}^*$ induces the (iso_T, T-coinitial sink) factorization in $\underline{A}^{op}$, a fact which may be proved (directly) from factorization properties.

2.5 Corollary. (cf. [10]). Let $\underline{A}$ be a category in which a class $\underline{\mathcal{E}}$ of bimorphisms induces a factorization of sources in $\underline{A}$ as well as A^{op}. Then there is an absolutely topological functor $T : \underline{A} \to \underline{X}$ with $\underline{\mathcal{E}} = \mathrm{iso}_T$.

Proof: Apply 2.1, with $\underline{\mathcal{E}} = \underline{\mathcal{E}}^*$, noting that the compatibility condition on $(\underline{\mathcal{E}}, \underline{\mathcal{E}}^*)$ holds (trivially), since $\underline{\mathcal{E}}$ induces a factorization on $\underline{A}^{op}$, and that every absolutely topological functor is dense.

2.6 Theorem. Suppose $\underline{A}$ is an $(\underline{\mathcal{E}}, \underline{\mathcal{M}})$-category and $\underline{\mathcal{E}}^*$ is a class of bimorphisms in $\underline{\mathcal{E}}$ inducing a factorization $(\underline{\mathcal{E}}^*, \underline{\mathcal{M}}^*)$ of sources in $\underline{A}$, as well as a factorization of sources in $\underline{A}^{op}$. Further, suppose that $e \cdot u \in \underline{\mathcal{E}}$ and $e \in \underline{\mathcal{E}}^*$ implies $u \in \underline{\mathcal{E}}$, and that $\underline{\mathcal{E}} \cap \underline{\mathcal{M}}^*$ is germinal. Then there exists an $(\underline{E}, \underline{M})$-category $\underline{X}$ and an $(\underline{E}, \underline{M})$-universally topological functor $T : \underline{A} \to \underline{X}$ with $(\underline{\mathcal{E}}, \underline{\mathcal{M}}) = (\underline{E}_T, \underline{M}_T)$ and $\underline{\mathcal{E}}^* = \mathrm{iso}_T$.

Proof: By Corollary 2.5 above, there exists an absolutely topological functor $T : \underline{A} \to \underline{X}$, with $\underline{\mathcal{E}}^* = \mathrm{iso}_T$. Further, $\underline{X}$ is an $(\underline{E}, \underline{M})$-category, by a similar proof to that of 2.3.4; and as $\underline{\mathcal{E}} \subset \underline{\mathcal{E}}^*$, $\underline{\mathcal{M}} = \underline{M}_T$ and (hence) $(\underline{\mathcal{E}}, \underline{\mathcal{M}}) = (\underline{E}_T, \underline{M}_T)$. Clearly, $\underline{\mathcal{E}} \cap \underline{\mathcal{M}}^* = E^T$, and hence $T : \underline{A} \to \underline{X}$ is an $(\underline{E}, \underline{M})$-universally topological functor.

2.7 Proposition. Let $T : \underline{A} \to \underline{X}$ and $U : \underline{A} \to \underline{Y}$ be dense right adjoint functors. Then the following are equivalent:

(1) $\mathrm{ISO}_T = \mathrm{ISO}_U$

(2) There exist equivalences $V : \underline{X} \to \underline{Y}$ and $W : \underline{Y} \to \underline{X}$ such that

$$V \cdot W \simeq \mathrm{Id}_{\underline{Y}}, \quad W \cdot V \simeq \mathrm{Id}_{\underline{X}}, \quad T \simeq V \cdot S \quad \text{and} \quad S \simeq W \cdot T$$

Proof: (2) implies (1) easily. For the converse, let K T and L U. Then $V = U \cdot K$ and $W = T \cdot L$ suffice.

2.8 Theorem. Let $(\underline{\mathcal{E}}, \underline{\mathcal{E}}^*)$ be an intrinsic topological structure on a category $\underline{A}$ and $U : \underline{A} \to \underline{Y}$ be a dense right adjoint functor with $\underline{\mathcal{E}}^* = \mathrm{ISO}_U$. Then $\underline{Y}$ is an $(\underline{E}, \underline{M})$-category, for some $\underline{E}$, and $U : \underline{A} \to \underline{Y}$ is an $(\underline{E}, \underline{M})$-topological functor with $\underline{\mathcal{E}} = \underline{E}_U$.

Proof: Apply Proposition 2.7 above with $T : \underline{A} \to \underline{X}$ the prototype functor constructed in Theorem 2.3 and $(\underline{E}, \underline{M})$ naturally induced on $\underline{Y}$ by equivalence to $\underline{X}$.

2.9 Corollary. Let $U : \underline{A} \to \underline{X}$ be a dense right adjoint functor such that Iso_U induces a factorization on $\underline{A}$ and $\underline{A}^{op}$. Then U is an absolutely topological functor.

2.10 Remarks.

(1) A further similar corollary may be obtained from Theorem 2.6 and Proposition 2.7.

(2) Item (2) of Theorem 1.3 proves useful in applications.

For instance, we have the following result.

2.11 Theorem [1,2] The forgetful functor $U : \underline{Cont} \to \underline{Prox}$, from the category of contiguity spaces to the category of generalized proximity spaces, is absolutely topological.

Proof: U is a dense right adjoint functor, as well as being a left adjoint functor. (cf. [2]). Hence, it preserves limits and colimits, so that Iso_U induces a factorization in $\underline{Cont}$ and $\underline{Cont}^{op}$.

2.12 Definition. (cf. [3, 4, 7]). A category $\underline{A}$ is autotrivial if each equivalence $F : \underline{A} \to \underline{A}$ is natural isomorphic to the identity functor on $\underline{A}$.

2.13 Examples.

(1) Examples of autotrivial categories : Set, Top, $\underline{Top}_o$, $\underline{Top}_1$, $\underline{Top}_2$ (cf. [3]), where $\underline{Top}_n$ is the full subcategory of Top consisting of T_n-spaces, and all concrete finitary algebraic categories $\underline{A}$ with a "distinguishable algebraic separator" i.e. a regular-projective regular-separator S such that $FS \simeq S$ for each equivalence F on $\underline{A}$ (cf. [7]).

(2) Some counter-examples are Cat and Preord.

Uniqueness results for topological functors with autotrivial 'ground categories' (codomains) follow from the following corollary of Proposition 2.7.

2.14 Proposition. Let $T,U : \underline{A} \to \underline{X}$ be dense right adjoint functors, where $\underline{X}$ is an autotrivial category. Then the following are equivalent:

(1) $Iso_T = Iso_U$

(2) $T \simeq U$.

REFERENCES

[1] Bentley, H.L. The role of nearness spaces in topology. Proc.Conf. Categorical Topology, Mannheim 1975, pp 1-22, LNM 540, Springer-Verlag, Berlin 1976.

[2] Bentley, H.L. and Herrlich, H. The forgetful functor Cont → Prox is topological. *Questiones Math.* 2 (1977), 44 - 57

[3] Freyd, P. Abelian Categories : and introduction to the theory of functors, Harper and Row, New York 1964.

[4] Give'on, Y. Transparent categories and categories of transition systems. Proc.Conf. Categorical Algebra, La Jolla 1965, pp 317-330, Springer, Berlin 1966.

[5] Harvey, J.M. T_0 separation in topological categories. *Questiones Math.* 2 (1977), 177 - 190.

[6] ———— Aspects of semi-topological functors. Preprint.

[7] ———— Autotrivial algebraic categories. Preprint.

[8] Herrlich, H. Topological functors. *Gen. Topology Appl.* 4 (1974), 125 - 142.

[9] Hoffmann, R.E. (E,M)-universally topological functors. Habilitationsschrift. Univ.Dusseldorf 1974.

[10] ———— Topological functors and factorizations, *Arch. Math. (Bassel)* 26 (1975), 1 - 7.

[11] ———— Topological functors admitting generalized Cauchy completions. in LNM 540, pp 286-344.

[12] ———— Factorization of cones. Preprint.

[13] Nel, L.D. Initially structured categories and contesian closedness. *Canad. J.Math.* 27 (1975), 1361-1377.

Department of Mathematics,
University of Rhodesia,
P. O. Box MP 167,
Mount Pleasant,
Salisbury.
Rhodesia.

Groupoids and classification sequences

Philip R. Heath, St. John's
Klaus Heiner Kamps, Hagen

0. Introduction.

It is known that R. Brown's exact sequence of a fibration of groupoids ([RB1], 4.3) can be used to deduce many standard long exact sequences of homotopy theory for example the homotopy sequence of a cofibre map - the Puppe sequence - and their duals, even in abstract homotopy theory (see [RB1], [He], [K2], [K3]).
It is our feeling and indeed experience ([K4]) that other multiple situations can be covered using homotopy theory of groupoids, which is, as a rule, much simpler than the corresponding theory of topological spaces.

In this work we derive a classification sequence in the theory of groupoids and use it to deduce classification sequences in obstruction theory and in localization in homotopy theory. Our approach is motivated by two considerations; firstly to use groupoids to separate the geometric from the algebraic ingredients in such a way as to allow for maximum simplicity; secondly to search for unifying factors in localization theory (particularly in the many and varied five lemma type of arguments) that will at the same time define the precise algebraic conditions the topology is required to reflect.

In this preliminary report we deal only with setting up the sequence - the study of naturality will be left until a later time. It will be shown elsewhere that a variation of the ideas considered here solve the problem of the lack of duality mentioned in section 2 below and in [HMRS].

Finally we have not aimed, here, at the maximum generality but rather indicated that the proofs given in the topological situation do indeed allow a formulation in abstract homotopy theory as it has been developed in [K1] and [HK].

1. A classification sequence for groupoids.

In this section let

(1.1)
$$\begin{array}{ccc} A & \xrightarrow{\bar{f}} & D \\ {\scriptstyle \bar{p}}\downarrow & & \downarrow{\scriptstyle p} \\ C & \xrightarrow[f]{} & B \end{array}$$

be a pullback diagram in the category *Gd* of groupoids such that

(1) C is 1-connected, i.e. $Ob(C) \neq \emptyset$ and $C(x,y)$ has exactly one element for all $x,y \in Ob(C)$,

(2) p or f is a fibration.

Diagram (1.1) gives rise to a commutative diagram of sets

(1.2)
$$\begin{array}{ccc} \pi_o A & \xrightarrow{\bar{f}} & \pi_o D \\ {\scriptstyle \bar{p}}\downarrow & & \downarrow{\scriptstyle p} \\ \pi_o C & \xrightarrow[f]{} & \pi_o B \end{array} ,$$

where π_o denotes the set of components. $\pi_o C$ is a singleton with unique element $*$. Thus $\pi_o B$ has a canonical base point $* = f(*)$.

If we choose an arbitrary object a of A, then *all* the maps in (1.2) are pointed, the base points of $\pi_o A$ and $\pi_o D$ being $\tilde{a}$ and $\tilde{d}$, where $d = \bar{f}(a)$ and $\tilde{}$ denotes the component.

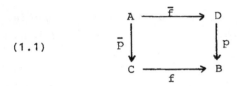

1.3. Proposition. *For each* $a \in Ob(A)$ *there exists an operation of* $B\{b\}$ *on* $\pi_o A$,

$$B\{b\} \times \pi_o A \xrightarrow{\bullet} \pi_o A, \quad (\beta,\tilde{a}_1) \mapsto \beta\bullet\tilde{a}_1,$$

and a sequence of homomorphisms and pointed maps

$$S(a): D\{d\} \xrightarrow{p_a} B\{b\} \xrightarrow{\partial_a} \pi_o A \xrightarrow{\bar{f}} \pi_o D \xrightarrow{p} \pi_o B,$$

where $b = p\bar{f}(a)$, $d = \bar{f}(a)$, $\partial_a(\beta) = \beta\bullet\tilde{a}$ *and* p_a *is the homomorphism of object groups induced by* p, *which is exact in the following sense:*

(1) $\operatorname{Im}\bar{f} = p^{-1}(*)$.

(2) *Let* $\beta_1,\beta_2 \in B\{b\}$. *Then* $\partial_a(\beta_1) = \partial_a(\beta_2)$ *if and only if* $\beta_1^{-1}\beta_2 \in \operatorname{Im} p_a$.

(3) *Let* $\tilde{a}_1, \tilde{a}_2 \in \pi_o A$. *Then* $\bar{f}(\tilde{a}_1) = \bar{f}(\tilde{a}_2)$ *if and only if there exists* $\beta \in B\{b\}$ *such that* $\beta\bullet\tilde{a}_1 = \tilde{a}_2$.

Proposition 1.3 gives us a factorization of ∂_a as

$$B\{b\} \twoheadrightarrow B\{b\}/\operatorname{Im} p_a \rightarrowtail \pi_o A.$$

Thus, as we vary a, we embed the various left cosets $B\{p\bar{f}(a)\}/\operatorname{Im} p_a$ in $\pi_o A$ allowing the following classification of $\pi_o A$, which is analogous to [RB1], 4.4.

1.4. Corollary. *There is a bijection*

$$\bigsqcup_{a\in R} B\{p\bar{f}(a)\}/\operatorname{Im} p_a \cong \pi_o A \qquad ,$$

where the disjoint union of the left hand side is taken over any set R *of objects of* A *which is mapped bijectively by* $\bar{f}: A \to D$ *on* $\bar{f}(R)$, *and such that* $\bar{f}(R)$ *is a complete set of representatives of the subset* $p^{-1}(*)$ *of* $\pi_o D$.

<u>Remark.</u> It is not hard to show the existence of such a set R using the assumption that f or p is a fibration.

<u>Proof of 1.3.</u> The proof is based on R. Brown's exact sequence [RB1], 4.3 and standard arguments of homotopy theory, but in the category of groupoids (see [K2], 5, [K1]).

<u>First case:</u> p is a fibration.
Consider the diagram

$$\begin{array}{ccccc} F & \xrightarrow{\bar{j}} & A & \xrightarrow{\bar{f}} & D \\ \downarrow & & \downarrow{\scriptstyle \bar{p}} & & \downarrow{\scriptstyle p} \\ * & \xrightarrow[j]{} & C & \xrightarrow[f]{} & B \end{array},$$

where $j(*) = \bar{p}(a)$ and the left hand square and hence the whole diagram is a pullback. Thus F is isomorphic to the fibre $p^{-1}(b)$ of p over $b = f\bar{p}(a)$. We have R. Brown's exact sequence for the fibration p and the object d of D:

$$D\{d\} \xrightarrow{p_a} B\{b\} \longrightarrow \pi_0 F \longrightarrow \pi_0 D \xrightarrow{p} \pi_0 B.$$

Since pullbacks preserve fibrations, $\bar{p}$ is a fibration. j is a homotopy equivalence, since C is 1-connected. Thus $\bar{j}$ is a homotopy equivalence and hence induces a bijection $\bar{j} : \pi_0 F \to \pi_0 A$.

<u>Second case:</u> f is a fibration.
We factorize $p : D \longrightarrow B$ as a homotopy equivalence h, followed by a fibration p'. We have a diagram with three pullbacks

$$\begin{array}{ccc} A & \xrightarrow{\bar{f}} & D \\ {\scriptstyle \bar{h}}\downarrow & & \downarrow{\scriptstyle h} \\ A' & \xrightarrow{f'} & D' \\ {\scriptstyle \bar{p}'}\downarrow & & \downarrow{\scriptstyle p'} \\ C & \xrightarrow[f]{} & B \end{array}.$$

Apply the first case to the lower square and the object $\bar{h}(a)$ of A'. We obtain an exact sequence

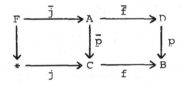

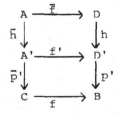

$$D'\{h(d)\} \xrightarrow{p'} B\{b\} \xrightarrow{\partial'} \pi_o A' \xrightarrow{f'} \pi_o D' \xrightarrow{p'} \pi_o B.$$

Since f is a fibration, f' is a fibration. Since h is a homotopy equivalence, $\bar{h}$ is a homotopy equivalence. We have a commutative diagram with respectively isomorphisms and bijections in the vertical rows.

$$\begin{array}{ccccccccc}
D'\{h(d)\} & \xrightarrow{p'} & B\{b\} & \xrightarrow{\partial'} & \pi_o A' & \xrightarrow{f'} & \pi_o D' & \xrightarrow{p'} & \pi_o B \\
h \uparrow \cong & & \| & & \bar{h} \uparrow \cong & & h \uparrow \cong & & \| \\
D\{d\} & \xrightarrow{p_a} & B\{b\} & \xrightarrow{\partial_a} & \pi_o A & \xrightarrow{\bar{f}} & \pi_o D & \xrightarrow{p} & \pi_o B
\end{array} ,$$

where $\partial_a := \bar{h}^{-1} \circ \partial'$. Thus the lower row is exact.□

1.5. Remark. For the applications it is more convenient to have a modified version of Corollary 1.4 where the cosets are formed with respect to a single group.

For this purpose let O be a given distinguished object of C, and O = f(O) the corresponding object of B.

For each $a \in Ob(A)$ we have a unique morphism $\gamma_a : O \to \bar{p}(a)$ of C, since C is 1-connected. Using the conjugation isomorphism

$$f(\gamma_a)_{\#} : B\{O\} \longrightarrow B\{p\bar{f}(a)\}$$

given by $\beta \longmapsto f(\gamma_a) \circ \beta \circ f(\gamma_a)^{-1}$ we make the following replacements in 1.3: replace $B\{b\} = B\{p\bar{f}(a)\}$ by $B\{O\}$, p_a by $p'_a = f(\gamma_a)_{\#}^{-1} \circ p_a$, and ∂_a by $\partial'_a = \partial_a \circ f(\gamma_a)_{\#}$. We thus obtain an exact sequence S'(a) and a classification of $\pi_o A$ by left cosets $B\{O\}/\mathrm{Im}\ p'_a$.

2. Applications.

For ease of exposition we work throughout this section in a convenient category *Top* of topological spaces (cf. for example the category of compactly generated Hausdorff spaces as used in [HK]). As remarked in the introduction the essential arguments of this section allow a formulation in abstract homotopy theory.

In order to give applications of Proposition 1.3 and its

corollary we need to know how to make the transition from topology to groupoids. Let π denote the fundamental groupoid functor. If $p : E \to B$ is a fibration in *Top* then $\pi p : \pi E \to \pi B$ is a fibration of groupoids (see [RB1], 6.1). Let Y^X and XY denote the based, respectively, free function spaces of continuous maps from X to Y, then $\pi_o \pi Y^X = [X,Y]$, the set of based homotopy classes, while $\pi_o \pi XY = (X,Y)$ the set of free homotopy classes (This notation is borrowed from [HMRS].). Further if $f : X \to Y$ then $\pi Y^X\{f\} = \pi_1(Y^X,f)$ the fundamental group of Y^X at f. If f is the constant map $\bullet$ into the base point then $\pi Y^X\{\bullet\} \cong [\Sigma X,Y]$ (see [RB1], 6).

In this context perhaps the subtlest interpretation is that of $\pi_o A$ in Proposition 1.3.

Consider the diagrams

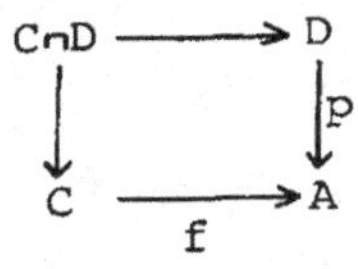

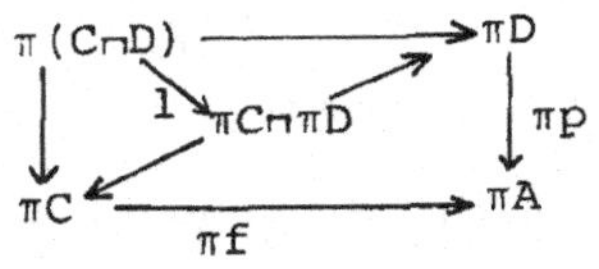

,

(2.1) (2.2)

where ⊓ denotes pullback, (2.1) is in *Top*, (2.2) in *Gd*, and 1 is the unique map.

<u>2.3. Proposition.</u> *If* p *or* f *is a fibration then*

$$\pi_o 1 : \pi_o \pi(C \sqcap D) \to \pi_o(\pi C \sqcap \pi D)$$

is a bijection, natural with respect to maps of (2.1).

<u>Proof.</u> The surjectivity of $\pi_o 1$ is trivial. Assume p is the fibration. Let $\pi_o 1(c,d)^\sim = \pi_o 1(c',d')^\sim$, where $^\sim$again denotes the component. Then there exist paths $k : d \to d'$ in D and $h : c \to c$ in C such that pk is equivalent to fh in πA. We deduce the existence of a homotopy $K : pk \simeq fh : I \times I \to A$ rel end maps. Consider the diagram

$$\begin{array}{ccccc} t & \quad & I & \xrightarrow{k} & D \\ \downarrow & & \downarrow & & \downarrow{\scriptstyle p} \\ (t,0) & & I \times I & \xrightarrow[K]{} & A \end{array} \qquad .$$

Now K lifts to $\tilde{K}$ and $\bar{K} = -\tilde{K}|1\times I + \tilde{K}|I\times 1 + \tilde{K}|0\times I$ is a lift of $0 + pk + 0$. The pair $0 + k + 0$ and $\bar{K}$ defines the required path $(c,d) \to (c',d')$. □

We remark that Proposition 2.3 corrects a stronger statement in [RB2], 4.

Consider the diagrams of based spaces

$$\begin{array}{ccccc} A & \xrightarrow{i} & CA & & \\ {\scriptstyle f}\downarrow & & \downarrow & & \\ X & \xrightarrow[j]{} & Y & \xrightarrow[g]{} & W \end{array} \qquad\qquad \begin{array}{ccccc} W & \xrightarrow{g} & E & \longrightarrow & PB \\ & & {\scriptstyle p}\downarrow & & \downarrow{\scriptstyle \varepsilon} \\ & & X & \xrightarrow[f]{} & B \end{array} \quad ,$$

(2.4) (2.5)

where the square in (2.4) is a pushout, i the inclusion of A into the cone on A, and where the square in (2.5) is a pullback, PB the paths on B eminating from the base point and ε evaluating each path at its end point.

Diagram (2.5) gives rise to further pullbacks

$$\begin{array}{ccc} E^W & \longrightarrow & PB^W \\ {\scriptstyle p^W}\downarrow & & \downarrow{\scriptstyle \varepsilon^W} \\ X^W & \longrightarrow & B^W \end{array} \qquad\qquad \begin{array}{ccc} WE & \longrightarrow & WPB \\ {\scriptstyle Wp}\downarrow & & \downarrow{\scriptstyle W\varepsilon} \\ WX & \longrightarrow & WB \end{array}$$

(2.6) (2.7)

in the based and free cases. The maps ε^W and $W\varepsilon$ are fibrations (see [HK], 5.4), moreover both PB^W and WPB are contractible (see [HK], 4.1) hence their fundamental groupoids are 1-connected. We deduce from 1.5 and 2.3

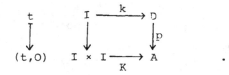

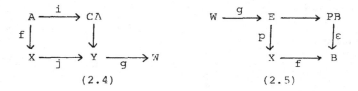

__2.8. Proposition.__ *In the situation of* (2.5) *there are exact sequences*

$$\text{(A)}\qquad \pi_1(X^W,pg) \xrightarrow{\varphi_g} [W,\Omega B] \xrightarrow{\partial} [W,E] \xrightarrow{p_*} [W,X] \xrightarrow{f_*} [W,B],$$

$$\text{(B)}\qquad \pi_1(WX,pg) \xrightarrow{\psi_g} \pi_1(WB,\bullet) \xrightarrow{\partial} (W,E) \xrightarrow{p_*} (W,X) \xrightarrow{f_*} (W,B).$$

Moreover $[W,\Omega B]$ *operates on* $[W,E]$, $\pi_1(WB,\bullet)$ *operates on* (W,E), *and* $[W,E]$ *and* (W,E) *are classified as the disjoint union of left cosets of* $[W,\Omega B]$ *modulo* Im φ_g, *respectively*, $\pi_1(WB,\bullet)$ *modulo* Im ψ_g *according* to 1.5.

We note that there has been replacements of $[W,\Omega B] \cong \pi B^W\{\bullet\} \cong \pi B^W\{fpg\}$ etc.

Baues derives sequence (A) in [B], (2.5). Hilton, Mislin, Roitberg and Steiner derive similar sequences in [HMRS] in order to compare the classifications of [W,E] and (W,E) using naturality and the evaluation fibrations $E^W \rightarrowtail WE \xrightarrow{ev} E$. Our methods will be extended elsewhere to discuss the comparison in the dual case - where duality is seen to break down (cf. [HMRS] and the remark below).

In the dual case we have

__2.9. Proposition.__ *In the situation of* (2.4) *there is an exact sequence*

$$(A^*)\quad \pi_1(W^X,gj) \xrightarrow{\varphi_g} [\Sigma A,W] \xrightarrow{\partial} [Y,W] \xrightarrow{j^*} [X,W] \xrightarrow{f^*} [A,W].$$

Moreover $[\Sigma A,W]$ *operates on* $[Y,W]$ *and the latter set is classified as the disjoint union of left cosets of* $[\Sigma A,W]$ *modulo* Im φ_g *according to* 1.5.

Baues derives sequence (A^*) in [B], (2.4), and Hilton derives it in [Hi], 5 in case A is a sphere.

Proposition 2.9 arises by raising W to the power of diagram (2.4). The free analogue does not work in this case since although W^{CA} is 1-connected for all W, in the free case $CAW \simeq *W \cong W$ is 1-connected only when W is. This bears out the lack of duality discussed in [HMRS].

Finally the relative classification sequences (A) p. 151 and (A') p. 161 of [B] also follow from our considerations - these we leave to the reader.

References

[B] H.J. Baues, Obstruction Theory on Homotopy Classification of Maps. Lecture Notes in Math. 628. Springer. Berlin 1977.

[RB1] R. Brown, Fibrations of groupoids. J. Algebra 15, 103-132 (1970).

[RB2] ________, Groupoids and the Mayer-Vietoris sequence. Manuscript. 1972.

[He] P.R. Heath, An Introduction to Homotopy Theory via Groupoids and Universal Constructions. Queen's Papers in Pure and Applied Mathematics (to appear).

[HK] P.R. Heath, K.H. Kamps, Induced homotopy in structured categories. Rendiconti di Matem. 9, 71-84 (1976).

[Hi] P. Hilton, Localization in group theory and homotopy theory. Jber. Deutsch. Math.-Verein. 79, 70-78 (1977).

[HMRS] P. Hilton, G. Mislin, J. Roitberg, R. Steiner, On free maps and free homotopies into nilpotent spaces. Algebraic Topology: Proceedings, Vancouver B.C. 1977. Lecture Notes in Math. 673. Springer. Berlin.

[K1] K.H. Kamps, Kan-Bedingungen und abstrakte Homotopietheorie. Math. Z. 124, 215-236 (1972).

[K2] __________, Zur Homotopietheorie von Gruppoiden. Arch. Math. 23, 610-618 (1972).

[K3] __________, On exact sequences in homotopy theory. Topol. Appl. Sympos. Budva (Yugoslavia) 1972, 136-141 (1973).

[K4] __________, On a sequence of K.A. Hardie. Cahiers de Top. et Géom. Diff. (to appear).

Philip R. Heath
Department of Mathematics
Memorial University of Newfoundland
St. John's, Newfoundland,
Canada, A1B 3X7

Klaus Heiner Kamps
Fachbereich Mathematik
Fernuniversität
Postfach 940
D - 5800 Hagen

Concentrated Nearness Spaces

N.C.Heldermann, Berlin

Abstract. The category R_0Top of topological R_0-spaces and continuous maps can be embedded as a nice subcategory into *Near*, the category of nearness spaces and nearness preserving maps. The resulting subcategory *TNear* of *Near* has the property that products and subspaces of *TNear*-objects taken in *Near* are generally different from those taken in *TNear*. This led H.Herrlich - who introduced nearness spaces - to the problem to characterize those spaces belonging to the epireflective hull EH(*TNear*) of *TNear* in *Near* internally, which is still open.
In this paper we introduce the property of being "concentrated" for nearness spaces, which gives rise to a subcategory of *Near* that contains all subtopological nearness spaces of Bentley and contributes to the problem mentioned above in the sense that it is the largest subcategory of EH(*TNear*) known so far for which an internal characterization exists. We discuss properties of these spaces and show that they may be helpful in solving Herrlich's problem, but are also interesting in their own right.

AMS Subject Classification: Primary 54E15, Secondary 54B05, 54B10, 54B15, 54B25.

The category R_0Top of symmetric topological spaces*) can nicely be embedded into *Near*, the category of nearness spaces and nearness preserving maps [13] [14]. The resulting subcategory, isomorphic to R_0Top, is denoted *TNear* and its objects are called topological nearness spaces.

The main concern of this paper is to investigate those nearness spaces that are obtained by taking subspaces of products of topological nearness spaces in the category *Near* of nearness spaces, i.e. to characterize the objects of the epireflective hull EH(*TNear*) of *TNear* in *Near*. The problem how to give an internal characterization of these spaces - posed explicitely by Herrlich in [12] and [16] - has already attracted the attention of a number of researchers: The conjecture that every nearness space can be embedded into a topological nearness space was independently disproved by Naimpally and Whitfield [23], and Bentley [1]. Furthermore, Bentley characterized subtopological nearness spaces, i.e. subspaces of topological nearness spaces, internally by means of grills. In [9], [10] and [11] analogous results were proved for more general categories than *Near*.

A negative answer to the conjecture that every nearness space can be emded-ded into a product of topological nearness spaces was given by Bentley and Herrlich [2], implying that EH(*TNear*) is a proper subcategory of *Near*.

It is the aim of this paper to "approximate" EH(*TNear*) from below, i.e. to specify a subcategory *Con* of EH(*TNear*) which is large enough to contain all subtopological spaces, but can still be characterized internally. Since topological nearness spaces are distinguished by the property that every near collection of subsets can be located at a certain point of the underlying set, it is not surprising, that the defining property for *Con*-objects still embodies a property of "local concentration": Every near collection of sub sets can be located within a suitable subset of the underlying set.

We derive properties of concentrated nearness spaces and give examples to show that the class of subtopological nearness spaces is properly contained in *Con*, and that *Con* is not all of EH(*TNear*).

---

*) Topological R_0-*spaces*, also called *symmetric* [19] or *essentially* T_1 [27] are defined by the property $x \in cl\{y\}$ whenever $y \in cl\{x\}$. These spaces were introduced by Shanin [26] and further investigated in [6] [7] [8] [18] and [22].

I am indebted to H.Herrlich for the kind permission to include example 2.3, and to H.Brandenburg and H.Pust for valuable comments.

1. Notations and preliminaries.

Nearness spaces were introduced by Herrlich [13] [14], and we adopt in essence the terminology used there.

A pair (X,ξ) consisting of a set X and a nonempty family ξ of nonempty covers of X is called a *nearness space* (*N-space*) provided the following axioms hold:

(N1) Every cover $\mathcal{B}$ of X which is refined by a member $\mathcal{A}$ of ξ belongs to ξ ;

(N2) if $\mathcal{A}, \mathcal{B} \in \xi$, then $\mathcal{A} \wedge \mathcal{B} = \{ A \cap B \mid A \in \mathcal{A} , B \in \mathcal{B} \} \in \xi$;

(N3) $\mathcal{A} \in \xi$ implies $\mathrm{int}_\xi \mathcal{A} = \{ \mathrm{int}_\xi A \mid A \in \mathcal{A} \} \in \xi$, where $x \in \mathrm{int}_\xi A$ iff $\{ A , X \setminus \{x\} \} \in \xi$.

Members of ξ are called *ξ-covers*. There is an equivalent axiomatization of N-spaces (X,ξ) by specifying conditions for "near collections" of subsets of X . The elements of the family $\bar{\xi}$ of all near collections stand in one-to-one correspondance to the non-ξ-covers: $\mathcal{A} \in \bar{\xi}$ iff $\{ X \setminus A \mid A \in \mathcal{A} \} \notin \xi$, or equivalently, iff for every $\mathcal{B} \in \xi$ there exists $B \in \mathcal{B}$ such that $A \cap B \neq \emptyset$ for all $A \in \mathcal{A}$. The last equivalence gives an intuitively appealing motivation why elements of $\bar{\xi}$ are called "near collections".

If (X,ξ) and (Y,η) are N-spaces, and if $f: X \longrightarrow Y$ is a map, then $f: (X,\xi) \longrightarrow (Y,\eta)$ is called a *nearness preserving map* or *N-map* , provided that $\mathcal{A} \in \eta$ implies $f^{-1}\mathcal{A} = \{ f^{-1}[A] \mid A \in \mathcal{A} \} \in \xi$ (equivalently: $\mathcal{A} \in \bar{\xi}$ implies $f\mathcal{A} = \{ f[A] \mid A \in \mathcal{A} \} \in \bar{\eta}$). The category of N-spaces and N-maps is denoted by *Near* .

N-spaces (X,ξ) that satisfy the condition $x, y \in X$, $x \neq y$ $\Rightarrow$ $\{ X \setminus \{x\} , X \setminus \{y\} \} \in \xi$, are called *N1-spaces* . They constitute the full subcategory *Near-1* of *Near* .

Subcategories are always assumed to be full and isomorphism-closed. Therefore we do not distinguish between classes of objects and subcategories.

We call a N-space (X,ξ) *topological* , iff $\mathcal{A} \in \xi$ whenever $\mathrm{int}_\xi \mathcal{A}$ is a cover of X (equivalently: $\mathcal{A} \in \bar{\xi}$ implies $\bigcap \mathrm{cl}_\xi \mathcal{A} \neq \emptyset$, where $\mathrm{cl}_\xi A =$

$\{ x \in X \mid \{A,\{x\}\} \in \bar{\xi} \}$ for $A \subset X$). *TNear* , the subcategory of topological N-spaces is contained bicoreflectively in *Near* . The coreflection for an arbitrary N-space (X,ξ) is given by $1_X \colon (X,\xi_t) \longrightarrow (X,\xi)$, where $\xi_t =$ $= \{ A \subset P(X) \mid \text{int}_\xi A \text{ covers } X \}$. *TNear* is not closed under the formation of products or subspaces [13], where these operations are assumed to be taken in *Near* (as is always done in the sequel).

If X is a set , then $G \subset P(X)$ is called a *grill* [5], if $\emptyset \notin G$ and if for subsets A, B of X the condition $A \cup B \in G$ is equivalent to $A \in G$ or $B \in G$.

1.1 Theorem 1 . For a N-space (X,ξ) the following conditions are equivalent: (1) (X,ξ) is a subspace of some topological N-space;
(2) for every $\mathcal{A} \in \bar{\xi}$ there exists a grill $G \in \xi$ with $\mathcal{A} \subset G$.

N-spaces satisfying these properties are called *subtopological* [14]. They constitute the subcategory *SubTNear* (originally called *Bun*), which is contained bicoreflectively in *Near* [1].

For any subcategory $\underline{B}$ of *Near* we denote the epireflective hull of $\underline{B}$ in *Near* , i.e. the subcategory consisting of all subspaces of products of $\underline{B}$-objects by $EH(\underline{B})$. It is evident from 1.1 that $EH(TNear) = EH(SubTNear)$. From a general result of Marny [20] [21] on topological categories it follows that the epireflective hull and the bireflective hull in *Near* of any class $\underline{B} \subset Near$ coincide if $\underline{B}$ contains non-N1-spaces.

1.2 Proposition. For a N-space (X,ξ) the following are equivalent:

(1) $(X,\xi) \in EH(TNear)$;

(2) ξ is the initial N-structure on X with respect to the family of all N-maps from (X,ξ) to subtopological N-spaces.

2. Concentrated nearness spaces.

We call a subset A of a N-space (X,ξ) a *point closure*, if there is $x \in X$ with $A = \text{cl}_\xi\{x\}$. A subset A of X is called *point closure finite* (***pc*-**

-finite) if it is contained in a union of finitely many pointclosures, and *point closure infinite (pc-infinite)* if it is not pc-finite.
Obviously the terms point closure, pc-finite and pc-infinite coincide with point, finite and infinite, respectively, if (X,ξ) is a N1-space.
A N-space (X,ξ) is called *concentrated*, if for each $\mathcal{A}\in\bar{\xi}$ with $\bigcap \mathrm{cl}_\xi \mathcal{A} = \emptyset$ and for each $\mathcal{B}\in\xi$ there exists $B\in\mathcal{B}$ such that $A\cap B$ is pc-infinite for all $A\in\mathcal{A}$. We denote the subcategory of *Near* consisting of all concentrated N-spaces by *Con*.

2.1 Lemma. If (X,ξ) is a N-space, $\mathcal{A}\in\xi$ and $A\in\mathcal{A}$ such that A is pc-finite, then there is $x\in A$ with $x\in\bigcap \mathrm{cl}_\xi \mathcal{A}$.

Proof. Choose $x_1,\ldots,x_n\in A$ such that $A\subset \mathrm{cl}_\xi\{x_1,\ldots,x_n\}$. Since $(\mathcal{A}\cup\{\mathrm{cl}_\xi\{x_1\}\})\vee\ldots\vee(\mathcal{A}\cup\{\mathrm{cl}_\xi\{x_n\}\})$ corefines*) $\mathcal{A}$, there is $i\in\{1,\ldots,n\}$ such that $\mathcal{A}\cup\{\mathrm{cl}_\xi\{x_i\}\}\in\xi$. This implies $x_i\in\bigcap \mathrm{cl}_\xi \mathcal{A}$.

2.2 Theorem. Every subtopological N-space is concentrated.

Proof. Let (X,ξ) be subtopological. We choose $\mathcal{A}\in\bar{\xi}$ and $\mathcal{B}\in\xi$ arbitrarily and assume $\bigcap \mathrm{cl}_\xi \mathcal{A} = \emptyset$. Since (X,ξ) is subtopological there is a grill $\mathcal{G}\in\bar{\xi}$ with $\mathcal{A}\subset\mathcal{G}$. From the characteristic property of near collections we infer the existence of some $B\in\mathcal{B}$ such that $B\cap G\neq\emptyset$ for all $G\in\mathcal{G}$. We assume now that there is $G\in\mathcal{G}$ such that $B\cap G$ is pc-finite. Since $G=(B\cap G)\cup(G\setminus B)$ we obtain from the properties of a grill that $B\cap G\in\mathcal{G}$ or $G\setminus B\in\mathcal{G}$. But $G\setminus B\in\mathcal{G}$ is impossible since $(G\setminus B)\cap B=\emptyset$. Hence the pc-finite set $B\cap G$ is an element of $\mathcal{G}$. Now 2.1 implies $\bigcap \mathrm{cl}_\xi \mathcal{G}\neq\emptyset$, hence $\bigcap \mathrm{cl}_\xi \mathcal{A}\neq\emptyset$ - a contradiction.

The implication of 2.2 is not reversable as the following example shows.

2.3 Example (Herrlich). Let $X=\mathbb{N}\times\mathbb{Z}$, i.e. the cartesian product of the natural numbers $\mathbb{N}=\{0,1,\ldots\}$ with the set of integers. We define

---

*) We say that a setsystem $\mathcal{A}$ *corefines* $\mathcal{B}$ if for all $B\in\mathcal{B}$ there exists $A\in\mathcal{A}$ with $A\subset B$.

$\mu \subset P^2(X)$ by $\mathcal{B} \in \mu :\Leftrightarrow \mathcal{B}$ covers X and there exists $n \in \mathbb{N}$ such that for all $m \geq n$ there is $k(m) \in \mathbb{N}$ and $B(m) \in \mathcal{B}$ with

$$\{m\} \times ( ]k(m),\infty[\cup]-\infty,-k(m)[ ) \subset B(m) .$$

We assert

(1) (X,μ) is a zerodimensional*), hence uniform N1-space with discrete underlying topology.

(2) (X,μ) is concentrated.

(3) (X,μ) is not subtopological.

ad 1) We shall first take a look at the interior operation defined by μ. More precisely, we show that for any $A \subset X$ $\mathrm{int}_\mu A = A$: Since $\mathrm{int}_\mu A \subset A$ is trivially fulfilled it remains to show that for any $x \in A$ $\{ X\setminus\{x\} , A \} \in \mu$. Assume $x = (i,j)$, $i \in \mathbb{N}$, $j \in \mathbb{Z}$. For $n := i+1$ and for all $m \geq n$ set $k(m) := 0$ and $B(m) := X\setminus\{x\}$. Then we obtain for all $m \geq n$

$$\{m\} \times ( ]k(m),\infty[\cup]-\infty,-k(m)[ ) = \{m\} \times]-\infty,\infty[\subset X\setminus\{x\} = B(m) ,$$

proving that $\{ X\setminus\{x\} , A \} \in \mu$.

It can now easily be seen that axioms (N1)-(N3) hold, and since $\{ X\setminus\{x\} , X\setminus\{y\} \} \in \mu$ for $x, y \in X$, $x \neq y$ is immediate, we proved that (X,μ) is a N1-space with discrete underlying topology. To show that (X,μ) is zerodimensional let $\mathcal{B} \in \mu$ be given. Then there exists $n \in \mathbb{N}$ such that for all $m \geq n$ there is $k(m) \in \mathbb{N}$ and $B(m) \in \mathcal{B}$ with

$$\{m\} \times ( ]k(m),\infty[\cup]-\infty,-k(m)[ ) \subset B(m) .$$

Define

$$\mathcal{C} := \{ \{m\} \times ( ]k(m),\infty[\cup]-\infty,-k(m)[ ) \mid m \geq n \} \cup$$
$$\cup \{ \{ (i,j) \} \mid (i,j) \in X , i<n \text{ or: } i \geq n \text{ and } -k(m) \leq j \leq k(m) \} .$$

It is easy to see that $\mathcal{C} \in \mu$ is a partition of X with $\mathcal{C} < \mathcal{B}$.

ad 2) Let $\mathcal{A} \in \bar{\mu}$ and $\mathcal{B} \in \mu$ be given with $\bigcap \mathrm{cl}_\mu \mathcal{A} = \emptyset$. This implies $\bigcap \mathcal{A} = \emptyset$, and following 1) we may assume that $\mathcal{B}$ consists of just two types of sets: sets of the form $\{x\}$ and sets of the form

$$B(m) = \{m\} \times ( ]k(m),\infty[\cup]-\infty,-k(m)[ )$$

where $m \geq n$ for some $n \in \mathbb{N}$. We assume now that (X,μ) is not concentrated. This implies that for all $B(m)$, $m \geq n$, there exists $A(m) \in \mathcal{A}$ such that $B(m) \cap A(m)$ is at most finite. Choose $k'(m)$ for each $m \geq n$ such that

---

*) A N-space (X,μ) is called *zerodimensional* iff for each $\mathcal{B} \in \mu$ there is a partition $\mathcal{C}$ of X with $\mathcal{C} \in \mu$ and $\mathcal{C} < \mathcal{B}$ [10] [15] [24] [25].

$(\,]k'(m),\infty[\,\cup\,]-\infty,-k'(m)[\,)\cap A(m) = \emptyset$ and define
$\mathcal{C} := \{\,]k'(m),\infty[\,\cup\,]-\infty,-k'(m)[\;|\; m \geq n\,\} \cup \{\,\{(i,j)\}\;|\;(i,j)\in X$,
$i<n$ or: $i\geq n$ and $-k'(m)\leq j\leq k(m)\,\}$.
Then $\mathcal{C}\in\mu$, but for each $C\in\mathcal{C}$ there exists $A\in\mathcal{A}$ such that $A\cap C=\emptyset$.
This implies $\mathcal{A}\notin\bar{\mu}$ - a contradiction.

ad 3) We assume that (X,μ) is subtopological. Define $A := \mathbb{N}\times\mathbb{N}$ and $B := \mathbb{N}\times]-\infty,-1]$. Then $\mathcal{A} := \{A , B\}\in\bar{\mu}$ since $\{X\backslash A , X\backslash B\}\notin\mu$ which implies that there exists a grill $\mathcal{G}\in\bar{\mu}$ with $\mathcal{A}\subset\mathcal{G}$. Then for every $n\in\mathbb{N}$ $A = (\{n\}\times\mathbb{N}) \cup ((\mathbb{N}\backslash\{n\})\times\mathbb{N})\in\mathcal{G}$ and it follows from the properties of a grill that $(\{n\}\times\mathbb{N})\in\mathcal{G}$ or $(\mathbb{N}\backslash\{n\})\times\mathbb{N}\in\mathcal{G}$. If $(\{n\}\times\mathbb{N})\in\mathcal{G}$ was true, we would obtain $\{\{n\}\times\mathbb{N}, B\}\in\bar{\mu}$ from $\{\{n\}\times\mathbb{N} , B\}\subset\mathcal{G}$, which is obviously false. Hence $(\mathbb{N}\backslash\{n\})\times\mathbb{N}\in\mathcal{G}$ for all $n\in\mathbb{N}$ which implies $\{B\}\cup\{(\mathbb{N}\backslash\{n\})\times\mathbb{N}\;|\;n\in\mathbb{N}\}\in\bar{\mu}$ - a contradiction.

2.4 Theorem. Every concentrated N-space is an element of the epireflective hull of *TNear* .

Proof. Following 1.2 it suffices to show that for an arbitrary concentrated N-space (X,ξ) ξ is initial with respect to the family of all N-maps from (X,ξ) to subtopological N-spaces. This will be done by proving that for any ξ-cover $\mathcal{A}$ of X there is a subtopological N-space (Y,η) , a η-cover $\mathcal{B}$ of Y and a N-map $f : (X,\xi)\longrightarrow(Y,\eta)$ with $f^{-1}\mathcal{B}<\mathcal{A}$. We will even prove that Y can be chosen equal to X and $f = 1_X$.
Let $\mathcal{A}\in\xi$ be given. Define
$\beta := \{\{\mathrm{int}_\xi(X\backslash\{x\}) , \mathrm{int}_\xi C\}\;|\;C\subset X , x\in\mathrm{int}_\xi C\}\cup\{\mathrm{int}_\xi\mathcal{A}\}$, and
$\kappa := \{\mathcal{C}\subset P(X)\;|\;$There are $\mathcal{B}_1,\ldots,\mathcal{B}_n\in\beta$ with $\mathcal{B}_1\wedge\ldots\wedge\mathcal{B}_n<\mathcal{C}\}$.
Since (X,κ) fulfills axioms (N1) and (N2) , it suffices to prove (N3) to show that (X,κ) is a N-space. Let $A\subset X$ and $x\in\mathrm{int}_\xi A$ be given. Then $\{\mathrm{int}_\xi(X\backslash\{x\}) , \mathrm{int}_\xi A\}\in\beta$ which implies $\{X\backslash\{x\} , A\}\in\kappa$ and hence $x\in\mathrm{int}_\kappa A$. This conclusion is reversable : if $x\in\mathrm{int}_\kappa A$, then $\{X\backslash\{x\} , A\}\in\kappa$. Consequentley there exist $\mathcal{B}_1,\ldots,\mathcal{B}_n\in\beta$ with $\mathcal{B}_1\wedge\ldots\wedge\mathcal{B}_n<\{X\backslash\{x\} , A\}$. For every $i\in N$, where $N = \{1,\ldots,n\}$ we define $C_i := \bigcup\{D\in\mathcal{B}_i\;|\;x\in D\}$ and $\mathcal{B}'_i := \{X\backslash\{x\} , C_i\}$. Since for all $i\in N$ $\mathcal{B}_i<\mathcal{B}'_i$, $\mathcal{B}_1\wedge\ldots\wedge\mathcal{B}_n<\mathcal{B}'_1\wedge\ldots\wedge\mathcal{B}'_n$ obviously holds. We show $\mathcal{B}'_1\wedge\ldots\wedge\mathcal{B}'_n<\{X\backslash\{x\} , A\}$ which amounts to prove $C_1\cap\ldots\cap C_n\subset A$. Let $y\in C_1\cap\ldots\cap C_n$ be given . Then for

every $i \in N$ there is $D_i \in \mathcal{B}_i$ with $y \in D_i$ and $x \in D_i$. This implies $y \in D_1 \cap \ldots \cap D_n \subset A$. Since the elements of members of β are open in (X,ξ) by definition, $C_1 \cap \ldots \cap C_n$ is an open set in (X,ξ) with $x \in C_1 \cap \ldots \cap C_n \subset A$, which implies $x \in \mathrm{int}_\xi A$.

We prove (N3) : If $\mathcal{C} \in \kappa$ then there are $\mathcal{B}_1, \ldots, \mathcal{B}_n \in \beta$ with $\mathcal{B}_1 \wedge \ldots \wedge \mathcal{B}_n < \mathcal{C}$. Since $\mathrm{int}_\xi \mathcal{B}_i = \mathcal{B}_i = \mathrm{int}_\kappa \mathcal{B}_i$ for all $i \in \{1,\ldots,n\}$ it follows that $\mathcal{B}_1 \wedge \ldots \wedge \mathcal{B}_n < \mathrm{int}_\kappa \mathcal{A}$ and thus $\mathrm{int}_\kappa \mathcal{A} \in \kappa$.

Since $\kappa \subset \xi$ $1_X : (X,\xi) \longrightarrow (X,\kappa)$ is a N-map from the concentrated N-space (X,ξ) to the N-space (X,κ), which was constructed in dependence of an arbitrary $\mathcal{A} \in \xi$. Since *SubTNear* is a bicoreflective subcategory of *Near*, there is a bicoreflection for an arbitrary N-space (Z,ζ) which can be described by $1_X : (Z,\zeta_g) \longrightarrow (Z,\zeta)$, where $\mathcal{A} \in \bar{\zeta}_g$ iff there is a grill $\mathcal{G} \in \bar{\zeta}$ with $\mathcal{A} \subset \mathcal{G}$ [1].

We proceed in the proof by considering $\eta := \kappa_g$. Since $1_X : (X,\eta) \longrightarrow (X,\kappa)$ is a N-map, i.e. $\kappa \subset \eta$, we have $\mathcal{A} \in \eta$. To complete the proof it suffices to show that $1_X : (X,\xi) \longrightarrow (X,\eta)$ is a N-map, i.e. $\bar{\xi} \subset \bar{\eta}$.

Take $\mathcal{B} \in \bar{\xi}$. Since $\bar{\xi} \subset \bar{\kappa}$ we obtain $\mathcal{B} \in \bar{\kappa}$, and to show $\mathcal{B} \in \bar{\eta}$ it remains to prove the existence of a grill $\mathcal{G} \in \bar{\kappa}$ with $\mathcal{B} \subset \mathcal{G}$.

Case 1: $\bigcap \mathrm{cl}_\xi \mathcal{B} = \emptyset$. Then there exists an $A \in \mathrm{int}_\xi \mathcal{A}$ such that $A \cap B$ is pc-infinite for all $B \in \mathcal{B}$ since $\mathrm{int}_\xi \mathcal{A} \in \xi$ and (X,ξ) is concentrated. Define $\mathcal{G} := \{ G \subset X \mid G \cap A$ is pc-infinite $\}$. It is easy to see that $\mathcal{G}$ is a grill containing $\mathcal{B}$. We remind the reader that $\mathcal{G} \in \bar{\kappa}$ iff for any $\mathcal{C} \in \kappa$ there is $C \in \mathcal{C}$ with $C \cap G \neq \emptyset$ for all $G \in \mathcal{G}$. Let $\mathcal{C} \in \kappa$ be given. Then there exist subsets $D_i \subset X$ and points $x_i \in \mathrm{int}_\xi D_i$ for all $i \in \{1,\ldots,n\}$ and some $n \in \mathbb{N}$ such that $\mathrm{int}_\xi \mathcal{A} \wedge \mathcal{B}_1 \wedge \ldots \wedge \mathcal{B}_n < \mathcal{C}$, where $\mathcal{B}_i := \{\mathrm{int}_\xi (X \setminus \{x_i\}), \mathrm{int}_\xi D_i\}$. If we define $A' := A \cap \mathrm{int}_\xi (X\setminus\{x_1\}) \cap \ldots \cap \mathrm{int}_\xi (X\setminus\{x_n\}) = A \cap (X \setminus \mathrm{cl}_\xi \{x_1\}) \cap \ldots \cap (X \setminus \mathrm{cl}_\xi\{x_n\})$, then there is $C \in \mathcal{C}$ with $A' \subset C$ since $A' \in \mathrm{int}_\xi \mathcal{A} \wedge \mathcal{B}_1 \wedge \ldots \wedge \mathcal{B}_n$. We know that for every element G of $\mathcal{G}$ the intersection $G \cap A$ is pc-infinite, and since A' is obtained from A by removing finitely many point closures, $G \cap A'$ must be pc-infinite, too. Thus $C \cap G$ is nonempty for every $G \in \mathcal{G}$ which implies $\mathcal{G} \in \bar{\kappa}$.

<u>Case 2</u> : $\bigcap cl_\xi \mathcal{B} \neq \emptyset$. Take $x \in \bigcap cl_\xi \mathcal{B}$ and define $\mathcal{G} := \{ G \subset X \mid x \in cl_\xi G \}$. Then $\mathcal{G}$ is a grill containing $\mathcal{B}$. Since $\bigcap cl_\xi \mathcal{G} \neq \emptyset$, i.e. $\bigcup \{ int_\xi (X \setminus G) \mid G \in \mathcal{G} \} = \bigcup \{ int_\kappa (X \setminus G) \mid G \in \mathcal{G} \} \neq X$, axiom (N3) implies $\mathcal{G} \in \bar{\kappa}$, and the proof is complete.

We have actually proved the following :

2.5 Corollary. If (X,ξ) is a concentrated N-space and $\mathcal{A} \in \xi$ then there exists $\eta \subset P^2(X)$ such that

(1) (X,η) is a subtopological N-space inducing the same topology as (X,ξ) ;

(2) $1_X : (X,\xi) \longrightarrow (X,\eta)$ is a N-map ;

(3) $\mathcal{A} \in \eta$.

It should be mentioned that in the proof of 2.4 the assertion that (X,κ) is a N-space could have been proved more elegantly if we had used the concept of "kernel-normal families" as introduced by Brandenburg [3] [4]. For reasons of simplicity, however, we choose the more elementary way.

The inclusion stated in theorem 2.4 is proper. An example of an object belonging to the epireflective hull of *TNear* without being concentrated is given in 2.10(1).

2.6 Theorem. Coproducts of concentrated N-spaces are again concentrated.

Proof. If (Y,η) is the coproduct of a family $(X_i,\xi_i)_{i \in I}$ of concentrated N-spaces with injections $j_i : (X_i,\xi_i) \longrightarrow (Y,\eta)$ then $\mathcal{A} \in \bar{\eta}$ and $\mathcal{B} \in \eta$ implies the existence of an $i \in I$ with $j_i^{-1}\mathcal{A} \in \bar{\xi}_i$ and $j_i^{-1}\mathcal{B} \in \xi_i$. Hence there is $B \in \mathcal{B}$ such that $j_i^{-1}[B] \cap j_i^{-1}[A]$ is pc-infinite for every $A \in \mathcal{A}$. This implies that $B \cap A$ is pc-infinite for every $A \in \mathcal{A}$.

2.7 Theorem. The inverse image of a concentrated N-space is concentrated.

Proof. Let X be a set, (Y,η) a N-space and $f : X \longrightarrow Y$ a mapping. The initial N-structure ξ on X has the properties $\mathcal{A}\in\xi$ iff there exists $\mathcal{B}\in\eta$ with $f^{-1}\mathcal{B}<\mathcal{A}$ and $\mathcal{A}\in\bar{\xi}$ iff $f\mathcal{A}\in\bar{\eta}$ [14;8.4]. Let $\mathcal{A}\in\bar{\xi}$ with $\bigcap cl_{\xi}\mathcal{A} = \emptyset$ and $\mathcal{B}\in\xi$ be given. Then $f\mathcal{A}\in\bar{\eta}$ and there exists $\mathcal{C}\in\eta$ with $f^{-1}\mathcal{C}<\mathcal{B}$.

If $\bigcap cl_{\eta} f\mathcal{A} = \emptyset$, then there exists $C\in\mathcal{C}$ with $f[A]\cap C$ pc-infinite for all $A\in\mathcal{A}$, i.e. for all finite subsets D of Y $f[A]\cap C\not\subset cl_{\eta}D$. For a fixed finite subset E of X we thus obtain for every $A\in\mathcal{A}$ a point $x_A\in A\cap f^{-1}[C]$ with $f(x_A)\notin cl_{\eta}f[E]$ which implies that for every $A\in\mathcal{A}$ $x_A\notin cl_{\xi}E$ and hence $A\cap f^{-1}[C]$ is pc-infinite for every $A\in\mathcal{A}$. Since $f^{-1}\mathcal{C}<\mathcal{B}$ there is a $B\in\mathcal{B}$ with $A\cap B$ pc-infinite for all $A\in\mathcal{A}$.

If $\bigcap cl_{\eta} f\mathcal{A}\neq\emptyset$ there exists $y\in Y\setminus f[X]$ with $y\in\bigcap cl_{\eta}f\mathcal{A}$. First we assume that there exists $A\in\mathcal{A}$ and $x\in A$ with $y\in cl_{\eta}\{f(x)\}$. This implies $f(x)\in cl_{\eta}\{y\}$, and since $cl_{\eta}\{y\}\subset\bigcap cl_{\eta}f\mathcal{A}$ we obtain $f(x)\in\bigcap cl_{\eta}f\mathcal{A}$ and hence $x\in\bigcap cl_{\xi}\mathcal{A}$ - a contradiction. Consequentley $y\notin cl_{\eta}f[D]$ for each $A\in\mathcal{A}$ and each finite subset D of A.

Using (N3) we obtain $int_{\eta}\mathcal{C}\in\eta$, so that there is $C\in\mathcal{C}$ with $y\in int_{\eta}C$. Since $y\in cl_{\eta}f[A]$ for each $A\in\mathcal{A}$, $(int_{\eta}C)\cap f[A]$ is nonempty. If we assume $(int_{\eta}C)\cap f[A]$ to be pc-finite for some $A\in\mathcal{A}$, we can choose a finite subset D of A with $(int_{\eta}C)\cap f[A]\subset cl_{\eta}f[D]$, fulfilling also $y\notin cl_{\eta}f[D]$ as proved above. But then $(int_{\eta}C)\setminus cl_{\eta}f[D]$ is an open neighborhood of y not intersecting $f[A]$, which is impossible. Consequently $f[A]\cap C$ is pc--infinite for each $A\in\mathcal{A}$, and following the argumentation in the case $\bigcap cl_{\eta}f\mathcal{A} = \emptyset$ we obtain a $B\in\mathcal{B}$ such that $A\cap B$ is pc-infinite for all $A\in\mathcal{A}$. Hence (X,ξ) is concentrated.

2.8 Corollary. Subspaces of concentrated N-spaces are again concentrated.

We turn now to the analysis of products of concentrated N-spaces.

2.9 Theorem. If (X,ξ) is a N1-space which has a closed discrete subspace A and a non-closed subspace B of equal cardinality, then $(X,\xi)\times(X,\xi)$ is not concentrated.

Proof. Denote the N-structure of the product by η, i.e. $(X^2,\eta) = (X,\xi)\times(X,\xi)$. We choose an index set I for A and B so that $A = \{a_i \mid i\in I\}$ and $B = \{b_i \mid i\in I\}$. Since B is not closed, there exists $x\in X$ with $x\in(\mathrm{cl}_\xi B)\setminus B$. Define $C := \{(a_i,x) \mid i\in I\}$ and $D := \{(a_i,b_i) \mid i\in I\}$. We show that D is closed in (X^2,η).
Since A is a discrete subspace of (X,ξ) there exists $\mathcal{E}\in\xi$ such that $\mathrm{card}(E\cap A) = 0$ or 1 for every $E\in\mathcal{E}$. Let $(x,y)\in X^2\setminus D$ be given. If $x \neq a_i$ for every $i\in I$ then $(X\setminus A)\times X$ is an open neighborhood of (x,y) not intersecting D. If $x = a_i$ for some $i\in I$ but $y\neq b_i$ then choose $E\in\mathcal{E}$ with $x\in E$. Clearly $(\mathrm{int}_\xi E)\times X\setminus\{b_i\}$ is an open neighbourhood of (x,y) not intersecting D. To show that C is closed one proceeds analogously.
Define $\mathcal{A} := \{C, D\}$. We show that $\mathcal{A}\in\bar{\eta}$, i.e. for all $\mathcal{B}\in\eta$ there exists $G\in\mathcal{B}$ such that for all $A\in\mathcal{A}$ $A\cap G\neq\emptyset$. Let $\mathcal{B}\in\eta$ be given. Then there are $\mathcal{D}\in\xi$ and $\mathcal{F}\in\xi$ with $p_1^{-1}\mathcal{D}\wedge p_2^{-1}\mathcal{F}<\mathcal{B}$, where p_1 and p_2 denote the projections. Since $\mathrm{int}_\xi\mathcal{F}$ is a cover of X there exists $F\in\mathcal{F}$ with $x\in\mathrm{int}_\xi F$. This implies $F\cap B\neq\emptyset$. Let $b_i\in F\cap B$. If $a_i\in H$, $H\in\mathcal{D}$, then $(a_i,x)\in H\times F$ and $(a_i,b_i)\in H\times F$. Since there exists $G\in\mathcal{B}$ with $H\times F\subset G$ we obtain $C\cap G\neq\emptyset$ and $D\cap G\neq\emptyset$ as required. If we now define $\mathcal{B} := p_1^{-1}\mathcal{E}$ where $\mathcal{E}$ was defined above, then the following holds: $\mathcal{B}\in\xi$, $\mathcal{A}\in\bar{\xi}$ with $\bigcap \mathrm{cl}_\eta\mathcal{A} = \emptyset$, and $C\cap G$ and $D\cap G$ is finite for every $G\in\mathcal{B}$, which proves that (X^2,η) is not concentrated.

This result generalizes proposition 3.7 in [16]. The following examples show that products (even finite products) of concentrated N-spaces need not be concentrated. Moreover, example (1) illustrates that *Con* is properly contained in EH(*TNear*). Example (2) shows that a quotient of a concentrated N-space needs no longer be concentrated.

2.10 Examples. (1) By $(\mathbb{R},\rho)$ we denote the reals supplied with the usual topological N-structure, i.e. $\mathcal{A}\in\rho$ iff $\mathcal{A}$ is refined by a cover of $\mathbb{R}$ which is open in the usual topology. Obviously $(\mathbb{R},\rho)$ is a concentrated N1-space. Define subsets A and B of $\mathbb{R}$ by $A := \{1,2,\ldots\}$ and $B := \{1,1/2,1/3,\ldots\}$, then the assumptions of theorem 2.9 are fulfilled.

(2) We reconsider the concentrated N1-space (X,ξ) of example 2.3. Let $Y = \mathbb{N}\times\{1,2\}$ and define $f : X \longrightarrow Y$ by

$$f(n,k) := \begin{cases} (n,1) & , \text{ if } k\leq 0 , \\ (n,2) & , \text{ if } k>0 . \end{cases}$$

Let $\eta\subset P^2(Y)$ be given by $\mathcal{A}\in\eta$ iff $f^{-1}\mathcal{A}\in\xi$. Then $\mathcal{A}\in\eta$ is equivalent to $\mathcal{A}$ covers Y and there is $n\in\mathbb{N}$ such that for all $m\geq n$ there is $A(m)\in\mathcal{A}$ with $\{(m,1),(m,2)\}\subset A(m)$. Since for all $A\subset Y$ $cl_\eta A = A$, it follows easily that (Y,η) is a N1-space with discrete underlying topology. In fact, (Y,η) is final with respect to (X,ξ) and f : Let $g: Y\longrightarrow Z$ be a map from Y to the underlying set Z of a N-space (Z,ζ) such that $g\circ f$ is a N-map. If $\mathcal{A}\in\zeta$, then $f^{-1}g^{-1}\mathcal{A}\in\xi$, hence $g^{-1}\mathcal{A}\in\eta$, proving that g is a N-map.
In analogy to the preceeding example we define $A := \{(n,1) \mid n\in\mathbb{N}\}$, $B := \{(n,2) \mid n\in\mathbb{N}\}$, and $\mathcal{A} := \{A , B\}$. Then $\mathcal{A}\in\bar{\eta}$, $\bigcap cl_\eta\mathcal{A} = \emptyset$, $\mathcal{B} = \{\{(n,1),(n,2)\} \mid n\in\mathbb{N}\}\in\eta$, and for all $D\in\mathcal{B}$ $D\cap A$ and $D\cap B$ are finite or empty. Hence (Y,η) is not concentrated, though being the quotient of a concentrated N1-space.

If (X,ξ) is a N-space we denote by ξ_ω the coarsest N-structure on X which induces the same topology as ξ on X . This N-space always exists and is given in the following way [17] : $\mathcal{A}\in\bar{\xi}_\omega$ iff $\mathcal{A}\in\bar{\xi}$ or A is pc-infinite for all $A\in\mathcal{A}$. An easy computation yields : $\mathcal{A}\in\xi_\omega$ iff $\mathcal{A}\in\xi$ and there exists $A\in\mathcal{A}$ such that $X\setminus A$ is pc-finite.

2.11 Proposition. If (X,ξ) is a N-space, then ξ_ω is a concentrated N-structure.

Proof. Let $\mathcal{A}\in\bar{\xi}_\omega$ with $\bigcap cl_\xi\mathcal{A} = \emptyset$ and $\mathcal{B}\in\xi_\omega$ be given. From 2.1 we know that A is pc-infinite for every $A \in \mathcal{A}$. Since there is $B \in \mathcal{B}$ such that $X \setminus B$ is pc-finite, the intersection $A \cap B$ must be pc-infinite for every $A \in \mathcal{A}$, proving that (X,ξ_ω) is concentrated.

Concentrated N-spaces provide an approximation of the epireflective hull of *TNear* , and, furthermore, they may be helpful in the further search for a solution of the problem how to characterize EH(*TNear*) internally : From 1.2 and 2.7 it is obvious that a N-space (X,ξ) belongs to EH(*TNear*) iff is the infimum of all concentrated N-structures on X that are coarser than ξ . And if (X,ξ) is an arbitrary N-space then the reflector belonging to EH(*TNear*) assigns to (X,ξ) the N-space (X,η) , where η is the coarsest N-structure on X finer than any concentrated N-structure on X which is coarser than ξ . It follows from 2.11 that (X,η) has the same underlying topology as (X,ξ) , a fact that had already been known to Hastings [9].

There is some evidence that the approach to view objects of EH(*TNear*) as infima of certain sets of concentrated N-spaces may lead to an internal characterization of these objects. It would be desirable to pursue the study of concentrated N-spaces in this direction.

References

[1] Bentley, H.L., Nearness spaces and extensions of topological spaces, Studies in Topology, N.Stavrakas and K.Allen (eds.), New York 1975, 47 - 66.

[2] Bentley, H.L.; Herrlich, H., The reals and the reals, General Topol. Appl., to appear.

[3] Brandenburg, H., Hüllenbildungen für die Klasse der entwickelbaren topologischen Räume, Dissertation, Freie Universität Berlin 1978.

[4] Brandenburg, H., On a class of nearness spaces and the epireflective hull of developable topological spaces, preprint.

[5] Choquet, G., Sur les notions de filtre et de grille, C.r. Acad. Sci., Paris, Sér. A, 224 (1947) 171 - 173.

[6] Davis, A.S., Indexed systems of neighborhoods for general topological spaces, Amer. math. Monthly 68 (1961) 886 - 893.

[7] Dube, K.K., A note on R_0-toplogical spaces, Mat. Vestnik, n. Ser., 11 (26) (1974) 203 - 208.

[8] Hall, D.W.; Murphy, S.K.; Rozicky, J., On spaces which are essentially T_1, J. Austral. math. Soc. 12 (1971) 451 - 455.

[9] Hastings, M.S., Epireflective hulls in Near, Thesis, University of Toledo 1975.

[10] Heldermann, N.C., Hüllenbildung in Nearness-Kategorien, Dissertation, Freie Universität Berlin 1977/78.

[11] Heldermann, N.C., On topological prenearness spaces and convergence spaces, preprint.

[12] Herrlich, H., Remarks and problems concerning nearness structures, Problems in Categorical Topology, Categorical Topology Conference at Southern Illinois University, Carbondale 1973.

[13] Herrlich, H., A concept of nearness, General. Topol. Appl. 4 (1974) 191 - 212.

[14] Herrlich, H., Topological structures, Math. Centre Tracts 52, Amsterdam 1974, 59 - 122.

[15] Herrlich, H., Some topological theorems which fail to be true, Lect. Notes 540, Proc. of the Conference on Categorical Topology held at Mannheim 1975, E.Binz and H.Herrlich (eds.), Berlin et al. 1976, 265 - 285.

[16] Herrlich, H., Products in topology, Quaest. math. 2 (1977) 191 - 205.

[17] Hunsaker, W.N.; Sharma, P.L., Nearness structures compatible with a topological space, Arch. der Math. 25 (1974) 172 - 177.

[18] Lee, S.M., On T_0'-spaces, Kyungpook math. J. 16 (1976) 61 - 62.

[19] Lodato, M.W., On topologically induced generalized proximity relations, Proc. Amer. math. Soc. 15 (1964) 417 - 422.

[20] Marny, T., Rechts-Bikategoriestrukturen in topologischen Kategorien, Dissertation, Freie Universität Berlin 1973.

[21] Marny, T., On epireflective subcategories of topological categories, preprint.

[22] Naimpally, S.A., On R_0-topological spaces, Ann. Univ. Sci. Budapest. Rolando Eötvös, Sect. Math. 10 (1967) 53 - 54.

[23] Naimpally, S.A.; Whitfield, J.H.M., Not every near family is contained in a near clan, Proc. Amer. math. Soc. 47 (1975) 237 - 238.

[24] Pust, H., Normalität und Dimension für Nearness-Räume, Dissertation, Freie Universität Berlin 1977/78.

[25] Pust, H., Covering dimension for normal nearness spaces, preprint.

[26] Shanin, N.A., On separation in topological spaces, Doklady Akad. Nauk. SSSR 38 (1943) 110 - 113.

[27] Worrell, J.M. jun.; Wicke, H.H., Characterizations of developable topological spaces, Canadian J. Math. 17 (1965) 820 - 830.

N.C. Heldermann
Zentralblatt für Mathematik
Otto-Suhr-Allee 26
D - 1ooo Berlin 1o

Initial and final completions

Horst Herrlich

I Introduction.

The theory of initial and final completions of concrete categories has two roots:

(1) Completions of partially ordered sets. Cf. Mac Neille (1937), Birkhoff (1948), Bourbaki (1963), Banaschewski and Bruns (1967), Ringleb (1969) a.o.

(2) Cartesian closed topological extensions of Top and similar concrete categories over Set. Cf. Spanier (1963), Antoine (1966), Day (1972), Chartrelle (1972), Machado (1973), Wyler (1976), Bourdaud (1976) a.o.

In a more systematic investigation of initial completions the author (1976) demonstrated that several of the former constructions were illegitimate in a reasonable set theoretic framework and that several constructions (even if legitimate) don't have those universal properties, attached to them by several authors. These observations triggered a great deal of (mostly still unpublished) research on initial completions The aim of this paper is to review the present state of this theory. Most amazing, perhaps, are those results, which correlate the question of the existence of certain initial completions with recent investigations of suitable generalizations of topological and algebraic functors.

II Terminology

Let $\underline{X}$ be a fixed, non empty category, called base category. The objects of study are concrete categories over $\underline{X}$, i.e. pairs $(\underline{A},U)$ where $\underline{A}$ is a category and $U:\underline{A}\to\underline{X}$ is a faithful and amnestic functor, called the underlying functor of $(\underline{A},U)$. Functors between concrete categories are called concrete, provided they commute with the underlying functors. Concrete full embeddings are called extensions. Extensions are called essential, provided they are initially and finally dense.* Extensions $E:(\underline{A},U)\to(\underline{B},V)$ are called initial completions, provided $(\underline{B},V)$ is initially complete and E is initially dense. (Previous articles did not contain the last requirement). The dual concept is called a final completion.

* This definition is equivalent to the standard categorical one [18,37].

Initial and final completions can always be constructed in some higher universe, provided the existence of such higher universe is assumed. Life within a fixed universe requires a more careful analysis. As in [22] we will assume a set-theory, whose members are called conglomerates, with a fixed universe, whose elements are called sets and whose sub-conglomerates are called classes. Categories are supposed to be classes. Conglomerates, which need not be classes but otherwise behave like categories, are called quasicategories.

A conglomerate X is called small (resp. legitimate), provided it is codable by some set (resp. class) Y, i.e. provided there exists some injection $X \rightarrow Y$. Legitimate quasicategories can and will be treated like categories.

III Basic Constructions

Let $(\underline{A},U)$ be a concrete category over $\underline{X}$. Basic tools for the construction of initial completions of $(\underline{A},U)$ are the concepts of structured morphisms, i.e. pairs (f,A) where A is an $\underline{A}$-object and $f:X \rightarrow UA$ is an $\underline{X}$-morphism, and structured sources, i.e. pairs (X,S) where X is an $\underline{X}$-object and S is a class-indexed family* of structured morphisms with domain X. An $\underline{X}$-morphism $f:X \rightarrow Y$ is called a source map $f:(X,S) \rightarrow (Y,T)$, provided $(g,A) \in T$ implies $(gf,A) \in S$.

$$\underline{E^1 : (A,U) \rightarrow (A^1,U^1)}$$

$\underline{A}^1$ is the quasicategory, whose objects are all structured sources and whose morphisms are all source maps. The forgetful functor U^1 is defined by $U^1(X,S)=X$ and $U^1(f)=f$. The embedding E^1 is defined by $E^1(A) = (UA, \{(Uf,B) \mid f:A \rightarrow B \text{ is an } \underline{A}\text{-morphism}\})$ and $E^1(f) = Uf$. If $\overline{A^1}$ is legitimate, which is seldom the case, $E^1 : (\underline{A},U) \rightarrow (\underline{A}^1,U^1)$ is a finality preserving,

---

* For the present article we may identify S with its image, i.e. a class of structured morphisms.

nitially and hence finally complete extension. It need neither be initially or finally dense nor preserve initial structures.

$E^2:(A,U)\to(A^2,U^2)$

$\underline{A}^2$ is the full subquasicategory of $(\underline{A}^1,U^1)$ consisting of all weakly-closed sources, i.e. of those $\underline{A}^1$-objects (X,S), which satisfy the following condition:

If $(f,A)\in S$ and $g:A\to B$ is an $\underline{A}$morphism then $(Ug\cdot f,B)\in S$.

If $\underline{A}^2$ is legitimate, which still is seldom the case, the induced $E^2:(\underline{A},U)\to(\underline{A}^2,U^2)$ is an initial completion. It need neither be a final completion nor preserve initial structures. $(\underline{A}^2,U^2)$ is the initial hull of $(\underline{A},U)$ (more precisely: of $E^2(\underline{A})$) in $(\underline{A}^1,U^1)$. It is initially and finally closed in $(\underline{A}^1,U^1)$, hence simultaneously a reflective and a coreflective modification of $(\underline{A}^1,U^1)$.

$E^3: (A,U)\to(A^3,U^3)$

$\underline{A}^3$ is the full subquasicategory of $(\underline{A}^2,U^2)$ consisting of all semi-closed sources, i.e. of those $\underline{A}^2$-objects (X,S), which satisfy the following condition:

If (f,A) is a structured morphism and $(A\xrightarrow{f_i}A_i)_I$ is an initial source, such that every $(Uf\cdot f,A_i)$ belongs to S, then (f,A) belongs to S.

If $\underline{A}^3$ is legitimate, the induced E^3: $(\underline{A},U)\to(\underline{A}^3,U^3)$ is an initiality preserving initial completion. It need not be a final completion. $(\underline{A}^3,U^3)$ is finally closed in $(\underline{A}^2,U^2)$, hence a coreflective modification of $(\underline{A}^2,U^2)$.

$E^4: (A,U)\to(A^4,U^4)$

For any structured source (X,S) its opposite sink $(X,S)^{op}$ is defined to be the structured sink (T,X), consisting of those costructured morphisms $UB\xrightarrow{g}X$ with the property that for any $X\xrightarrow{f}UA$ in S the $\underline{X}$-morphism $UB\xrightarrow{f\cdot g}UA$ underlies some

$\underline{A}$-morphism $B \to A$. Likewise for structured sinks (T,X) the <u>opposite source</u> $(T,X)^{op}$ is defined.
$\underline{A}^4$ is the full subcategory of $(\underline{A}^3,U^3)$ consisting of all <u>closed sources</u>, i.e. of those $\underline{A}^I$-objects (X,S) with $(X,S)^{op\ op} = (X,S)$.
If $\underline{A}^4$ is legitimate, the induced $E^4:(\underline{A},U)\to(\underline{A}^4,U^4)$ is simultaneously an initial and final completion. $(\underline{A}^4,U^4)$ is finally closed in $(\underline{A}^3,U^3)$, hence a coreflective modification of $(\underline{A}^3,U^3)$.

<u>$E^{-n}: (A,U)\to(A^{-n},U^{-n})$</u>

Replacing structured sources by structured sinks, one obtains for n=1,2,3,4 the dual constructions.

IV <u>Largest initial completions</u>

4.1 <u>Proposition</u> [18] Every initially dense extension of $(\underline{A},U)$ is isomorphic to a subcategory of $(\underline{A}^2,U^2)$.

4.2 <u>Corollary</u> [5]: Every initial completion preserves final structures.

The conglomerate of all extensions of a concrete category $(\underline{A},U)$ is preordered by:

$E:(\underline{A},U)\to(\underline{B},V) \leq E':(\underline{A},U)\to(\underline{B}',V')$ iff there exists an extension $F:(\underline{B},V)\to(\underline{B}',V')$ with $F\cdot E = E'$.

An extension $E:(\underline{A},U)\to(\underline{B},V)$ with some property P is called a <u>largest</u> extension with property P, provided the following conditions hold:

(1) If an extension $E':(\underline{A},U)\to(\underline{B}'V')$ has property P, then $E' \leq E$

(2) If an extension $E':(\underline{B},V)\to(\underline{B}',V')$ is such that $E'\cdot E$ has property P, then E' is an isomorphism.

Largest extensions with property P, if they exist, are determined uniquely up to isomorphism.

4.3 Theorem [18,1]: $(\underline{A},U)$ has a largest initial completion if and only if $\underline{A}^2$ is legitimate. In this case, $E^2:(\underline{A},U)\rightarrow(\underline{A}^2,U^2)$ is the largest initial completion of $(\underline{A},U)$.

Largest initial completions are uncommon. Every concrete category $(\underline{A},U)$ has a proper initially dense extension. In particular, no concrete category $(\underline{A},U)$ ever equals its largest initial completion (since empty sources give rise to objects in $\underline{A}^2$, which are not in $E^2(\underline{A})$). The process of forming largest initial completions, if it can be performed at all, is not idempotent.

This unpleasant situation is caused by the fact that the conglomerate of initially dense extensions is not closed under composition. If we restrict attention to initiality preserving extensions the picture becomes much more pleasant.

Universal initial completions

5.1 Theorem [18]: Every initiality preserving initially dense extension of $(\underline{A},U)$ is isomorphic to a subcategory of $(\underline{A}^3,U^3)$.

An initial completion $E:(\underline{A},U)\rightarrow(\underline{B},V)$ is called universal, provided the following conditions hold:

(1) E preserves initiality

(2) For every initially complete category $(\underline{C},W)$ and every initiality preserving concrete functor $F:(\underline{A},U)\rightarrow(\underline{C},W)$ there exists a unique initiality preserving concrete functor $G:(\underline{B},V)\rightarrow(\underline{C},W)$ with $F=G\cdot E$.

A universal initial completion, if it exists, is uniquely determined up to isomorphism.

5.2 Theorem [18,1]: Equivalent are:

(a) $(\underline{A},U)$ has a largest initiality preserving initial completion

(b) $(\underline{A},U)$ has a universal initial completion

(c) $\underline{A}^3$ is legitimate.

In this case, $E^3:(\underline{A},U)\rightarrow(\underline{A}^3,U^3)$ has the desired properties

For practically all "everyday" categories $(\underline{A},U)$, $\underline{A}^3$ is legitimate. For concrete examples see in particular [46,23].

VI Mac Neille completions

An extension, which is simultaneously an initial and final completion, is called a Mac Neille completion.
A Mac Neille completion, if it exists, is uniquely determined up to isomorphism.

6.1 Theorem [5,18,1] : Equivalent are:

(1) $(\underline{A},U)$ has some initial completion
(2) $(\underline{A},U)$ has some final completion
(3) $(\underline{A},U)$ has a Mac Neille completion
(4) $(\underline{A},U)$ has a smallest initial completion
(5) $(\underline{A},U)$ has a smallest final completion
(6) $\underline{A}^4$ is legitimate.

In this case $E^4:(\underline{A},U)\to(\underline{A}^4,U^4)$ has the desired properties.

In the quasicategory of all concrete categories over $\underline{X}$ and all concrete functors over $\underline{X}$, the initially complete categories are precisely the injective objects with respect to extensions [8,18,37]. Hence:

6.2 Theorem [8,18,37,1]: Equivalent are:

(1) $(\underline{A},U)$ has a smallest injective extension
(2) $(\underline{A},U)$ has a largest essential extension
(3) $(\underline{A},U)$ has an injective hull, i.e. an essential injective extension
(4) $\underline{A}^4$ is legitimate.

In this case, $E^4:(\underline{A},U)\to(\underline{A}^4,U^4)$ has the desired properties.

VII Final completions

For the final completions $E^{-n}:(\underline{A},U)\to(\underline{A}^{-n},U^{-n})$ corresponding results hold. In particular, E^{-2} is the largest final completion, E^{-3} the universal final completion, and E^{-4} the smallest final completion, provided these extensions are legitimate. Obviously the extensions E^4 and E^{-4} are isomorphic. Even for partially ordered sets (= concrete categories over the terminal base category), the remaining extensions may all be different [39,18].

III Cartesian closed topological extensions

In this section, we restrict attention to concrete categories $(\underline{A},U)$ over $\underline{Set}$, satisfying the following conditions:

(1) $(\underline{A},U)$ is fibre-small

(2) $(\underline{A},U)$ has finite concrete products

(3) Constant maps between $\underline{A}$-objects are $\underline{A}$-morphisms.

(4) $\emptyset$ carries precisely one $\underline{A}$-structure.

In this context, an extension $E:(\underline{A},U)\to(\underline{B},V)$ is called a <u>catesian closed topological extension,</u> provided $(\underline{B},V)$ is cartesian closed and initially complete and E preserves finite products. Since initial completions preserve final sinks, they are improper candidates for such extensions. In fact, if $(\underline{A},U)$ is initially complete, then $(\underline{A},U)$ has some cartesian closed topological extension, which is simultaneously an initial completion, if and only if $(\underline{A},U)$ is cartesian closed itself [17]. Final completions are better candidates. The largest final completion E^{-2} violates the above requirement (3). But if we remove the "lowest layer", i.e. consider the full subcategory $\underline{A}^{-2\cdot 1}$ of $\underline{A}^{-2}$, whose objects are those A^{-2}-objects (S,X) with $S \neq \emptyset$, then the induced extension $E^{-2\cdot 1}:(\underline{A},U)\to(\underline{A}^{-2\cdot 1},U^{-2\cdot 1})$ is a cartesian closed topological extension, provided it is legitimate.

8.1 <u>Theorem</u> [1]: $(\underline{A},U)$ has a largest cartesian closed final completion if and only if $\underline{A}^{-2\cdot 1}$ is legitimate. In this case $E^{-2\cdot 1}:(\underline{A},U)\to(\underline{A}^{-2\cdot 1},U^{-2\cdot 1})$ is the largest cartesian closed final completion of $(\underline{A},U)$.

$\underline{A}^{-2\cdot 1}$ is rarely legitimate. Even in the case of compact Hausdorff spaces it is illegitimate [1].

8.2 <u>Theorem</u> [1,3,4,5,21]: Equivalent are:

(1) $(\underline{A},U)$ has a cartesian closed topological extension

(2) $(\underline{A},U)$ has a cartesian closed final completion

(3) $(\underline{A},U)$ has a smallest cartesian closed final completion.

(4) $(\underline{A},U)$ is strictly fibre-small in the sense of [3,4].

In this case the smallest cartesian closed final completion of $(\underline{A},U)$ can be construced as the initial hull of the conglomerate of all powers of $\underline{A}$-objects in $(\underline{A}^{-2\cdot 1},U^{-2\cdot 1})$. It can be characterized as the unique cartesian closed final completion of $(\underline{A},U)$ in which powers of $\underline{A}$-objects are initially dense, and will be called the cartesian closed topological hull (CCTH) of $(\underline{A},U)$.

IX Reflective initial completions

An extension $E:(\underline{A},U)\to(\underline{B},V)$ is called reflective, provided E has a (not necessarily concrete) left adjoint.

9.1 Theorem: Equivalent are:

(1) $(\underline{A},U)$ has a reflective initial completion
(2) $(\underline{A},U)$ has a reflective final completion
(3) $(\underline{A},U)$ has a reflective Mac Neille completion.
(4) $\underline{A}^4$ is legitimate and every final completion of $(\underline{A},U)$ is reflective
(5) Every essential extension of $(\underline{A},U)$ is reflective.

The observation, leading to the equivalence (3) $\Longleftrightarrow$ (5) above and to the following results, is the following: $(\underline{A},U)$ has a reflective smallest (universal, largest) initial completion if and only if each closed (semi-closed, weakly closed) source (X,S) can be represented by a single structured epimorphism (f,A), i.e.

$$S = \{(Ug\cdot f,B) \mid g:A\to B \text{ is an A-morphism}\}.$$

Since empty sources obviously don't have such a representation, we conclude

9.2 Theorem [38]: Largest initial completions are never reflective.

A semi-closed source can be represented if and only if it has a (structured epi, initial)-factorization. If every semi-closed source has such a factorization, then so does every structured source. The structured morphisms, appearing in these factorizations, are characterized by the diagonalization property with respect to initial sources. They are called semi-universal maps and automatically structured epimorphisms [23].

9.3 Theorem [23,11]: Equivalent are:

(1) ($\underline{A}$,U) has a reflective universal initial completion

(2) Every initiality preserving, initially dense extension of ($\underline{A}$,U) is reflective

(3) U is topologically-algebraic (in the sense of Y.H. and S.S.Hong [29-31]), i.e. every structured source has a (structured epi,initial)-factorization

(4) U is a (semi-universal, initial)-functor (in the sense of [23]).

A closed source can be represented if and only if it has a semi-initial factorization [24,41], equivalently: if and only if its opposite sink has a semi-final solution (=semi-identifying lift) [24,41,44]. If such factorizations (resp. solutions) exist for all closed sources (sinks), then they exist for all structured sources (sinks). Underlying functors for such concrete categories are called semi-topological [24,41,43,44].

9.4 Theorem [41,28,38]: Equivalent are:

(1) ($\underline{A}$,U) has a reflective Mac Neille completion

(2) Every essential extension of ($\underline{A}$,U) is reflective

(3) U is semi-topological.

Reflective Mac Neille and even universal initial completions exist for any decent category:

9.5 Theorem [25,47,20,12]: Consider the following conditions:

(a) U is an (epi,-)functor in the sense of [23]

(b) ($\underline{A}$,U) has a reflective universal initial completion

(c) ($\underline{A}$,U) has a reflective Mac Neille completion

(d) U has a left adjoint

Then:

(1) Each of the above conditions implies all the following ones

(2) If $\underline{A}$ is an (epi,-) category (in particular: if $\underline{A}$ is cocomplete and cowellpowered), then all the above

conditions are equivalent.

(3) If A is complete, well-powered and cowellpowered, then the conditions (b), (c) and (d) are equivalent.

(4) If A is finite, the conditions (b) and (c) are equivalent.

X Counterexamples

Consider the following diagram:

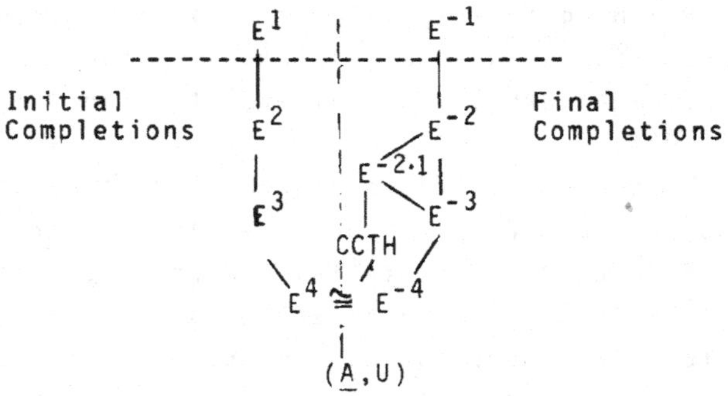

In general, all the above extensions are different [18]. If any of them is legitimate, none of the larger need to be so, even under strong additional requirements.

1o.1 Example: Initial completions need not exist [32,18]. The simplest example is the concrete category over Set of all sets and all bijections and constants between sets of equal cardinality.

1o.2 Example: Reflective Mac Neille completions of fibre-small categories may exist, but not be fibre-small [19].

1o.3 Example: Universal initial completions need not exist, even for concrete categories over Set, which have a fibre-small, reflective Mac Neille completion [1o].

1o.4 Example: Universal initial completions may exist but not be reflective, even though the Mac Neille completions is reflective and A is countable [2o,9].

1o.5 Example: Largest initial completions, final completions and cartesian closed final completions, rarely exist [1].

1o.6 Example: Cartesian closed topological extensions needn't exist, even for initially complete categories satisfying conditions (1) - (4) of section 6 [3,4].

References

1 J.ADÁMEK, H.HERRLICH, G.E.STRECKER: Least and largest initial completions. Preprint.

2 -,-,- : The structure of initial completions. Preprint.

3 J.ADÁMEK, Y.KOUBEK: What to embed into a cartesion closed topological category. Comment.Math.Univ. Carolinae 18, 817-821 (1977).

4 -.- : Cartesian closed fibre-completions. Preprint.

5 P.ANTOINE: Étude élémentaire des catégories d'ensembles structurés. Bull.Soc.Math.Belgique 18, 142-164 (1966).

6 - : Extension minimale de la catégorie des espaces topologiques. C.R.Acad.Sc., Paris A262, 1389-1392 (1966).

7 B.BANASCHEWSKI, G.BRUNS: Categorical characterization of the Mac Neille completion. Archiv Math. 18, 369-377 (1967).

8 G.C.L.BRÜMMER, R.-E.HOFFMANN: An external characterization of topological functors. Springer Lecture Notes Math. 54o, 136-151 (1976).

9 R.BÖRGER: Semitopologisch ≠ topologisch algebraisch. Preprint.

1o - : Universal topological completions of semi-topological functors over Ens need not exist. Preprint.

11 - : Legitimacy of certain topological completions. These Proceedings.

12 R.BÖRGER, W.THOLEN: Remarks on topologically algebraic functors Preprint.

13 G.Bourdaud: Some cartesian closed topological categories of convergence spaces. Springer Lecture Notes Math. 540, 93-108 (1976)

14 N.BOURBAKI: Théorie des ensembles Ch.3 Ensembles ordonnés. Paris: Hermann 1963.

15 M.CHARTRELLE: Constructions de catégories auto-dominées. C.R.Acad. Sci., Paris A.B 274, 388-391 (1972).

16 B.DAY: A reflection theorem for closed categories. J.pure appl. Algebra 2, 1-11 (1972).

17 H.HERRLICH: Cartesian closed topological categories. Math. Colloq. Univ. Cape Town 9, 1-16 (1974).

18 - : Initial completions. Math.Z. 15o, 1o1-11o (1976).

19 - : Reflective Mac Neille completions of fibre-small categories need not be fibre-small. Comment.Math.Univ.Carolinae 19, 147-149 (1978).

2o H.HERRLICH, R.NAKAGAWA, G.E.STRECKER, T.TITCOMB: Equivalence of semi-topological and topologically-algebraic functors. Canad.J.Math.

21 H.HERRLICH, L.D.NEL: Cartesian closed topological hulls. Proc. Amer. Math. Soc. 62, 215-222 (1977).

22 H.HERRLICH, G.E.STRECKER: Category Theory. Allyn and Bacon, Boston 1973.

23 -, - : Semi-universal maps and universal initial completions. Pacific J.Math.

24 R.-E.HOFFMANN: Semi-identifying lifts and a generalization of the duality theorem for topological functors. Math. Nachr. 74, 295-3o7 (1976).

25 - : Topological functors admitting generalized Cauchy-completions. Springer Lecture Notes Math. 54o, 286-344 (1976).

26 - : Topological completion of faithful functors. Kategorienseminar 1, Hagen, 26-37 (1976).

27 - : Full reflective restrictions of topological functors. Math. Colloq. Univ. Cape Town 11, 65-88 (1977).

28 - : Note on semi-topological functors. Math.Z. 16o, 69-74 (1978).

29 S.S.HONG: Categories in which every mono-source is initial. Kyungpook Math.J. 15, 133-139 (1975).

30 Y.H.HONG: Studies on categories of universal topological algebras. Thesis, Mc Master Univ. 1974.

31 - : On initially structured functors. J.Korean Math.Soc. 14, 159-165 (1978).

32 L.KUČERA, A.PULTR: On a mechanism of defining morphisms in concrete categories. Cahiers Topol. Géom. Diff. 13, 397-41o (1972).

33 A.MACHADO: Espaces d'Antoine et pseudo-topologies, Cahiers Topol. Géom. Diff. 14, 3o9-327 (1973).

34 E.G.MANES: Algebraic Theories, Springer Verlag 1975.

35 L.D.NEL: Initially structured categories and cartesian closedness. Canad.J.Math. 27, 1361-1377 (1975).

36 - : Cartesian closed topological categories. Springer Lecture Notes Math. 54o, 439-451 (1976).

37 H.E.PORST: Characterization of Mac Neille completions and topological functors. Bull.Austral.Math.Soc. 18, 2o1-21o (1978).

38 Y.T.RHINEGHOST: Global completions. Preprint.

39 P.RINGLEB: Untersuchungen über die Kategorie der geordneten Mengen. Thesis, Free Univ. Berlin 1969.

40 E.SPANIER: Quasi-topologies. Duke Math.J. 3o, 1-14 (1963).

41 W.THOLEN: Semi-topological functors. J.Pure Appl. Algebra

42 - : Konkrete Funktoren. Habilitationsschrift, Hagen 1978.

43 V.TRNKOVÁ: Automata and categories. Springer Lecture Notes Computer Sci. 32, 138-152 (1975).

44 M. WISCHNEWSKY: A lifting theorem for right adjoints. Cahiers Topol. Gêom. Diff.

45 O. WYLER: Are there topoi in topology? Springer Lecture Notes Math. 54o, 699-719 (1976).

46 R.-E. HOFFMANN: Note on universal topological completion. Preprint.

47 W. THOLEN: On Wyler's taut lifting theorem. Gen. Topol. Appl. 8, 197-206 (1978).

ALGEBRA $\cup$ TOPOLOGY

by

Horst Herrlich and George E. Strecker

Abstract: The smallest collection of functors that is closed under composition and contains all algebraic and all topological functors is characterized as the collection of those semi-topological functors that preserve regular epimorphisms.

AMS(MOS) subject classifications (1973) Primary: 18C10, 18A99, 18A20.
Key Words and Phrases: (regular) monadic functors, regular functors, algebraic functors, topological functors, topologically-algebraic functors, semi-topological functors; regular epimorphisms.

§0. Introduction: The question of what is algebraic versus what is topological makes more sense for (forgetful) functors than for categories. Using the rule that nice collections of functors ought to be closed under composition, the question of what algebraic and topological functors have in common can best be made precise by asking: what is the smallest collection of functors that is closed under composition and contains all algebraic and all topological functors?

To avoid unneccessary complications (due partially to slightly different definitions by different authors) let us assume that all functors in question are faithful and that all categories in question are regular; i.e., have (regular epi, mono-source)-factorizations. Topological and algebraic functors each come in two versions: a more rigid (= transportable) one and a more flexible (= equivalence-invariant) one. In this paper we will adopt the more flexible versions.

Whereas the concept of "topological" functor is well settled by now, the concept of "algebraic" functor is still a subject of discussion. For some time many categorists believed that monadic functors provide the answer. However monadic functors not only violate our basic rule of being closed under composition, they also may fail to preserve regular epimorphisms [3] and may fail to detect colimits [1]. In fact, they seem to behave reasonably well only for very Set-like base categories; e.g., those for which regular epimorphisms are automatically retracts. Monadic functors with such codomains must be regular. More reasonable candidates are the regular monadic functors introduced by E. Manes [11], the algebraic functors [5], and the regular functors [3]. Since the latter two concepts coincide (for regular categories) and, as has been shown by H. E. Porst [12], form the compositive hull of all regular monadic functors, we choose them as "algebraic" functors in this paper.

Categorical concepts, which simultaneously generalize the concepts of algebraic and topological functors, have been proposed before. Notable examples are the concepts of topologically-algebraic functors, introduced by Y. H. Hong [10] (see also [9] and [13]) and of semi-topological functors, introduced by V. Trnková [15], R.-E. Hoffmann [7], M. Wischnewsky [16], and W. Tholen [14]. As has been discovered independently in [2] and [6], the latter form the compositive hull of the former. As has been shown by T. Titcomb (Example 2.3 below) they need not preserve regular epimorphisms. The collection of those semi-topological functors that preserve regular epimorphisms, forms the answer to our problem.

§1. Definitions:

Throughout we assume that all categories in question are regular, and that all functors in question are faithful.

1.1 A functor is called algebraic, provided that it has a left-adjoint and preserves and reflects regular epimorphisms.

1.2 A functor $U:\underline{A} \to \underline{X}$ is called topologically-algebraic, provided that for any U-source $(X \xrightarrow{f_i} UA_i)_{i \in I}$ there exists a U-epimorphism $X \xrightarrow{e} UA$ and a U-initial source $(A \xrightarrow{g_i} A_i)_{i \in I}$ with $f_i = Ug_i \cdot e$ for each $i \in I$.

1.3 A functor $U:\underline{A} \to \underline{X}$ is called semi-topological, provided any U-sink $(UA_i \xrightarrow{f_i} X)_{i \in I}$ has a semi-final solution, i.e. a U-morphism $e:X \to UA$ such that:

(1) each $UA_i \xrightarrow{e \cdot f_i} UA$ is an $\underline{A}$-morphism

(2) for any $e':X \to UA'$ such that each $UA_i \xrightarrow{e' \cdot f_i} UA'$ is an $\underline{A}$-morphism there exists a unique $\underline{A}$-morphism $f:A \to A'$ with $e' = Uf \cdot e$.

1.4 A functor $U:\underline{A} \to \underline{X}$ is called topological, provided it is semi-topological and all semi-final solutions $e:X \to UA$ are $\underline{X}$-isomorphisms.

§2. Results:

2.1 Theorem: For any functor $U:\underline{A} \to \underline{X}$, the following conditions are equivalent:

(1) U is semi-topological and preserves regular epimorphisms.

(2) U belongs to the compositive hull of all regular monadic and all topological functors.

(3) U belongs to the compositive hull of all algebraic and all topological functors.

Proof: (1) $\Longrightarrow$ (2). Let $\underline{B}$ be the category whose objects are the semi-final solutions $X \xrightarrow{e} UA$ of all U-sinks $(UA_i \xrightarrow{f_i} X)_{i \in I}$, whose morphisms from $X \xrightarrow{e} UA$ to $X' \xrightarrow{e'} UA'$ are all pairs (f,g), where $f: X \to X'$ is an $\underline{X}$-morphism and $g: A \to A'$ is an $\underline{A}$-morphism with $e' \cdot f = Ug \cdot e$, and whose composition is defined coordinatewise. Let $E: \underline{A} \to \underline{B}$ be defined by $EA = UA \xrightarrow{1_{UA}} UA$ and $Ef = (Uf, f)$. Let $V: \underline{B} \to \underline{X}$ be defined by $V(X \xrightarrow{e} UA) = X$ and $V(f,k) = f$. If U is semi-topological, then $\underline{B}$ is regular, V is topological, and E is a full embedding, which reflects regular epimorphisms and has a left-adjoint (see [8] or [14]). If U preserves regular epimorhpisms, so does E. Hence (1) implies that U is the composition $U = V \circ E$ of the regular monadic functor E and the topological functor V.

(2) $\Longrightarrow$ (3) $\Longrightarrow$ (1). Straightforward.

2.2 Hence we have the following strict implications among functors.

① regular monadic

$\Downarrow$

② regular
= algebraic
= compositive hull of ①

③ topological

$\Downarrow$

(② $\Longrightarrow$ ④, ③ $\Longrightarrow$ ④)

④ topologically-algebraic and preserving regular epimorphisms

$\Downarrow$

⑤ semi-topological and preserving regular epimorphisms = compositive hull of ①∪③

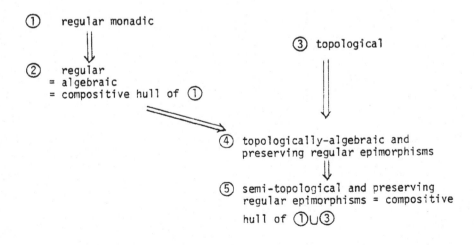

2.3 The following example of a topologically-algebriac (and hence semi-topological) functor $U:\underline{A} \to \underline{X}$, which does not preserve regular epimorphisms, has been discovered by Tim Titcomb. Consider the regular category $\underline{X}$:

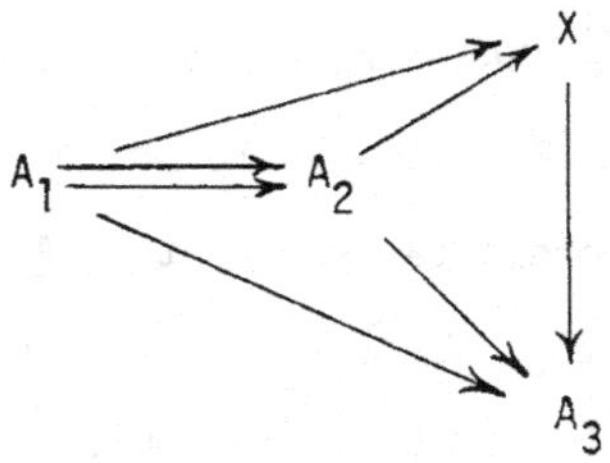

and the regular category $\underline{A}$, which is the full subcategory of $\underline{X}$ determined by A_1, A_2 and A_3. The embedding functor $U:\underline{A} \to \underline{X}$ has the desired properties.

REFERENCES

1. Adámek, J.: Colimits of algebras revisited. Bull. Austral. Math. Soc. 17, 433-450 (1977).

2. Börger, R. and W. Tholen: Is any semi-topological functor topologically algebraic? Preprint.

3. Herrlich, H: Regular categories and regular functors. Canad. J. Math. 26, 709-720 (1974).

4. Herrlich, H.; R. Nakagawa; G. E. Strecker; T. Titcomb: Semi-topological and topologically-algebraic functors (are and are not equivalent). Preprint.

5. Herrlich, H. and G. E. Strecker: Category Theory, Allyn and Bacon, Boston, 1973.

6. Herrlich, H. and G. E. Strecker: Semi-universal maps and universal initial completions. Preprint.

7. Hoffmann, R.-E.: Semi-identifying lifts and a generalization of the duality theorem for topological functors. Math. Nachr. 74, 295-307 (1976).

8. Hoffmann, R.-E.: Note on semi-topological functors. Preprint.

9. Hong, S. S.: Categories in which every mono-source is initial. Kyungpook Math. J. 15, 133-139 (1975).

10. Hong, Y. H.: Studies on categories of universal topological algebras. Thesis, McMaster University 1974.

11. Manes, E.: A triple miscellany, Thesis, Wesleyan University, 1967.

12. Porst, H.-E.: On underlying functors in general and topological algebra. Manuscripta Math. 20, 209-225 (1977).

13. Tholen, W.: On Wyler's taut lift theorem. General Topol. Appl. (to appear).

14. Tholen, W.: Semi-topological functors I. Preprint.

15. Trnková, V.: Automata and categories. Lecture Notes Computer Sci. 32, 138-152 (1975).

16. Wischnewsky, M. B.: A lifting theorem for right adjoints, Cahiers Topol. Geom. Diff. (to appear).

H. Herrlich
F. S. Mathematik
Universität Bremen
28 Bremen
Fed. Rep. Germany

G. E. Strecker
Department of Mathematics
Kansas State University
Manhattan, Kansas 66502
U.S.A.

TOPOLOGICAL SPACES ADMITTING A "DUAL" *) **)

Rudolf-E.Hoffmann

Recall that a topological spaces is A(lexandrov)-discrete iff every intersection of open sets is open. To an A-discrete space X there is associated a space Y such that the lattice $\underline{O}(X)$ of open subsets of X (ordered by inclusion) is isomorphic to the lattice $\underline{A}(X)$ of closed subsets of Y; Y may be considered as a "dual" of X (for Y take the same points as X and the closed sets of X as the open sets of Y) - cf.[1].

It is the purpose of this note to describe a method of producing a class of spaces X admitting a dual which are not lattice-equivalent to an A-discrete space Z, i.e. not $\underline{O}(X) \cong \underline{O}(Z)$ for any such Z. Our method consists in a comparison of two product theorems (1.3,1.7) which are deduced from a general product theorem (2.2). This approach also seems to be useful in similar situations (cf.2.3).

The existence of these (counter-)examples has a more significant interpretation in purely lattice-theoretic terms, see 1.9 below.

At first glance, it seems that spaces admitting a dual very rarely occur. A closer inspection, however, reveals that every injective T_0-space, i.e. every continuous

*) The present paper replaces the elaborate version (45 pp.) of the results of the talk "Projective sober spaces" given at the conference.

**) Conversations with A.Batbedat, K.H.Hofmann, J.D.Lawson, and M.Mislove during the workshop II "Continuous lattices at TH Darmstadt, July 1978, are gratefully acknowledged.

lattice supplied with the Scott topology [13,15], has this property.
Our terminology usually is compatible with [15].

We recall a basic result from the literature which was obtained first by J.R.Büchi [5] §2 (in a more general form).

o.1 THEOREM (cf.[5] §2, [11] prop.1, [8] 4.1, [4] I p.199, [14] Thm.3.1): Suppose L is a complete lattice, then the following are equivalent:

(i) L is distributive and every element of L is a meet of meet-irreducible elements $a \in L$ (i.e. whenever $a = \inf\{b,c\}$, then $a \in \{b,c\}$).

(ii) Every element of L is a meet of meet-prime elements $a \in L$ (i.e. whenever $a \geq \inf\{b,c\}$, then $b \leq a$ or $c \leq a$).

(iii) There is a topological space X with $L \cong \underline{O}(X)$.

o.2 LEMMA. For an open subset M of a space Y the following are equivalent:

(i) M is join-irreducible in $\underline{O}(Y)$.

(ii) Whenever $M \subseteq K \cup N$ for open subsets K,N of Y, then $M \subseteq K$ or $M \subseteq N$.

(iii) Every (relatively) closed subset of M is connected.

(iv) The induced pre-order on M (i.e. $x \leq y$ iff $x \in cl\{y\}$) is downward directed.

The spaces M satisfying (ii) and/or (iii) are called strongly connected in [1o]. The equivalence of (ii) and (iv) is observed in [2].

§1

1.1 DEFINITION (N.LEVINE [1o]). A space X is locally strongly connected ("strongly locally connected") iff for every $x \in X$ and for every neighborhood V of x in X there is an open, strongly connected neighborhood U of x with $U \subset V$.

A continuous lattice in its Scott-topology has a basis of Scott-open filters ([15] II.1.13.2), hence is locally strongly connected.

1.2 THEOREM. A space X is locally strongly connected iff X has a "dual" Y.

In order to produce the desired counterexamples, we use the following two "product theorems" which are analogous to the product theorem for local connectedness.

1.3 THEOREM [9] (4.12). $\prod_I X_i$ for a family $\{X_i\}_{i\in I}$ of non-empty spaces is locally strongly connected iff (1) and (2) are satisfied:

(1) Every X_i is locally strongly connected.

(2) $\{i\in I \mid X_i$ is not strongly connected$\}$ is finite.

1.4 A non-empty space X is called "monogeneous" [3] iff every open cover of X necessarily contains X, i.e. iff every filter on X converges, i.e. iff the induced pre-order of X contains a (=at least one) smallest element o (of which X is the smallest neighborhood).

1.5 LEMMA. A space X has an open basis of monogeneous sets iff every element of $\underline{O}(X)$ is the join of completely join-irreducible elements (i.e. those which are no proper joins).

1.6 LEMMA. A space X is lattice-equivalent to an A-discrete space iff it has an open basis of "monogeneous" sets.

Proof:

(i) A space Y is A-discrete iff every point $a \in Y$ has a smallest neighborhood U_a. U_a is clearly monogeneous with smallest point a. The U_a's clearly form an open basis.

(ii) Let Y denote the set of all those points of X (with X having an open "monogeneous" basis) which have a smallest neighborhood in X. Then Y is clearly A-discrete. Let W,V be open in X with $W \nsubseteq V$, say $z \in W-V$. Now let U be a monogeneous open set with $z \in U \subseteq W$ and let b denote a smallest point of $U = U_b$, then $b \notin V$ (otherwise $U_b \leq V$), hence $W \cap Y \neq V \cap Y$, i.e. the embedding $Y \hookrightarrow X$ induces an isomorphism $\underline{O}(X) \to \underline{O}(Y)$.

In view of 1.5, 1.6 is another consequence of Büchi's theorem in [5] §2.

1.6 is easily extended to say that X is lattice-equivalent to an v-semi-lattice in its A-discrete (=upper end) topology iff it has a monogeneous open basis which is stable under binary intersections (the presence of a smallest element 0 then corresponds to the monogeneity of X itself).

1.7 <u>THEOREM.</u> $\prod_I X_i$ for a family of non-empty spaces is lattice-equivalent to an A-discrete space iff (1) and (2) are satisfied:

(1) Every X_i is lattice-equivalent to an A-discrete space.

(2) $\{i \in I \mid X_i$ is not monogeneous$\}$ is finite.

1.8 <u>EXAMPLE.</u> Let D be a downward directed non-empty partially ordered set endowed with the A-discrete topology (i.e. the upper sets of D are the open sets) and let I be an infinite set. Now D^I has a dual (1.3), but is lattice-equivalent to an A-discrete space iff D contains a smallest element (cf. 1.7(2)). - We note that single examples of this sort are easily obtained, e.g. the set $\mathbb{R}$ of real numbers with open sets (r,∞) $(r \in \mathbb{R})$ and $\emptyset$ and $\mathbb{R}$. Note, on the other hand, that every power of the Sierpinski space $S = \{o,1\}$ with open sets $\emptyset, \{1\}$, and $\{o,1\}$ is - surprisingly - lattice-equivalent to an A-discrete space.

1.9 Let us reword the role of these examples (1.8) in lattice-theoretic terms (which make their relevance more evident). We consider the follwoing conditions (1), (1^*), (2), (2^*), $D_\wedge$, $D_\vee$ for <u>distributive</u> complete lattices $(L,\leq)$:

(1) Every element of L is a meet of meet-irreducible elements;

(1*) every element of L is a join of join-irreducible elements;

(2) every element of L is a meet of completely meet-irreducible elements;

(2*) every element of L is a join of completely join-irreducible elements;

$D_\wedge$ $a \wedge \bigvee_I b_i = \bigvee_I (a \wedge b_i)$, for every $a, b_i \in L$, $i \in I$;

$D_\vee$ $a \vee \bigwedge_I b_i = \bigwedge_I (a \vee b_i)$ for every $a, b_i \in L$, $i \in I$

for arbitrary sets I.

Then it is immediate from the preceding discussion that under the given hypotheses on $(L, \leq)$:

(2) implies (1);	(2*) implies (1*);
(1) implies $D_\wedge$;	(1*) implies $D_\vee$;
(1) $\wedge$ (2*) implies (2) by 1.5 and 1.6;	(1*) $\wedge$ (2) implies (2*).

It is shown in [7a] that $(L, \leq)$ satisfies (2*) iff there is a (necessarily: unique) T_D-space X(cf. [14]; = $T_{1/2}$-space, [4] IIp.7) with $\underline{A}(X) \cong (L, \leq)$. Thus a result of [7b] gives that

$(2^*) \wedge D_\wedge$ implies (2); $(2) \wedge D_\vee$ implies (2*).

Examples demonstrating the difference between (1) (1*), (2), (2*) now are easily available. It is 1.8 that guarantees that (1) $\wedge$ (1*) does imply neither (2) nor (2*).

§2

We formulate a general product theorem (2.2) which covers both 1.3 and 1.6.

2.1 Let $\underline{P}$ be a class of topological spaces which is closed under the formation of

(i) products and

(ii) continuous images.

Then a space X is said to be $\underline{P}$-local iff it has an open basis consisting of members of P.

2.2 THEOREM. A product $\prod_I X_i$ of non-empty spaces X_i is $\underline{P}$-local iff (1) and (2) are satisfied:

(1) Every X_i is $\underline{P}$-local.

(2) $\{i \in I \mid X_i \notin \underline{P}\}$ is finite.

Proof:

(a) Suppose $\prod_I X_i \neq \emptyset$ is $\underline{P}$-local. Let O be open in X_j $(j \in I)$ and $b \in O$, then $O \times \prod_{I-\{j\}} X_i$ is open in $\prod_I X_i$. Let $(b_i)_I \in \prod_I X_i$ with $b_j = b$, then there is an open subspace P of $O \times \prod_{I-\{j\}} X_i$ with $(b_i)_I \in P \in \underline{P}$. Since the j-th projection $pr_j[P]$ is open, X_j is seen to be $\underline{P}$-local. Let $V \in \underline{P}$ be an open non-empty subspace of $\prod_I X_i$. Then $O_1 \times \ldots \times O_n \times \prod_{I'} X_i \subseteq V \subseteq \prod_I X_i$ for some non-empty open sets O_k of X_k - with $k = 1,\ldots,n \in I$ and $I' = I - \{1,\ldots,n\}$. As a consequence, $X_i = pr_i[V] \in \underline{P}$ for every $i \in I'$.

(b) Suppose $\{X_i\}_{i\in I}$ satisfies (1) and (2). Let V be open in $\prod_I X_i$ and let $(x_i)_{i\in I} \in V$. Then $(x_i)_I \in O_1 \times \ldots \times O_n \times \prod_{I'} X_i \subseteq V$ with $1,\ldots,n \in I$ and $I' = I - \{1,\ldots,n\}$ and O_i open in X_i $(i = 1,\ldots,n)$. Since every X_i is $\underline{P}$-local, there are open subspaces $V_1,\ldots,V_n \in \underline{P}$ of $O_1,\ldots$ and, resp., O_n with $x_1 \in V_1,\ldots,x_n \in V_n$. As a consequence, $V_1 \times \ldots \times V_n \times \prod_{I'} X_i$ is a member of $\underline{P}$ (since $\underline{P}$ is productive), an open neighborhood of $(x_i)_I$ and is contained in V.

We apply 2.2 to the following classes $\underline{P}$ of spaces:

(1) monogeneous spaces,

(2) strongly connected spaces,

(3) connected spaces,

i.e. to the following classes of $\underline{P}$-local spaces:

(1*) spaces which are lattice-equivalent to an A-discrete space,

(2*) spaces which admit a dual,

(3*) locally connected spaces.

Note that the verification of the requirements 2.1(i) and (ii) is easier in cases (1) and (2) than in case (3):

Since (1) and (2) can be defined in terms of the associated pre-order $x \leq y$ iff $x \in cl\{y\}$ (1.4, o.2(iv)), it suffices for 2.1(i) to show that these properties are inherited by the pre-order of the product (=the componentwise pre-order).

2.3 EXAMPLE: Let $\underline{P} = \{$quasi-compact spaces$\}$. Then a space X is $\underline{P}$-localiff $\underline{O}(X)$ is an algebraic lattice. For an A-discrete space Y, the U_a's ($a \in Y$) form an open quasi-compact basis.

(1) Let Y be an A-discrete space with a smallest element (i.e. Y is monogeneous), then Y is quasi-compact and has a basis of quasi-compact open sets. As a consequence, this is also true for $X = Y \amalg Y$ (sum) and - by 2.2 - for an infinite power X^I. X^I does not have a dual, since $X = Y \amalg Y$ is not co nnected.

(2) An A-discrete space Z is quasi-compact*) iff (i) the number of minimal elements of $(Z,\leq)$ (=closed points) is finite, and (ii) for every $a \in Z$ there is a minimal element $b \in Z$ with $b \leq a$ (i.e. every point closure contains a closed point, cf. [7a] §2). Let $\mathbb{N}$ denote the set of natural numbers, inversely ordered in its A-discrete topology, then an infinite power $\mathbb{N}^I$ is locally strongly connected, but $\underline{O}(\mathbb{N}^I)$ is not algebraic.

---

*) This is easily verified; it corrects the claim in [6] prop.8 and MR 41, # 1614 that an A-discrete space Y is quasi-compact iff it is monogeneous. The space X of 2.3(1) shows that this is wrong.

2.4 REMARKS:

a) The above theorem 2.2 was obtained by G.Preuß [12] 5.3.12. However, his setting in [12] 5.3.2 is too restrictive to cover the cases 1.3 and 1.7, since {monogeneous spaces} do not form a "class of components" in the sense of [12] 5.3.2(a), and {strongly connected spaces} form a "class of components" which does not satisfy the additivity requirement of [12] 5.3.2(b) (a 3-element counterexample suffices).

b) Clearly, the product theorem 2.2 has an obvious analogue with "P-local" replaced by the following property: "Every point has a neighborhood basis consisting of members of P".
(For P = {quasi-compact spaces}, this is one the versions of "locally quasi-compact").

c) G. Grimeisen, Math.Ann. 173, (1967) 241-252 has coined the term "supercompact" to the "monogeneous" spaces of [3] (not excluding the empty space). At present, the term "supercompact" is reserved to a notion introduced by J.De Groot, in: Contributions to extension theory of topological structures, Berlin 1969, pp.89-9o: "There exists an open subbasis such that every cover of the space consisting of members of this subbasis admits a 2-element subcover". Replacing 2 by an arbitrary natural number n, we obtain the concept of n-supercompactness*). In particular, we have 1-supercompact = supercompact (a la Grimeisen), 2-supercompact = supercompact (a la De Groot).

d) The class of P-local spaces is clearly stable under the formation of arbitrary sums. It need not in general be stable under quotients. Is the quotient of a locally strongly connected space again locally strongly connected? (cf. [1o] lemma 1 and [12] 5.3.8). This links with the

---

*) E.Wattel has kindly informed me that there is some work on this topic (for n=3,4) on the M.Sc. thesis level by the school of De Groot.

question of determining the "lattice-invariant bi-co-reflective hull" of all A-discrete spaces, and, resp., of the Sierpinski space (cf. the present appendix of [7b]).

e) We intend to prepare a paper*) where we shall assign to every topological space a (natural) "dual space". Then, the spaces of 1.1, 1.2 above will be precisely those which are (not homeomorphic, but) lattice-equivalent to their bi-dual ("reflexive topological spaces").

REFERENCES

1. Alexandrov, P.S., Diskrete Räume. Mat. Sb. 2(1937), 5o1-519

2. Andima, Susan J. and W.J.Thron, Order-induced topological properties. Pacific J.Math.75(1978), 297-318.

3. Batbedat, A., Le langage des schêmas pour les demi-groupes commutatifs. Semigroup Forum.

4. Bruns, G., Darstellungen und Erweiterungen geordneter Mengen I. und II. J.reine angew.Math.2o9 (1962), 167-2oo, and, resp., 21o (1962), 1-23.

5. Büchi, J.R., Representation of complete lattices by sets. Portugaliae Math.11(1952), 151-167.

6. Georgescu, G., and B.Lungulescu, Sur les propriêté topologiques des structures ordonnés. Rev.Roumaine Math. Pures Appl. 14(1969), 1453-1456.

7. Hoffmann, R.-E., Sobrification of partially ordered sets. Semigroup Forum.

7a. - - , On the sobrification remainder ${}^{s}X-X$. Pacific J.Math.

7b. - - , Essentially complete T_0-spaces. Preprint.

8. Kowalsky, H.J., Verbandstheoretische Kennzeichnung topologischer Räume. Math.Nachr.21(196o), 297-318.

---

*) This paper will continue both the investigations of the present paper and those reported in the talk given at the conference.

9. Leuschen,J.E. and B.T.Sims, Stronger forms of connectivity. Rend.Circ.Mat.Palermo (II)21(1972),255-266.

1o. Levine,N., Strongly connected sets in topology. Amer. Math.Monthly 72(1965), 1o98-11o1.

11. Papert,S., Which distributive lattices are lattices of closed sets? Proc. Cambridge Phil.Soc.55 (1959), 172-176.

12. Preuß,G., Allgemeine Topologie. Springer Verlag: Berlin-Heidelberg-New York 1972.

13. Scott,D., Continuous lattices. In:Springer Lect.Notes in Math.274(1972), 97-136.

14. Thron,W.J., Lattice-equivalence of topological spaces. Duke Math.J. 29(1962), 671-679.

15. SCS (Seminar on Continuity in Semi-Lattices), A compendium of continuous lattices. Part I (preliminary version). By K.H.Hofmann, J.Lawson, G.Gierz (TH Darmstadt 1978).

Universität Bremen
Fachbereich Mathematik
D-28 Bremen
German Federal Republic

SPECIAL CLASSES OF COMPACT SPACES

M.Hušek,Praha

The main aim of this talk is to acquaint you with two new subclasses of compact Hausdorff spaces,namely with the spaces uniformly generated by well-ordered nets and the spaces with κ-inaccessible diagonal.

The first class was constructed as an interesting example in connection with the study of productive coreflective subclasses of *Unif* in [HR] and the second one as a generalizetion of spaces with G_δ-diagonal in factorization theorems (see $[H_1]$, $[H_2]$).Here we shall mention their properties,relations to other classes of compact Hausdorff spaces,and open problems concerning this subject.

First a discussion of general basic properties of subclasses of the category *Comp* of compact Hausdorff spaces.From categorical point of view,the epireflective or coreflective subcategories seem to be the best subclasses.That means in *Comp* those subclasses which are either productive and closed-hereditary or closed under continuous images and contain all βD,D discrete.Clearly,there are only trivial coreflective classes in *Comp* (i.e.,($\emptyset$),*Comp*).We shall obtain more subclasses if we want them to be traces of coreflective subcategories in a topological category *K* containing *Comp* (e.g.compact sequential spaces, compact locally connected spaces,finite spaces).In that case the property about βD must be replaced by the following one:if a compact space X is inductively generated in *K* by subspaces belonging to the subclass,then X also belongs to the subclass.

If *C* is epireflective in *Comp* and contains a two-point space,then *C* contains all βD,D discrete (hence *C* = *Comp* provided *C* is closed under continuous images),and the ordered space $\omega+1$ (hence *C* = *Comp* provided *C* is closed under *K*-generation by subspaces from *C*,since *C* contains then the closed unit interval).

Classes satisfying the remaining two combinations with three properties (of those four described above) are traces of coreflective subcategories of *K* which are either productive or closed-hereditary. There are many known coreflective subcategories of *K* the traces of which to *Comp* are closed-hereditary,e.g.all compact spaces inductively generated in *K* by subspaces of cardinality less than a given cardinal (these classes may be different for different *K*: $[0,1]^{\omega_1}$ is not generated by countable subspaces in *Top* but it is in the category of all completely regular spaces).

I do not know how to prove purely topologically that there is a coreflective subcategory of *Top* the trace of which to *Comp* is productive and nontrivial. I do not know whether there is a nontrivial productive coreflective subcategory of *Top*; if there is any, then there is the least one which can be constructed as follows: put C_0 to be all discrete spaces, and C_α to be the coreflective hull in *Top* of all products from $\bigcup\{C_\beta \mid \beta\in\alpha\}$; then $C = \bigcup C_\alpha$ is the least productive coreflective subcategory of *Top* different from $(\emptyset)$ and I conjecture that $\beta\omega \notin C$.

If we use uniform spaces instead of topological ones we are able to construct many productive subclasses of *Comp* being traces of coreflective subcategories of *Unif* at the same time. We shall outline ideas and procedures of $[H_3]$, [HR] where one can find most of the details.

Two nets $\{x_a \mid a\in A\}$, $\{y_a \mid a\in A\}$ in a uniform space X are said to be adjacent if the net $\{(x_a,y_a) \mid a\in A\}$ of pairs lies eventually in each uniform neighborhood of the diagonal Δ_X (i.e., it converges "uniformly" to Δ_X). We shall make use of well-ordered nets only. The concept of adjacent nets leads to the following

DEFINITION 1. Let κ be an infinite regular cardinal. We shall denote by L_κ a "pair of adjacent well-ordered nets" of length κ:
the underlying set of L_κ is the product $(0,1)\times\kappa$,
a base of the uniformity on L_κ is formed by the covers
$\{(0,\beta) \mid \beta\in\alpha\} \cup \{(1,\beta) \mid \beta\in\alpha\} \cup \{((0,\beta),(1,\beta)) \mid \beta\in\kappa-\alpha\}$, $\alpha\in\kappa$.

The adjacent nets are used in *Unif* in the same way as converging nets in *Top*. The topological spaces X generated by well-ordered nets are called chain-net spaces (i.e., $A\subset X$ is closed provided every well-ordered net in A has all its limits in A). If we denote by C_κ, κ an infinite regular cardinal, the compact spaces generated by well-ordered nets of length at most κ and by C all compact chain-net spaces ($C = \bigcup C_\kappa$), then it is known that

(a) The classes C and all C_κ are closed-hereditary, closed under continuous images in *Comp* and under inductive generation in *Top*, and are not productive.

(b) If $\kappa\in\lambda$ are infinite regular, then $C_\kappa \subsetneq C_\lambda \subsetneq C \subsetneq$ *Comp*.

The class C_ω of compact sequential spaces is finitely productive and it is not known whether it is countably productive and whether C_κ for uncountable regular κ are finitely productive. The class C contains all compact orderable spaces.

The similar procedure in *Unif* gives us "uniformly chain-net" spaces, i.e. the uniform spaces inductively generated by well-ordered adjacent nets, or equivalently, the spaces belonging to the coreflective hull in *Unif* of all spaces L_κ. The corresponding definition of spaces generated by nets of length at most κ is not exactly what we want because productivity of such spaces depends on set-theoretical assumptions.

DEFINITION 2. Let κ be an infinite regular cardinal. Denote by L_κ the class of compact spaces inductively generated in *Unif* by all spaces $L_\lambda, \lambda\in\kappa$ and by all powers of uniformly discrete spaces. Call the spaces from $L = \bigcup L_\kappa$ uniformly chain-net compact spaces.

A cardinal κ is said to be uniformly sequential (an analogue) of Mazur's sequential cardinal) if there is a non-continuous real function on the Cantor space 2^κ which is uniformly sequentially continuous (i.e. preserves adjacent sequences). One may suppose that no uniformly sequential cardinal exists. In that case one may omit powers of uniformly discrete spaces in the definition of L_κ: L_κ consists of compact spaces generated by all $L_\lambda, \lambda\in\kappa$, hence L_ω consists of compact spaces inductively generated by metric spaces (in *Unif*).

THEOREM 1. (a) *L and all L_κ are productive, closed under continuous images in Comp and under inductive generation in Unif, and are not closed-hereditary.*

(b) *If $\kappa\in\lambda$ are regular infinite cardinals, then $L_\kappa \subsetneq L_\lambda \subsetneq L$, $C_\kappa \subsetneq L_\kappa$ and $C \subsetneq L$.*

Proof. To prove productivity, one must use the fact (Theorem 1 in [HR]) that each product of uniform quotient mappings is again uniform quotient (all products in this proof are taken in *Unif*). If D is a uniformly discrete space, then $L_\lambda\times D$ is inductively generated in *Unif* by L_λ, so that $X\times D$ belongs to L_κ provided $X\in L_\kappa$. Since $X\times Y$ is the least upper bound of $X\times D, D_1\times Y$ for convenient uniformly discrete D, D_1, it follows that L_κ is finitely productive. Suppose now that $X_i\in L_\kappa, i\in I$, f is a mapping on ΠX_i into a metric space (M,d) such that any composition of f with a uniformly continuous map on D^α or L_λ into ΠX_i is uniformly continuous; we must prove that f is uniformly continuous, i.e. that f can be uniformly approximated by uniformly continuous maps. Let $\varepsilon>0$. We know that f is uniformly continuous on ΠD_i, where D_i is the set X_i with the discrete uniformity. Thus there is a finite set $F\subset I$ such that $d(fx,fy)<\varepsilon$ provided $pr_F x = pr_F y$. If we denote by e a canonical embedding of $\Pi\{X_i\}_F$ onto a subspace $\Pi\{X_i\}_F \times \{a_i\}_{I-F}$ of ΠX_i, by r the canonical retraction of ΠX_i onto $\Pi\{X_i\}_F$ and $g=f\cdot e\cdot r$, then $d(f,g)<\varepsilon$ and g is uniformly continuous because $\Pi\{X_i\}_F \in L_\kappa$.

All the facts in (b), except perhaps $L_\kappa \neq L_\lambda$, follow from definitions and productivity of L_κ, L. If $\kappa\in\lambda$ and we put X to be the compact ordered space of all ordinals less or equal to κ^+, then $X\in L_\lambda - L_\kappa$. Indeed, any uniformly continuous map f on D^α, D a uniformly discrete space, into X depends on κ^+ coordinates so that we may suppose that $\alpha\leq\kappa^+$. But D^α is inductively generated by spaces L_β, $\beta\in\kappa$ (this fact can be proved by transfinite induction and using uniformly sequential cardinals-see Theorem 6 in [HR]). Since X is not inductively generated by L_β, $\beta\leq\kappa$, X does not belong to L_κ.

It follows from (b) that L contains all compact ordered spaces and all compact sequential spaces; thus we have the first part of the following Theorem (here ω-inaccessibility of diagonal means that there are no nontrivial adjacent sequences - see Definition 3):

THEOREM 2. (a) *The class L contains all compact sequential spaces, all compact ordered spaces, all dyadic (in fact, κ-adic) spaces, hence all compact topological groups.*

(b) *The class L does not contain infinite compact spaces with ω-inaccessible diagonal (thus e.g. infinite compact extremally disconnected spaces) and* βX *for non-pseudocompact* X.

Proof. Suppose that $X\in L$ contains a non-closed open set A. Then $(A,X-A)$ is not a uniform cover of X and, hence, there are adjacent well-ordered nets $\{x_\alpha | \alpha\in\lambda\}\subset A$, $\{y_\alpha|\alpha\in\lambda\}\subset X-A$. If A is an F_σ-set, then $\lambda=\omega$. Consequently, if an infinite X has ω-inaccessible diagonal, it has non nontrivial adjacent sequences and cannot belong to L because X contains a non-closed cozero set. Similarly if $X = \beta Y$, Y non-pseudocompact, then there is a cozero A such that $Y\subset A \subsetneq X$. We may suppose that the set $\{x_n\}\cup\{y_n\}$ from above is discrete; since $A\cup\{y_n\}$ is normal (as a σ-compact space), the subspace $\{x_n\}\cup\{y_n\}$ is C^*-embedded in X - a contradiction with adjacency.

We shall look more carefully at the spaces with κ-inaccessible diagonal mentioned in the previous theorem for $\kappa = \omega$. These spaces have their origin in factorization via subproducts of mappings defined on products [H$_1$].

DEFINITION 3. Let κ be an infinite regular cardinal. A compact space X is said to have κ-inaccessible diagonal if there are no nontrivial adjacent well-ordered nets of length κ in X. The class of such spaces is denoted by D_κ.

Clearly, a compact space X belongs to D_κ iff for every $A \subset X \times X$ such that $|A-\Delta_X| \geq \kappa$ there is an open $U \supset \Delta_X$ such that $|A-U| \geq \kappa$. If X has $G_{<\kappa}$-diagonal, then $X \in D_\kappa$. The most interesting cases are $\kappa=\omega$ and $\kappa=\omega_1$; the first case because D_ω is much bigger than the class of finite spaces (i.e. the class of compact spaces with $G_{<\omega}$-diagonal), and the second one because, unlike $\kappa=\omega$, the class D_{ω_1} almost coincides with the class of compact metrizable spaces (i.e. the class of compact spaces with G_δ-diagonal). First we shall repeat without proofs properties of D_κ, [H$_1$], [H$_2$].

PROPOSITION 1. *The class D_κ is λ-productive for all $\lambda<\kappa$, closed-hereditary, and not closed under inductive generation in Top. The class D_ω is not closed under continuous images.*

PROPOSITION 2. *A compact space X belongs to D_κ iff any continuous map on a κ-compact product into X depends on less than κ coordinates.*

PROPOSITION 3. *If any countable discrete subspace of a compact space X is C^*-embedded in X, then $X \in D_\omega$.*

The κ-compactness in Proposition 2 can be replaced by pseudo-κ-compactness. It follows from Proposition 3 that all compact F-spaces (thus all compact extremally disconnected spaces, all closed subspaces of βD, D discrete) and their finite products belong to D_ω. If $X \in D_\omega$ is embedded in a countable product, it may be embedded into a finite subproduct.

As to the class D_{ω_1}, it is not known yet whether there exists a nonmetrizable space from D_{ω_1}. In the next part, several conditions are found under which spaces from D_{ω_1} are metrizable.

THEOREM 3. *Let $X \in D_{\omega_1}$ be either scattered or orderable or Fréchet-Uryson. Then X is metrizable.*

Proof. Suppose that $X \in D_{\omega_1}$ is scattered and uncountable. If $A \subset X$, $|A|=\omega_1$, then there is a complete accumulation point of A and its closed neighborhood U such that U contains no other complete accumulation point of A. Then $U \cap A = \{x_\alpha | \alpha \in \omega_1\}$ converges to x and

$\{x_\alpha|\alpha\in\omega_1\}$ and $\{x|\alpha\in\omega_1\}$ are adjacent in X - a contradiction.

Suppose now that $X\in D_{\omega_1}$ is orderable. We may assume that $wX>\omega_1$ (if $wX=\omega$ then we are ready, if $wX=\omega_1$ then $\chi(\Delta_X, X\times X)=\omega_1$ and, hence, $X\notin D_{\omega_1}$). Then either there is an uncountable collection $\{(a_\alpha,b_\alpha)|\alpha\in\omega_1\}$ of disjoint open nonvoid intervals or there are uncountably many immediate neighbors $a_\alpha, b_\alpha, \alpha\in\omega_1$ (see e.g. [J]); in both cases, $\{a_\alpha|\alpha\in\omega_1\}, \{b_\alpha|\alpha\in\omega_1\}$ are adjacent in X - a contradiction.

Finally, let $X\in D_{\omega_1}$ be a Fréchet-Uryson space. Suppose first that $\chi(x)>\omega$ for an $x\in X$ and that U is a base of open neighborhoods of x in X with $|U| = \chi(x)$. There is a sequence S_0 in $X-(x)$ converging to x and a countable $U_0\subset U$ such that $\bigcap U_0\cap S_0=\emptyset$, $\bigcap U_0=\bigcap\overline{U_0}$. Again there is a sequence S_1 in $\bigcap U_0-(x)$ converging to x and a countable $U_1\subset U$ such that $U_0\subset U_1$ and $\bigcap U_1=\bigcap\overline{U_1}$, $\bigcap U_1\cap(S_0\cup S_1)=\emptyset$. We can go up to ω_1 and obtain sequences $\{S_\alpha|\alpha\in\omega_1\}$ in X and countable families $\{U_\alpha|\alpha\in\omega_1\}$ in U such that $U_\alpha\subset U_\beta$ if $\alpha\in\beta\in\omega_1$, each S_α converges to x, $S_\alpha\subset\bigcap_{\beta\in\alpha}\bigcap U_\beta-(x)$, $\bigcap U_\alpha\cap\bigcup\{S_\beta|\beta\in\alpha\}=\emptyset$, $\bigcap U_\alpha=\bigcap\overline{U_\alpha}$. Then $\chi(x,\overline{\bigcup\{S_\alpha|\alpha\in\omega_1\}})=\omega_1$ - a contradiction with $X\in D_{\omega_1}$. If $\chi X=\omega$, then the quotient Y of $X\times X$ along Δ_X is Fréchet-Uryson and, if nonmetrizable, $\chi(\Delta_X,Y)>\omega$. It follows now from the preceding part that $X\notin D_{\omega_1}$.

Under CH one can find other conditions such that the spaces $X\in D_{\omega_1}$ satisfying them are metrizable, e.g. if the tightness of X is countable or if X is small in the sense that $|C(X)|\leq 2^\omega$ (see $[H_2]$).

<u>PROPOSITION 4</u>. *Assume that P is a class of compact spaces having the property that $X\in P$ contains a nontrivial convergent well-ordered net of length ω_1 provided $\chi(X)>\omega$. Then every space from $P\in D_{\omega_1}$ is metrizable.*

<u>Proof</u>. Let $X\in P\cap D_{\omega_1}$. By our assumption on P, X must be first-contable and thus metrizable by Theorem 3.

It is not known whether $P=Comp$ satisfies the assumption of Proposition 4. Such classes P were investigated in $[H_4]$; Proposition 4 together with results oh $[H_4]$ imply e.g. that every dyadic space from D_{ω_1} is metrizable and, under CH, that every space $X\in D_{\omega_1}$ with $cX=\omega$ is metrizable.

The problem from $[H_2]$ whether $\beta(\omega)$ or ω^* belong to D_{ω_1} was answered in the negative independently by B.Balcar, P.Simon [BS] and by S.Shelah [S]. Their procedure can be generalized to get the following result.

THEOREM 4. *If* $X\in D_\omega\cap D_{\omega_1}$ *then* X *is finite.*

Proof. If $X\in D_\omega$, then it has no nontrivial convergent sequence. It follows now from [Š] that any such X can be mapped continuously onto $[0,1]^{\omega_1}$. Thus there is a compact $Y\subset X$ and an irreducible continuous f on Y onto $[0,1]^{\omega_1}$. Denote by y_α the point of $[0,1]^{\omega_1}$ with the coordinates 0 up to α and 1 from α to ω_1. The net $\{y_\alpha\}$ converges to 0 in $[0,1]^{\omega_1}$. Take an $x\in f^{-1}(0)$; if U is a neighborhood of x then $f(U)\subset \overline{\mathrm{Int} f(U)}$ and these sets depend on countably many coordinates so that $\overline{f(U)}$ contains all y_α but countably many. The family $\{f^{-1}(y_\alpha)\cap\overline{U} \mid U$ is a neighborhood of $x, \overline{f(U)}$ depends on the set $\alpha\}$ is centerd and its intersection contains an x_α. It is easy to show that $\{x_\alpha\}$ converges to x, which implies $X\notin D_{\omega_1}$.

COROLLARY. *No infinite extremally disconnected compact space (moreover compact F-space) belongs to* D_{ω_1}.

The last Corollary may be regarded as a generalization of the fact that no compact F-space which is infinite can be dyadic.

Using the same procedure as in the proof of Theorem 4 one can show more, namely existence of a closed set Z such that $\chi(x,Z)=\omega_1$ (also for higher cardinals), which in the case of $X=\beta(\omega)$ can be reformulated as follows: there is a free filter on ω which can be completed to a base of an ultrafilter by adding ω_1 sets. Among other consequences of the above procedure there are conditions under which a compact space can be mapped continuously onto $[0,1]^{\omega_1}$ (e.g. if $2^{cX}<wX$). For details see [H_4].

There are other interesting classes such their interrelations are not clear. We shall mention two of them: Eberlein compact spaces and supercompact spaces.

The class *E* of Eberlein compact spaces (i.e., homeomorphs of weakly compact subspaces of Banach spaces) has the following properties (see e.g. [Ro] for basic properties and references):

(a) *E* is countably productive, closed-hereditary, closed under continuous images [Ru] and not under inductive generation in *Top*.

(b) *E* contains all compact metrizable spaces, one-point compactifications of discrete spaces and no βX for noncompact X [PS]. *E* is contained in C_ω.

Thus there are Eberlein compact spaces X and continuous maps on ω_1-compact products into X which do not depend on countably many coordinates.

The class S of supercompact spaces was defined by J.de Groot: X is said to be supercompact if it has an open subbase such that each cover of X formed by members of the subbase contains a subcover of cardinality at most 2.For basic properties and references see e.g. [vM].

(a) S is productive,not closed-hereditary,not closed under continuous images [MvM] and under inductive generation in *Top*.

(b) S contains all compact metrizable spaces [SS] ,all compact orderable spaces,all compact groups [M] and does not contain infinite spaces with ω-inaccessible diagonal,βX for non-pseudocompact X and all first-countable spaces [vDvM].

It is not known whether S contains E (a generalization of the result from [SS]) or all dyadic spaces.

At the end we shall summarize the main open questions that appeared in this talk.

PROBLEM 1. *Is there a nontrivial productive coreflective subcategory of Top?*

PROBLEM 2. *Is every compact space with ω_1-inaccessible diagonal metrizable?*

PROBLEM 3. *Is every Eberlein compact space supercompact?*

PROBLEM 4. *Is every supercompact space a uniformly chain-net space?*

REFERENCES

[HR] Hušek M.,Rice M.D.:Productivity of coreflective subcategories of uniform spaces (to appear in Gen.Top.Appl.)

[H_1] Hušek M.:Continuous mappings on subspaces of products, Symp.Math.17(1976)25-41

[H_2] Hušek M.:Topological spaces without κ-accessible diagonal, Comment.Math.Univ.Carolinae 18(1977)777-788

[H_3] Hušek M.:Products of uniform spaces (to appear in Czech.Math.J.)

[H_4] Hušek M.:Convergence versus character in compact spaces (to appear in Proc.Top.Coll.Budapest 1978)

[J] Juhász I.:Cardinal functions in topology,Math.Center Tract 34 (Amsterdam 1971)

[vM] van Mill J.:Supercompactness and Wallman spaces,Math.Center Tract 85 (Amsterdam 1977)

[M] Mills C.F.:Compact topological groups are supercompact (to appear)

[MvM] Mills C.F.,van Mill J.:A nonsupercompact continuous image of a supercomact space (preprint 1978)

[PS] Preiss D.,Simon P.:A weakly pseudocompact subspace of Banach space is weakly compact,Comment.Math.Univ.Carolinae 15(1974)603-609

[Ro] Rosenthal H.P.:The heredity problem for weakly compactly generated Banach spaces,Comp.Math.28(1974)83-111

[Ru] Rudin M.E.:A narrow view of set theoretic topology, Proc.Top. Conf.Prague 1976,Lecture Notes Math.609(1977)190-195

[S] Shelah S.:On P-points,β(ω) and other results in general topology, Notices Amer.Math.Soc.25(1978)A365

[SS] Strok M.,Szymański A.:Compact metric spaces have binary bases, Fund.Math.89(1975)81-91

[Š] Šapirovskij B.E.:On hereditarily normal compact spaces (russian),Abstracts Top.Conf.Minsk 1977.

[BS] Balcar B.,Simon P.:Convergent nets in the spaces of uniform ultrafilters (preprint,January 1978)

[vDvM] van Douwen E.,van Mill J.:Supercompact spaces (preprint, April 1976)

Mathematical Institute of Charles University
Sokolovská 83
186 00 Praha
Czechoslovakia

PAIRS OF TOPOLOGIES WITH SAME FAMILY OF CONTINUOUS SELF-MAPS

by

V. Kannan, Hyderabad University; India

ABSTRACT: We characterize pairs of comparable topologies with same family of continuous self-maps, in terms of categorical notions. This requires a description of bireflection in the bireflective hull of a given family of topological spaces; this is obtained in the first section. The known characterizations of bireflective subcategories of TOP follow as easy consequences of this description of bireflection.

§ 1. A description of reflection

Notations: TOP denotes the category of all topological spaces, with continuous maps as morphisms. If $\underline{E}$ is a family of topological spaces and X is a topological space, then $C(X,\underline{E})$ denotes the class of all continuous maps from X into members of $\underline{E}$. We sometimes write $C(X,Y)$ for $C(X,\{Y\})$; when there is no confusion regarding the underlying set, we even write $C(\underline{t},\underline{E})$ instead of $C(X,E)$ where $\underline{t}$ is the topology of X; thus, $C(\underline{t}_1,\underline{t}_2)$ denotes $\{f: X_1 \to X_2 : V \in \underline{t}_2$ implies $f^{-1}(V) \in \underline{t}_1\}$ where $\underline{t}_1$ and $\underline{t}_2$ are topologies respectively on the sets X_1 and X_2.

Definitions: a) $\underline{E}$ is said to be bireflective in TOP, if $\underline{E}$ is reflective such that each reflection morphism is both one-to-one and onto.

(All the subcategories of TOP considered here, are full and replete; we do not distinguish sometimes a full replate subcategory from its class of objects.)

Though the bireflective subcategories of TOP have been elegently characterized in the works of Kennison [K], Herrlich and Strecker [HS], their results do not suffice for our purpose; we now need an 'internal' description of the feflection of a space X in the smallest bireflective subcategory containing a family $\underline{E}$ ('internal' in the sense that the description involves only $\underline{E}$ and X, but not the external notions such as reflective subcategories containing $\underline{E}$.) For this purpose, we prove the following theorem which subsumes the known characterizations of bireflective subcategories.

<u>Theorem 1:</u> Let $(X,\underline{t})$ be a topological space, let $\underline{E}$ be a family of topological spaces. Then the following are equivalent for a topology $\underline{t}^*$ on the set X :-

1) $\{f^{-1}(V) : f\in C(X,Y),\ Y\in\underline{E},\ V \text{ open in } Y\}$ is a subbase for $\underline{t}^*$.
2) $C(\underline{t}^*, \underline{E}) = C(\underline{t},\underline{E})$; further if $\underline{t}_1$ is any topology strictly smaller than $\underline{t}$, then $C(\underline{t}_1, \underline{E}) \neq C(\underline{t}, \underline{E})$.
3) The $\underline{t}^*$-closure of a subset A of X is given by $\bigcap_{f\in C(\underline{t},\underline{F})} f^{-1}\overline{(f(A))}$ where $\underline{F}$ is the family of all finite products of members of $\underline{E}$.
4) $(X,\underline{t}^*)$ is the reflection of $(X,\underline{t})$ in the smallest bireflective subcategory containing $\underline{E}$. (The existence of such a subcategory is also a part of this assertion.)
5) $\underline{t}^*$ is the supremum of all topologies $\underline{t}_1 \leq \underline{t}$ such that $(X,\underline{t}_1)$ belongs to every bireflective subcategory containing $\underline{E}$.
6) $\underline{t}^*$ is the infimum of all topologies of reflections of $(X,\underline{t})$ in the bireflective subcategories containing $\underline{E}$.

<u>Proof:</u> 1) implies 2): Every subbasic $\underline{t}^*$-open set is easily seen to be t-open; therefore $\underline{t}^* \leq \underline{t}$ and therefore $C(\underline{t}^*,\underline{E}) \subseteq C(\underline{t},\underline{E})$. Conversely, if $f\in C(\underline{t},\underline{E})$, then there is $Y\in\underline{E}$ such that $f\in C(\underline{t},Y)$; if V is open in Y, then $f^{-1}(V)$ is $\underline{t}^*$-open; therefore $f\in C(\underline{t}^*,Y)$. Thus $C(\underline{t}^*,\underline{E}) = C(t,\underline{E})$. Now if $\underline{t}_1$ is strictly smaller than $\underline{t}^*$, there is $Y\in\underline{E}$, $f\in C(X,Y)$ and an open subset V of Y such that $f^{-1}(V)$ is not in $\underline{t}_1$. This tells that $f\in C(\underline{t},\underline{E})\setminus C(\underline{t}_1,\underline{E})$.

2) is equivalent to 3): Let us denote $\bigcap_{f\in C(\underline{t},\underline{F})} f^{-1}(\overline{f(\bar{A})})$ by $V(A)$. Then we easily check that i) $V(\emptyset) = \emptyset$ and ii) $V(A)\supset A$. <u>Now for</u> any <u>fixed</u> $f\in C(\underline{t},\underline{F})$, we have $V(A)\subset f^{-1}(\overline{f(A)})$; therefore $\overline{f(v(A))} = \overline{f(A)}$; therefore $v(v(A)) = v(A)$. Lastly, if A and B are two subsets of X and if $v(A\cup B)$ is not contained in $v(A)\cup v(B)$, there exist $x\in X$ and $f,g\in C(\underline{t},\underline{F})$ such that $f(x)\notin\overline{f(A)}$, $g(x)\notin\overline{g(B)}$,

but $h(x) \in \overline{h(A \cup B)}$ for every $h \in C(\underline{t},\underline{F})$. If Y and Z are the codomains of f and g, then $Y \times Z \in \underline{F}$ because $\underline{F}$ is stable under the formation of finite products; define $h: X \to Y \times Z$ by the rule $h(x) = (f(x), g(x))$; then $h \in C(\underline{t},\underline{F})$; therefore $h(x) \in \overline{h(A \cup B)} = \overline{h(A)} \cup \overline{h(B)}$; if $h(x) \in \overline{h(A)}$, then composing h with the projection on Y, we get $f(x) \in \overline{f(A)}$, a contradiction; if $h(x) \in \overline{h(B)}$, we similerly get $g(x) \in \overline{g(B)}$, a contradiction. This proves that always $v(A \cup B) = v(A) \cup v(B)$. These properties of the operator v say that v is a Kuratowski - closure operation. Let $\underline{t}^1$ be the topology defined by v. Since we have $f(v(A)) \subset \overline{f(A)}$ for every f in $C(\underline{t}, \underline{F})$, we have $C(\underline{t},\underline{F}) \subset C(\underline{t}^1,\underline{F})$. If $\underline{t}_1$ is any topology such that $C(\underline{t}_1,\underline{F}) \supset C(\underline{t},\underline{F})$, we have the inequality $f(\underline{t}_1\text{- closure of } A) \subset \overline{f(A)}$ for every f in $C(\underline{t},\underline{F})$, which implies that the $\underline{t}_1$- closure of A is contained in $V(A)$. Thus $\underline{t}^1 \leq \underline{t}_1$. Thus $\underline{t}^1$ is the least topology with the property $C(\underline{t},\underline{F}) \subset C(\underline{t}^1, \underline{F})$. This implies that $C(\underline{t}, \underline{F}) = C(\underline{t}^1, \underline{F})$. The proposition "a function into a product topological space is continuous if and only if each of its projections is continuous" gives that for any topology $\underline{t}_1$, we have $C(\underline{t}_1, \underline{F}) = C(\underline{t},\underline{F})$ if and only if $C(\underline{t}_1,\underline{E}) = C(\underline{t},\underline{E})$. Thus $\underline{t}^1$ is the weakest topology on X such that $C(\underline{t}^1, \underline{E}) = C(\underline{t}, \underline{E})$. Now it is clear that 2) and 3) are equivalent.

2) implies 4):- Let $\underline{A}$ be the collection of all spaces X with the following property: If Y has the same underlying set as X and a strictly smaller topology, then $C(Y,\underline{E})$ is not equal to $C(X,\underline{E})$. We first claim that $\underline{A}$ is bireflective. For any topological space $(X,\underline{t})$, define $\underline{t}^*$ as the smallest topology $\underline{t}'$ on X such that $C(\underline{t},\underline{E}) = C(\underline{t}',\underline{E})$. (The existence of this is easy to prove). Then we see immediately that $(X, \underline{t}^*) \in \underline{A}$. If we take r_X as the identity map from $(X,\underline{t})$ to $(X,\underline{t}^*)$, then we can check that for every $f \in C(\underline{t},\underline{E})$, there is a unique $\tilde{f} \in C(\underline{t}^*, \underline{E})$ (namely $\tilde{f} = f$ itself) such that $f = \tilde{f} \circ r_X$. Since r_X is both one-to one and onto, it follows that $\underline{A}$ is a bireflective subcategory of TOP and that $\underline{t}^*$ is the reflection of $\underline{t}$ in $\underline{A}$. If $\underline{B}$ is any bireflective subcategory containing $\underline{E}$, then the reflection morphism (with respect to $\underline{B}$) is a bijection; therefore the reflection of any space may be taken to have the same underlying set with a smaller topology, (by composing with a suitable homeomorphism), so that the reflection morphisms are

identity maps. Now if $(X,\underline{t})$ is any topological space and if $(X,\underline{t}_*)$ is the reflection of it in $\underline{B}$, then we must have $C(\underline{t},\underline{B}) = C(\underline{t}_*,\underline{B})$. In particular, we have $C(\underline{t}^*,\underline{B}) = C((\underline{t}^*)_*, \underline{B})$. This implies, since $\underline{E}\subset\underline{B}$, that $C(\underline{t}^*, \underline{E}) = C((\underline{t}^*)_*, \underline{E})$. Thus $(\underline{t}^*)_*$ is a topology $\leq t^*$, having the same family of $\underline{E}$-valued continuous maps. Since $\underline{t}^* \in \underline{A}$, this is impossible unless $\underline{t}^* = (\underline{t}^*)_*$. Thus we have $(X,\underline{t}^*)\in \underline{B}$. This proves that $\underline{A}$ is the smallest bireflective subcategory containing $\underline{E}$.

4) implies 1): The description of the smallest bireflective subcategory $\underline{A}$ containing $\underline{E}$ obtained in the previous paragraph, can be used to prove this implication.

We have proved that the first four assertions are equivalent. We leave the proof of the equivalence of these with the last two, to readers.

<u>Corollary 2:</u> Let X be a topological space and $\underline{E}$ a family of topological spaces. Then the following are equivalent.

a) X belongs to every bireflective subcategory containing $\underline{E}$.

b) If Y is any space having the same underlying set as X, but a strictly smaller topology, then $C(Y,\underline{E}) \neq C(X,\underline{E})$.

c) There is a family of functions from X to members of $\underline{E}$ with respect to which X is having the weak topology.

d) $C(X,\underline{F})$ separates points and closed subsets of X, where $\underline{F}$ is the family of all finite products of members of $\underline{E}$.

e) X is homeomorphic to a subspace of YxZ where Y is a (possibly infinite) product of members of $\underline{E}$ and Z is an indiscrete space.

<u>Proof:</u> The statement 3) of Theorem 7 can be used to prove that a) implies d).

To prove d) implies e): Let S_1 be the set of all pairs (A,x) where A is a closed subset of X and $x\in X\setminus A$. For each (A,x) in S_1, choose an $f_{(a,x)}$ in $C(X,\underline{E})$ such that $f_{(A,x)}(x) \notin \overline{f_{(A,x)}(A)}$ (Here we used d)); Let $Y_{(A,x)}$ be the codomain of $f_{(A,x)}$. Now let S_2 be the set of all unorderdd pairs (x,y) of elements of X such that $f_{(A,t)}(x) = f_{(A,t)}(y)$ for each $f_{(A,t)}$ chosen above. For each (x,y) in S_2, let $A_{(x,y)}$ be an indiscrete space having just two elements. We now put $Y = \prod_{(A,x)\in S_1} Y_{(A,x)}$ and $Z = \prod_{(x,y)\in S_2} A_{(x,y)}$.

Define a map $\phi : X \to Y\times Z$ by defining its projections as follows:

$$\phi(A,x)^{(t)} = f_{(A,x)}^{(t)} \quad \text{for every } (A,x) \text{ in } S_1$$

$$\phi_{(x,y)}^{(t)} = \begin{cases} x \text{ if } t \neq y \\ y \text{ if } t = y \end{cases}$$

It is a routine matter to verify that ϕ is an one-to-one continuous map, that is a closed map onto its image; that is, ϕ is a homeomorphism onto its image.

To prove (e) impeies (c):- If A is a subspace of YxZ where Z is indiscrete and Y is a product $\prod_{\alpha\in J} E_\alpha$ of members of $\underline{E}$, then A receives weak topology from the following collection of maps: $\{$Projections from Y to E_α (restricted to A) for each α in J$\}$ $\cup\{C(Z,E) \circ p_z$: E is any fixed member of $\underline{E}$ and p_z is the projection from A to Z$\}$.

The proofs of other implications are straightforward applications of the previous Theorem.

Corollary ([K], [HS]): The following are equivalent for a family $\underline{A}$ of topological spaces:

a) $\underline{A}$ is bireflective.

b) $\underline{A}$ contains indiscrete spaces, and $\underline{A}$ is stable under products and subspaces.

c) $\underline{A}$ is stable under the formation of weak topologies. (that is, if $\{f_\alpha : \alpha\in J\}$ is a collection of maps from a set X into members of $\underline{A}$, then the set X with the smallest topology making each f_α continuous, is a member of $\underline{A}$).

d) $\underline{A} = \{X$: $C(X,\underline{A})$ separates points and closed subsets of X$\}$.

e) There is a family $\underline{E}$ of topological spaces such that $\underline{A} = \{X$: $C(X,\underline{E})$ separates points and closed subsets of X$\}$.

§ 2. Same family of continuous self-maps

'Coreflective subcategory', 'coreflection', 'coreflection morphism', 'bicoreflective', etc. are dual notions of 'relective subcategory', 'reflection', 'reflection morphism', 'bireflective' etc. defined in the previous section. In TOP, every coreflective subcategory is automatically bicoreflective, whereas not every reflective subcategory is bireflective. Theorems dual to Theorem 1, corollaries 2 and 3 of §1 can be obtained for an arbitrary coreflective subcategory. Vide [K1] for more details. Sometimes we attribute the adjectives 'reflective', 'coreflective' etc. to the topological properties, in the sense that the family of all spaces having that property deserves these adjectives. In this sense, Indiscreteness, regularity, complete regularity, zerodimensionality etc. are bireflective properties; being a k-space, sequentialness, local connectedness etc. are coreflective properties.

<u>Theorem 4:</u> Let $\underline{t}_1$ and $\underline{t}_2$ be two topologies on a set X such that $\underline{t}_1 \leq \underline{t}_2$. If P is a bireflective property and Q is a coreflective property such that $\underline{t}_1$ is the P-reflection of $\underline{t}_2$ and $\underline{t}_2$ is the Q-coreflection of $\underline{t}_1$ then $C(\underline{t}_1,\underline{t}_1) = C(\underline{t}_2,\underline{t}_2) = C(\underline{t}_2,\underline{t}_1)$.

Conversely if $C(\underline{t}_1,\underline{t}_1) = C(\underline{t}_2,\underline{t}_2) = C(\underline{t}_2,\underline{t}_1)$, there exist a bireflective property P and a coreflective property Q such that $\underline{t}_1$ is the P-reflection of $\underline{t}_2$ and $\underline{t}_2$ is the Q-coreflection of $\underline{t}_1$.

<u>Proof:</u> Let $\underline{t}_1$, $\underline{t}_2$, P and Q be as stated. Then

$C(\underline{t}_1,\underline{t}_1) \subset C(\underline{t}_2,\underline{t}_1)$ (because $\underline{t}_1 \leq \underline{t}_2$)

$C(\underline{t}_2,\underline{t}_1) \subset C(\underline{t}_2,\underline{t}_2)$ (by the property of coreflections)

$C(\underline{t}_2,\underline{t}_2) \subset C(\underline{t}_2,\underline{t}_1)$ (because $\underline{t}_1 \leq \underline{t}_2$)

$C(\underline{t}_2,\underline{t}_1) \subset C(\underline{t}_1,\underline{t}_1)$ (by the property of bireflections. See §1 .)

These inequalities together prove the first part.

For the converse, let $\underline{t}_1 \leq \underline{t}_2$ be such that $C(\underline{t}_1,\underline{t}_1) = C(\underline{t}_2,\underline{t}_2) = C(\underline{t}_2,\underline{t}_1)$.

Let P be the strongest bireflective property satisfied by $\underline{t}_1$. Then by Theorem 1, the P-reflection $\underline{t}_2^*$ of $\underline{t}_2$ is the weakest of topologies $\underline{t}_3$ such that $C(\underline{t}_3,\underline{t}_1) = C(\underline{t}_2,\underline{t}_1)$. It follows that $\underline{t}_2^* \leq \underline{t}_1$. Considering the identity map, it has to be in $C(\underline{t}_2^*,\underline{t}_1)$ since it is in $C(\underline{t}_2,\underline{t}_1)$. Therefore $\underline{t}_1 \leq \underline{t}_2^*$. We therefore get that $\underline{t}_1$ is the P-reflection of t_2. Similarly, if Q is the strongest coreflective property satisfied by $(X,\underline{t}_2)$, then $\underline{t}_2$ is the Q-coreflection of $\underline{t}_1$.

<u>Corollary 5:</u> Let $\underline{t}_1 \leq \underline{t}_2$ be two topologies on X. Let P be a bireflective property satisfied by $\underline{t}_1$; let Q be a coreflective property satisfied by $\underline{t}_2$. Then either there is an intermediate topology satisfying both P and Q or there is a pair of intermediate topologies having the same family of continuous self-maps.

b) Let P be any bireflective property, Q any coreflective property, $\underline{t}_1$ any topology having P whose Q-coreflection $\underline{t}_2$ does not have P. Then $\underline{t}_2$ and its P-reflection have the same family of continuous self-maps. (Remark: Such instances of P,Q and $\underline{t}_1$ have been provided by many authors. See e.g. [FR] . Our assertion is that every such instance gives rise to an interesting pair of topologies.)

Remarks: a) Some papers of Yu Lee Lee construct pairs of topologies with the same class of self-homeomorphisms. It is clear that what we have obtained, is something more. b) The first part of the above corollary has been announced in [K3]; the second part has been announced earlier.

<u>Theorem 6:</u> Let $\underline{t}_1$ and $\underline{t}_2$ be two topologies on a set X such that $C(\underline{t}_1 , \underline{t}_1) = C(\underline{t}_2 , \underline{t}_2) = C(\underline{t}_2 , \underline{t}_1)$. Let $\underline{t}_3$ be an intermediate topology. Then the following are equivalent.

1) $\underline{t}_3$ is the coreflection of $\underline{t}_1$ in some coreflective subcategory.
2) $\underline{t}_3$ is the reflection of $\underline{t}_2$ in some bireflective subcategory.
3) $C(\underline{t}_3 , \underline{t}_3) = C(\underline{t}_1 , \underline{t}_1) = C(\underline{t}_3 , \underline{t}_1)$
4) $C(\underline{t}_3 , \underline{t}_3) = C(\underline{t}_2 , \underline{t}_2) = C(\underline{t}_2 , \underline{t}_3)$
5) $\underline{t}_3$ is the $\underline{t}_3$- coreflection of $\underline{t}_1$.
6) $\underline{t}_3$ is the $\underline{t}_3$- bireflection of $\underline{t}_2$.
7) $C(\underline{t}_3 , \underline{t}_3) \supset C(\underline{t}_1 , \underline{t}_1)$.

<u>Proof:</u> We prove 7) $\Rightarrow$ 4) only and leave the rest to the readers. We first prove that $C(\underline{t}_2 , \underline{t}_3) = C(\underline{t}_2 , \underline{t}_2)$.

Since $\underline{t}_3 \leq \underline{t}_2$, we have $C(\underline{t}_2,\underline{t}_2) \subset C(\underline{t}_2,\underline{t}_3)$ -- -- (1). As $\underline{t}_3$ is intermediate, we have that $\underline{t}_2$ is the $\underline{t}_2$ -coreflection of $\underline{t}_3$ also. (Since $\underline{t}_2$ is the $\underline{t}_2$ -coreflection of $\underline{t}_1$ (by Theorem 4)).
Therefore by the property of coreflections, we have $C(\underline{t}_2, \underline{t}_3) \subset C(\underline{t}_2,\underline{t}_2)$

-- -- (2)

Now $C(\underline{t}_3, \underline{t}_3) \subset C(\underline{t}_2 , \underline{t}_3)$ since $\underline{t}_3 \leq \underline{t}_2$ -- -- (3)

(3) and (2) imply $C(\underline{t}_3 , \underline{t}_3) \subset C(\underline{t}_2 , \underline{t}_2)$

7) and the assumption imply $C(\underline{t}_3 , \underline{t}_3) \supset C(\underline{t}_2 , \underline{t}_2)$.

These together prove 4).

__Corollary 7:__ Let $\underline{t}_1$ and $\underline{t}_2$ be two topologies on a set X such that $\underline{t}_1 \leq \underline{t}_2$.

Then the following are equivalent:-

1) There is a bireflective property B and a coreflective property C such that in the interval $[\underline{t}_1 , \underline{t}_2]$, $\underline{t}_1$ is the unique topology having B, and $\underline{t}_2$ is the unique topology having C.
2) $C(\underline{t}_1 , \underline{t}_1) = C(\underline{t}_2 , \underline{t}_2) \supset C(\underline{t}_3 , \underline{t}_3)$ holds for every $\underline{t}_3$ in between $\underline{t}_1$ and $\underline{t}_2$; further $C(\underline{t}_2 , \underline{t}_1)$ is a union of semigroups (under composition).
3) $C(\underline{t}_1 , \underline{t}_1) = C(\underline{t}_2 , \underline{t}_2) = C(\underline{t}_2 , \underline{t}_1)$.
4) $\underline{t}_1$ is the $\underline{t}_1$- bireflection of $\underline{t}_2$, and $\underline{t}_2$ is the $\underline{t}_2$-coreflection of $\underline{t}_1$.

__Proof:__ 2) implies 3):- If 3) is not true, pick f in $C(\underline{t}_2,\underline{t}_1) \setminus C(\underline{t}_1 , \underline{t}_1)$ and let for each positive integer n and each V in $\underline{t}_1$, $V_n = \{x \in X: f^n(x) \in V\}$. Let $\underline{t}$ be the topology generated by $\{V_n: V \in \underline{t}_1, n = 0, 1, 2 \ldots\}$. Then $f \in C(\underline{t}_3 , \underline{t}_3)$, a contradiction.

__Remarks 8:__ a) There are topologies $\underline{t}_1$, $\underline{t}_2$ on X such that $\underline{t}_1 \leq \underline{t}_2$, $C(\underline{t}_1 , \underline{t}_1) = C(\underline{t}_2 , \underline{t}_2) \supset C(\underline{t}_3 , \underline{t}_3)$ for every intermediate $\underline{t}_3$, but still $C(\underline{t}_1 , \underline{t}_1) \neq C(\underline{t}_2 , \underline{t}_1)$. For example, consider $X = \{1,2,3\}$, $\underline{t}_1 = \{\emptyset, \{1,2\} , X\}$ and $\underline{t}_2 = \{\emptyset, \{1,2\}, \{3\}, X\}$.

b) Let $\underline{t}_1$ and $\underline{t}_2$ be two topologies on a finite set such that $\underline{t}_2$ strictly contains $\underline{t}_1$. Then $C(\underline{t}_2 , \underline{t}_1)$ strictly contains

$C(\underline{t}_2, \underline{t}_2)$ unless $\underline{t}_2$ is discrete. This can be proved from the known result that if a coreflective property does not imply discreteness, then it is satisfied by every finite space. Our results show that a similar situation does not hold for infinite spaces.

REFERENCES

[FR] S.P. Franklin and M. Rajagopalan, Some examples in Topology, Trans. Amer.Math. Soc. 155 (1971) 305-314

[HS] H. Herrlich and G.E. Strecker, Topologische Reflectionen and Coreflectionen, Springer Verlag Lecture Notes in Math. 78 (1969).

[K1] V. Kannan, Coreflexive Subcategories in Topology, Ph.D. Thesis, Madurai University (1970).

[K3] V. Kannan, Lattices of Topologies, Topology Seminar at Meerut (1977).

[K] J.F. Kennison, Reflective functors in General Topology and elsewhere, Trans.Amer.Math.Soc.(1965)118,303-315.

HYDERABAD UNIVERSITY
NAMPALLY STATION ROAD
HYDERABAD-500 001
INDIA

HEREDITARILY LOCALLY COMPACT SEPARABLE SPACES

BY

V. KANNAN AND M. RAJAGOPALAN*

ABSTRACT: We first obtain some neat characterizations of hereditarily locally compact separable spaces. The first section includes also some characterizations of minimal one among them. The second section describes intrinsically the one-point-compactifications of such spaces. It is also proved that a compact Hausdorff sequential space of type (2,1) fails to be Frechet if and only if it contains one such space as a subspace. Thus a good class of test spaces for Frechet property is obtained here in answer to a problem of Arhangelskii and Franklin. In constrast no this we like to mention that it was proved in [R1] that S_2 cannot be a test space for sequential spaces of order 2.

Key words:- Sequential space, type (2,1), test space, Ψ^*,

A.M.S. classification (1970): 5452, 5440, 5422.

* This author had an NSF contract MCS 77-22201 when this paper was written.

Throughout, N denotes the set of all natural numbers; ω_1 is the first uncountable ordinal number; 'p.a.d.' is the abbreviation for 'pairwise almost disjoint'. Two sets A and B are said to be almost disjoint if their intersection is finite; they are said to be almost equal if their summetric difference is finite; if $A \setminus B$ is finite, we say that A almost contains B. We use the word 'clopen' in the sense of 'both closed and open'. 'h.l.c.s.' is our abbreviation for 'hereditarily-locally compact and separable'. The h.l.c.s. spaces may contain monseparable subspaces.

Let $\underline{F}$ be a family of infinite, p.a.d., subsets of N. Let $\psi(\underline{F})$ denote the disjoint set union $N \cup \underline{F}$, topologized as bolow. If F is an element of $\underline{F}$, it is on one hand an element of $\psi(\underline{F})$, and on the other hand a subset of N and hence of $\psi(\underline{F})$, Since this may create some confusion in our later discussions, we shall denote by F^*, the element F of $\underline{F}$, when viewed as an element of $\psi(\underline{F})$. Thus we have

$$\psi(\underline{F}) = N \cup \{F^* : F \in \underline{F}\}.$$

Each element of N is declared to be isolated in $\psi(\underline{F})$. If $F \in \underline{F}$, a set containing F^* is defined to be a neighbourhood of F^* if and only if it almost contains F.

An easy application of Zorn's lemma shows that every such $\underline{F}$ is contained in a maximal family of infinite p.a.d. subsets of N. The spaces $\psi(\underline{F})$ for such maximal families $\underline{F}$, play a special role in this paper.

Remarks: These spaces were first introduced by J.R. Isbell and considered in [G-J, p. 79]; they appear in [F] to provide an example of a compact Hausdorff sequential non-Frechet space; [R1] contains some non-embeddability theorems concerning them; [K1] and [R2] have used them to construct compact Hausdorff spaces of higher sequential orders; [K2] speaks of them as the spaces that let to guess the result of that paper; [M] has utilised them in connection with rings of continuous functions; [S] has implicitly used them to answer a question of Semadeni,

With this background, we are ready for the main results. All spaces considered, are assumed to be Hausdorff.

§ 1

Theorem 1.1:

A) The following are equivalent for a separable space X:-

1) X is hereditarily locally compact.
2) X is locally compact and the set of accumulation points of X is discrete.
3) X is homeomorphic to $\psi(\underline{F})$ for some $\underline{F}$.

B) The following (4 through 7) are equivalent for a separable space X:-

4) X is pseudocompact and hereditarily locally compact.
5) X is a minimal h.l.c.s. space; that is, X is hereditarily locally compact and no strictly coarser Hausdorff topology is so.
6) X is homeomorphic to $\psi(\underline{F})$ for some maximal $\underline{F}$.
7) X is regular; the set of accumulation points of X is discrete; and every clopen discrete subset of X is finite.

Proof:

1) implies 2): Since X is separable, it has a countable dense subset Y; since X is h.l.c.s., this Y is locally compact. It follows by an easy application of Baire category theorem (or from known characterizations of countable locally compact spaces) that Y contains a discrete dense subset, say N. It is easily seen that every point of N has to be isolated in X and that there can be no other isolated points. Now suppose its complement $X \setminus N$ has an accumulation point x. Then the subspace $N \cup \{x\}$ is not open in its closure, namely X. Hence it cannot be locally compact. (For, in a locally compact space, a subspace is locally compact if and only if it is open in its closure). This contradicts 1).

2) implies 3): Let N be the set of all isolated points

of X. Then N must be contained in every dense subset of X. Since X is separable, it follows that N is countable. This N is dense in X, because the set $X \setminus N$ of accumulation points of X is assumed to be discrete. Now for each x in $X \setminus N$, let V_x be a compact neighbourhood of x such that $V_x \subset N \cup \{x\}$. We now claim that $\{V_x : x \in X \setminus N\}$ is a family of p.a.d. infinite subsets of N. For this, we first observe that each V_x is the set of a sequence in N converging to x. This is because every compact countable space with a unique accumulation point must be homeomorphic to $\omega+1$. If $V_x \cap V_y$ is infinite, then we get a sequence converging to both x and y; this is impossible in a Hausdorff space unless $x = y$. It is now clear that the topology of X is same as that of $\psi(\{V_x : x \in X \setminus N\})$.

3) implies 1): Clearly $(\underline{F})$ is locally compact. The pairwise almost disjointness of members of $\underline{F}$ assures the Hausdorffness of this space. If A is any subset of this space, the set $\bar{A} \setminus A$ is closed. (In fact, any subset of $X \setminus N$ is closed, since any subset containing N is open). Thus A is open in its closure and therefore locally compact.

4) implies 6): Since we have already proved that 1) implies 3), we may assume that X is $\psi(\underline{F})$ for some family $\underline{F}$ of p.a.d. infinite subsets of N. We have to deduce the maximality of $\underline{F}$ from the pseudocompactness of X. If $\underline{F}$ is not maximal, there is an infinite subset A of N. almost disjoint with every member of $\underline{F}$. Then every F^* (where $F \in \underline{F}$) has a neighbourhood disjoint with A. Thus A is infinite and clopen. Clearly then there is a continuous real function on X which maps A onto any countable set we please.

6) implies 7): X is locally compact and Hausdorff and therefore regular. The set of accumulation points of X is discrete, because every such point has a neighbourhood containing no other such point. If A is a clopen discrete subset of $\psi(\underline{F})$, then consider $A \cap N$; the maximality of $\underline{F}$ implies that either

$A \cap N$ is finite or $A \cap N \cap F$ is infinite for some F in $\underline{F}$; in the latter case F^* is a limit point of A, contrary to the assumption that F is closed and discrete; consequently, $A \cap N$ is finite; but since N is dense and A is open, we hawe $A \subset \overline{A \cap N}$, thereby proving that A its elf is finite.

7) implies 4): First, we shall prove that X is locally compact. At isolated points, the singletons form compact neighbourhoods. Let X be an accumulation on point. Then since the set of all accumulation points of X is discrete, there is a neighbourhood V_x of x such that x is the only accumulation point of V_x, inside V_x. By regularity, we may assume that this V_x is closed. If A is any infinite subset of $V_x \setminus \{x\}$, then A is open (since every point of A is isolated in X) and $A \cup \{x\}$ is closed (since $A \subset V_x = \overline{V}_x$ and since x is the only accumulation point of V_x); our assumption that everu clopen discrete subset of X is finite, therefore implies that x is a limit point of A. Thus x is a limit point of every infinite subset of V_x. Clearly then, V_x is compact.

Now the discreteness of the set of all accumulation points implies (as we have already seen) that every subset is locally closed (that is, closed in a bi-ger open set, or equivalenty open in its closure) and therefore locally compact.

In the space X, there is a dense subset (namely, the set of all isolated points) every infinite subset of which has a limit point in X (since every clopen discrete subset is finite). Hence if f is any real valued continuous function, the range $f(X)$ must contain a dense subset, every infinite subset of which has a limit point. In other words, some bounded subset is dense in $f(X)$. Clearly then $f(X)$ itself is bounded. Hence X is pseudo compact.

5) implies 6): By what we have proved already, 5) implies that X is $\psi(\underline{F})$ for some $\underline{F}$. If this $\underline{F}$ is not maximal, there is an infinite subset A of N which meats every member of $\underline{F}$

in a finite set. Now weaken the topology at the point F^* (for one fixed F in $\underline{F}$) by declaring that every neighbourhood by F^* must almost contain $F \cup A$. [That is, we consider $\psi(\underline{G})$, where $\underline{G} = (\underline{F} \setminus \{F\}) \cup \{F \cup A\}$] this is a coarser T_2 topology that is h.l.c.s.

6) implies 5): We claim that any coarser h.l.c.s. topology gives rise to a family $\underline{G}$ of p.a.d. infinite subsets of N and an one-to-one map θ from $\underline{G}$ to $\underline{F}$ such that G almost contains $\theta(G)$ for every G in $\underline{G}$. For each F in $\underline{F}$, take a compact neighbourhood G of F^* in the coarser topology, containing no other accumulation point in that topology; this is possible because that topology is also h.l.c.s.; $\underline{G}$ is the collection of all such G; the map θ is clear. If G and $\theta(G)$ are not almost equal for some G in $\underline{G}$, then $G \setminus \theta(G)$ is infinite, therefore meets some member F of $\underline{F}$ infinitely, therefore has F^* in its closure in both the topologies, therefore has both F^* and $(\theta(G))$ in its closure in the coarser topology, a contradiction to the choice of G.

Now $\underline{F}$ and $\underline{G}$ are two families of p.a.d. infinite subsets of N and θ is a bijection. (Reason: There cannot be non accumulation points in the coarser topology. For, in any h.l.c.s. space, the set of accumulation points is discrete; hence, if X is a new accumulation point and V_x is a compact neighbourhood of x containing no other accumulation point, the some infinite subset of V_x has some F^* in its closure, by maximality of $\underline{F}$; hence a contradiction) from $\underline{G}$ to $\underline{F}$ such that G and (G) are almost equal for each G in $\underline{G}$. It follows that the two topologies that we are considering, are identical.

Corollaries 1.2:

a) The following are equivalent for a separable space X:-

1) X is hereditarily locally compact, pseudo compact, but not compact.

2) X is minimal h.l.c.s. and uncountable.

3) X is homeomorphic to $\psi(\underline{F})$ for some maximal and

infinite $\underline{F}$.

b) A space X is a subspace of an h.l.c.s. space if and only if either X itself is h.l.c.s. or X is the sum of an h.l.c.s. space with an uncountable discrete space of cardinality $\leq c$.

c) Every hereditarily separable hereditarily locally compact space is countable.

d) The class of pseudo-compact h.l.c.s. spaces is stable under the formation of finite sums, clopen subspaces and quotients that are one-to-one except on a finite subset of the domain.

e) Every h.l.c.s. space has an h.l.c.s. pseudo compact extension.

Remarks 1.3:

a) The assumption of separability in Theorem 1.1. cannot be deleted. The one-point-compactification of an uncountable discrete space satisfies 1), 2), 4), 5) and 7), but not 3) and 6) of the Theorem. There may even exist (we do not know) spaces satisfying 1) but not 2).

b) The following can now be easily proved: A noncompact T_3 space is $\psi(\underline{F})$ for some maximal $\underline{F}$ if and only if there is a subset D such that i) D is countable, ii) D is open, iii) both D and its complement are discrete and iv) every sequence in D has a subsequence converging in X.

c) Proposition: A space X is a continuous image of a pseudo-compact h.l.c.s. space if and only if X contains a countable dense set D such that every sequence in D has a subsequence convergent in X.

To prove the 'if' part, take a maximal family $\underline{F}$ of p.a.d. infinite subsets of D that are sets of convergent sequences in X. Our asumption implies that if A is any infinite subset of D, then there is an infinite subset B of A that is the set of a convergent sequence; $B \cap F$ is infinite for some F in $\underline{F}$; hence $A \cap F$ is infinite. Thus $\underline{F}$ is a maximal family of p.a.d. infinite subsets of X (without any further condition on its members). Clearly X is the image of $\psi(\underline{F})$ under the abvious continuous map.

d) It can be proved that the following are equivalent for an h.l.c.s. space: i) -compactness, ii) Lindelofness, iii) hereditarily separability iv) metrizability v) the first countability of its one-point-compactification, vi) normality, vii) second countability and viii) countability.

§ 2

We say that a space is of type ψ is it is homeomorphic to $\psi(\underline{F})$ for some maximal $\underline{F}$; we say that it is of type ψ^* if it is homeomorphic to the one-point-compactification of some space of type ψ.

Theorem 2.1.: A topological space X is of type ψ^* if and only if it satisfies the following four conditions:-

i) X is compact Hausdorff.

ii) X is separable.

iii) The set of accumulation points of X has a unique accumulation point x_o.

and iv) No sequence of isolated points of X converges to x_o.

Proof: If X is a space of type ψ^*, let x_o be its point such that $X \setminus \{x_o\}$ is of type ψ. Then $X \setminus \{x_o\}$ satisfies the conditions of Theorem 1.1 and is in particular separable; hence X is separable. Clearly, X is compact and Hausdorff. To prove iii) we observe that the set of accumulation points of $X \setminus \{x_o\}$ is infinite and discrete and hence cannot be closed in X, whereas it is closed in $X \setminus \{x_o\}$. To prove iv) let B be the set of a sequence of distinct isolated points. Then by the maximality of

$\underline{F}$, (where $\underline{F}$ is the family such taht $X \setminus \{x_o\}$ is homeomorphic to $\psi(\underline{F})$, there is F in $\underline{F}$ such that $B \cap F$ is infinite. (A set is not distinguishad by us from its image under the above homeomorphism and hence this is meaningful.) Then F^* is in the closure of B. Hence the sequence that we started with, cannot converge to any point other then F^*. In particular, it can not converge to x_o.

Conversely, let X be a space satisfying these four conditions. Let Y be the subspace $X \setminus \{x_o\}$. Then clearly Y is locally compact, Hausdorff, non-compact and separable. Further, the set of all accumulation points of Y is discrete. If W is a clopen discrete infinite subset of Y, them $W \cup \{x_o\}$ will be closed in X and hence compact; since every point of W has to be isolated, this means that W is the set of a sequence of isolated points of X converging to x_o. This contradicts iv), Thereby proving that every clopen discrete subset of Y is finite. Thus Y satisfies the condition 7) of Theorem 1.1 and therefore is of type ψ. Thus X is of type ψ^*.

<u>Theorem 2.2</u>: Let X be any space such that

1) It is locally compact and Hausdorff.

2) It is scattered, with derived langth 2.

and 3) It is sequential, with sequential order 2.

Then X contains a subspace of type ψ^*. (Conversely, it is easy to prove that any space of type ψ^* satisfies these three conditions). (See [K] for the definitions of some new terms here).

<u>Proof</u>: Let x_o be a point in X with sequential order 2. Let W be a compact open neighbourhood of x_o containing no other point of derived length ≥ 2. Since the sequential order at x_o is 2, there is a subset A of W such that x_o is in $\overline{A}$, but no sequence from A converges to x_o. Let B the set of those points of A that are accumulation points of X. We claim that x_o is not in $\overline{B}$. We observe that $\overline{B} \cup \{x_o\}$ has at

most one accumulation point and hence Frechet. (Since it is already sequential). Therefore, if x_o were in $\overline{B}$, there would be a sequence from B (and hence from A) converging to x_o, contradicting our choice of A. Thus $x_o \notin \overline{B}$ and hence we may as well assume that every point of A is isolated in X, Since every sequential space is countably generated, we can choose a countable subset $\overline{C}$ of A such that $x_o \in \overline{C}$. Now let $Y = \overline{C}$. Then the space Y satisfies the four conditions of Theorem 2.1.; therefore it is of type ψ^*.

Remarks:

a) The spaces of type ψ^* are thus test spaces to verify Frechet property, among a fairly good class of spaces. A (locally) compact space is said to be of type (2,1) if there is a unique point in its second derived set. (This is a standard terminology, introduced by sierpinski). A restatement of our theorem reads like this: A compact Hausdorff sequential space of type (2,1) fails to be Frechet if and only if it contains an uncountable minimal h.l.c.s. subspace. In this connection, the equivalence of the following three assertions, is also an easy consequence of our results, for a separable compact Hausdorff space X of type (2,1).

i) X is not first-countable.

ii) X contains an uncountable discrete subspace.

iii) X contains an uncountable h.l.c.s. subspace.

b(It will be interesting to know whether 2) can be deleted in Theorem 2.1.

REFERENCES

[F] S.P. Franklin, Spaces in which sequences suffice II, Fund. Math. 61 (1967) 51-66.

[G-J] L. Grillman and M. Jerison, Rings of continuous functions, Princeton, Van Nostrand co, (1960).

[K] V. Kannan, Ordinal invariants in Topology, a monograph, To appear.

[K1] ___, Ordinal invariants in Topology I, on two questions of Arhangelskii and Franklin, general topology and applns. 5 (1975) 269-296.

[K2] ___, Every compact T_5 sequential space is Frechet, Fund. Math., To appear.

[M] S. Mrowka, Some set-theoretic constructions in topology, Fund. Math. XCIV (1977) 83-92.

[R1] M. Rajagopalan, Sequential order and spaces S_n, Proc. Amer. Math. Soc. 54 (1976) 433-438.

[R2] ___, Sequential order of compact spaces, In preparation.

[S] R.C. Solmon, A scattered space that is not- o-dimensional, Bull. Lond. Math. Soc; 8 (1976) 239-240.

1) Madurai University,
Madurai 625021,
INDIA.

2) Universidad de Los Andes
VENEZUELA.

VK/MR/codea

INJECTIVES IN TOPOI, I:

REPRESENTING COALGEBRAS AS ALGEBRAS

F.E.J. Linton (Wesleyan U.) and R. Paré (Dalhousie U.)

Call a cotriple $\mathbb{G}$ on a topos $\mathbb{E}$ with subobject classifier Ω <u>dually algebraic</u> if the composite $(\mathbb{E}_{\mathbb{G}})^{op} \xrightarrow[(V_{\mathbb{G}})^{op}]{} \mathbb{E}^{op} \xrightarrow[\Omega^{(-)}]{} \mathbb{E}$ is tripleable, $V_{\mathbb{G}}$ being the $\mathbb{E}$-valued "underlying" functor on the $\mathbb{G}$-coalgebras (cf. [ZTB]). During what has come to be known as the Lawvere-Tierney topos year 1969-70 at Dalhousie, J.R. Isbell raised the question whether all cotriples on the topos of sets are dually algebraic (see [IGF], p. 588, ℓ. 4*). By April, 1970, we had a composite tripleableness lemma ([LDC], Th. 2; see [MAT], Exer. 3.1.17, for a more polished version) informing us that they are ([LDC], Cor. 5).

In this lemma, as against the earlier ones of Barr and Beck, the crucial ingredient is a requirement (ZHD) on the middle category, roughly, that practically all objects be projective. In 1973, consequently, when it was clear that $\Omega^{(-)} : \mathbb{E}^{op} \longrightarrow \mathbb{E}$ is always tripleable ([PCT], p. 558), the same lemma automatically revealed, for any topos $\mathbb{E}$, that all cotriples on $\mathbb{E}$ must be dually algebraic -- <u>provided</u> (coZHD) each nonzero object of $\mathbb{E}$ is injective.

Here we give the latest full proof of that lemma. Moreover, for those content merely to know all <u>indexed</u> cotriples (à la [RAF]) on $\mathbb{E}$ are dually algebraic, we weaken the seemingly overrestrictive coZHD proviso: it suffices that each object X of $\mathbb{E}$ be internally injective over its own support σX, i.e., as an object of $\mathbb{E}|_{\sigma X}$. It suffices -- but it is also necessary; and, as a corollary, we establish that the coZHD proviso, sufficient for all cotriples on $\mathbb{E}$ to be dually algebraic, is likewise necessary as well. Thus:

<u>Theorem A</u>. <u>Necessary and sufficient for all cotriples on the topos</u> $\mathbb{E}$ <u>to be dually algebraic is that each nonzero object</u> $X \neq 0$ <u>of</u> $\mathbb{E}$ <u>be injective</u>.

<u>Theorem B</u>. <u>Necessary and sufficient for all indexed cotriples on</u> $\mathbb{E}$ <u>to be dually algebraic is that each object</u> X <u>of</u> $\mathbb{E}$ <u>be internally injective in the open subtopos</u> $\mathbb{E}|_{\sigma X}$, <u>where</u> σX <u>is the support of</u> X.

Over the years, the research embodied herein was supported, in part and at times, by the Izaak Walton Killam Trust, Dalhousie U., Wesleyan U., McGill U., the National Research Council of Canada (grant # A-8141), and the U.S. National Science Foundation (MCS 76-10615): we gratefully thank them all.

1. The ZHD-Lemma and Theorem A (sufficiency proof). The ZHD-Lemma below, of which the sufficiency assertion in Theorem A is a direct consequence, makes use of the following definitions.

(1.1) Definitions. (i) An object Q of a category $\mathcal{B}$ is an artificial terminal object ("Q is AT") if Q is terminal in $\mathcal{B}$ and every $\mathcal{B}$-morphism with domain Q is an isomorphism ("AT" is "isolated" in [MAT], Exer. 3.1.10).

(ii) The category $\mathcal{B}$ is ZHD (think "zero homological dimension") if all objects, save perhaps those that are AT, act projective when tested against coequalizers; if $\mathcal{B}^{op}$ is ZHD, we say $\mathcal{B}$ is coZHD.

[It is clear that a topos is ZHD if and only if it satisfies the axiom of choice (AC). H.-M. Meyer has observed (see Satz 6.4 in [MDT]) that a topos is well-pointed if and only if it is coZHD and Boolean. Sets, pointed sets, modules over a semisimple ring, and their full subcategories, all constitute examples of categories that are both ZHD and coZHD.]

(1.2) ZHD-Lemma. Let $\mathcal{A}$, $\mathcal{B}$, and $\mathcal{C}$ be categories, with $\mathcal{C}$ tripleable over $\mathcal{B}$ via $V: \mathcal{C} \longrightarrow \mathcal{B}$, and $\mathcal{B}$ tripleable over $\mathcal{A}$ via $U: \mathcal{B} \longrightarrow \mathcal{A}$. Assume $\mathcal{B}$ is ZHD. Then $\mathcal{C}$ is tripleable over $\mathcal{A}$ via the composite

$$U \circ V: \mathcal{C} \xrightarrow{V} \mathcal{B} \xrightarrow{U} \mathcal{A}.$$

Proof. $U \circ V$ reflects isomorphisms because U and V do. By the absolute version ([PAC], Th. 7.3) of Beck's tripleableness theorem, we need only show that $\mathcal{C}$ has and $U \circ V$ preserves coequalizers of $(U \circ V)$-absolute pairs. If

(1.2.1) $$D \overset{x}{\underset{y}{\rightrightarrows}} W \qquad (\text{in } \mathcal{C})$$

(1.2.2) $$UVD \rightrightarrows UVW \xrightarrow{p} P \qquad (\text{in } \mathcal{A})$$

depict such a pair and accompanying absolute coequalizer data, find a map q in $\mathcal{B}$, with $Uq \cong p$, coequalizing the U-absolute pair (Vx, Vy):

(1.2.3) $$VD \rightrightarrows VW \xrightarrow{q} Q \qquad (\text{in } \mathcal{B}).$$

Where T is (the functor component of) the triple associated with the tripleable functor V, the lemma below will assure that both T and $T \circ T$ preserve the coequalizer diagram (1.2.3). It follows (see [LCA], Prop. 3) that $\mathcal{C}$ will have a map r coequalizing (1.2.1) and satisfying $Vr \cong q$; the inference $UVr \cong p$ then being immediate, the proof will be complete.

(1.3) Lemma. Let $\mathbb{T} = (T, \eta, \mu)$ be a triple on the ZHD category $\mathcal{B}$, and let $U: \mathcal{B} \longrightarrow \mathcal{A}$ be a functor reflecting coequalizers of U-absolute pairs and having a left adjoint $F: \mathcal{A} \longrightarrow \mathcal{B}$, with counit $\epsilon: FU \longrightarrow id_{\mathcal{B}}$.

Then not only is every diagram

(1.3.1) $$E \rightrightarrows X \longrightarrow Q \quad (\text{in } \mathcal{B})$$

whose transform under U *is an absolute coequalizer diagram in* $\mathcal{A}$ *already a coequalizer diagram in* $\mathcal{B}$, *but so are its transforms under* T *and* $T \circ T$.

Proof. We distinguish two cases, according as Q is or is not AT. If Q is AT, merely apply the following observation to T, $T \circ T$, and (1.3.1): *If* $D\colon \mathcal{D} \longrightarrow \mathcal{B}$ *is any diagram with AT colimit* Q, *then any functor* $S\colon \mathcal{B} \longrightarrow \mathcal{B}$ *admitting a natural transformation* $\lambda\colon \mathrm{id}_{\mathcal{B}} \longrightarrow S$ *satisfies* $SQ \cong Q$ *and preserves the colimit of* D. Indeed, λ_Q is an isomorphism; moreover, for any cone $S \circ D \circ\!\circ\!\circ\!\!> B$, the cone $D \circ\!\circ\!\circ\!\!> B$ induced by composition with λ factors uniquely through Q, whence $B \cong Q$ and $SQ \cong \operatorname{colim}(S \circ D)$.

If instead Q is not AT, neither is X or E, and all three are projective; it will then turn out that (1.3.1) is an *absolute* coequalizer diagram, which amply fulfills our requirement. Writing $G = FU$, consider the beginnings of a G-resolution of (1.3.1):

(1.3.2) $$\begin{array}{ccccc}
GGE & \rightrightarrows & GGX & \longrightarrow & GGQ \\
G\epsilon_E \downarrow\downarrow \epsilon_{GE} & & G\epsilon_X \downarrow\downarrow \epsilon_{GX} & & G\epsilon_Q \downarrow\downarrow \epsilon_{GQ} \\
GE & \rightrightarrows & GX & \longrightarrow & GQ \\
\epsilon_E \downarrow & & \epsilon_X \downarrow & & \epsilon_Q \downarrow \\
E & \rightrightarrows & X & \longrightarrow & Q
\end{array} .$$

Using the unit of adjunction $\mathrm{id}_{\mathcal{A}} \longrightarrow UF$, note that U transforms each column of (1.3.2) into a split coequalizer diagram in $\mathcal{A}$ (compare the discussion around display formula (5) in [McL], Ch. VI, §7). As each column is then a coequalizer diagram with projective coequalizer, it follows that

(1.3.3) the maps ϵ_E, ϵ_X, and ϵ_Q are split epimorphisms.

Moreover, as the transform of (1.3.1) under U is absolute, we know that

(1.3.4) G transforms (1.3.1) into an absolute coequalizer diagram.

The following observation, used again in §2, now concludes the proof:

(1.4) *ABS-Lemma*. *Let* G *be any endofunctor on a category* $\mathcal{B}$, *and* $\epsilon\colon G \longrightarrow \mathrm{id}_{\mathcal{B}}$ *any natural transformation*. *Then* (1.3.1) *is an absolute coequalizer diagram if conditions* (1.3.3) *and* (1.3.4) *hold*.

Proof. Referring to (1.3.2), the upper two rows are obviously absolute coequalizer diagrams by (1.3.4). But the columns are absolute coequalizer diagrams, too -- indeed, they are split: writing B for any one of E, X, or Q, and choosing a section $s: B \longrightarrow GB$ for ϵ_B (available by (1.3.3)), we have a split coequalizer diagram

$$GGB \underset{\epsilon_{GB}}{\overset{G\epsilon_B}{\rightrightarrows}} GB \xrightarrow{\epsilon_B} B, \qquad Gs: GB \to GGB,\quad s: B \to GB$$

as is verified by recording the section equation, $\epsilon_B \circ s = \mathrm{id}_B$, and applying G to obtain $G\epsilon_B \circ Gs = \mathrm{id}_{GB}$; the remaining two splitting equations are but instances of the naturality of ϵ: $\epsilon_{GB} \circ Gs = s \circ \epsilon_B$ and $\epsilon_B \circ G\epsilon_B = \epsilon_B \circ \epsilon_{GB}$. Applying any functor to (1.3.2), therefore, we obtain a similar 3×3 diagram in which all columns and the upper two rows are coequalizers. But then the 3×3 lemma (= Noether Isomorphism Theorem -- the special case of the Fubini Theorem (cf. [McL], p. 227) asserting that coequalizers commute with coequalizers) assures that the bottom row is a coequalizer, too, whence the lemma.

(1.5) Corollary (Theorem A -- sufficiency). *If the topos $\mathbb{E}$ is coZHD, then every cotriple $\mathbb{G}$ on $\mathbb{E}$ is dually algebraic.*

Proof. It is obvious that $(V_{\mathbb{G}})^{op}: (\mathbb{E}_{\mathbb{G}})^{op} \longrightarrow \mathbb{E}^{op}$ is tripleable, and it is known ([PCT], §2, Th'm) that $\Omega^{(-)}: \mathbb{E}^{op} \longrightarrow \mathbb{E}$ is tripleable, too. Now just apply the ZHD-Lemma (1.2).

(1.6) Corollary ([LDC]). *Every category cotripleable over the category $\mathbb{S}$ of sets and functions is the dual of a variety.*

Proof. By (1.5), cotriples on the coZHD topos $\mathbb{S}$ are dually algebraic.

2. *Internal injectives and Theorem B (sufficiency proof).* Properly to understand the basic facts concerning internal injectives, it helps to bear in mind, by way of comparison, that an object X of a topos $\mathbb{E}$ is injective (in the usual sense, that every extension problem

(2.1.1)
$$\begin{array}{ccc} A & \overset{m}{\rightarrowtail} & B \\ & \searrow^{\varphi} & \\ & & X \end{array}$$

with m a monomorphism has a solution $\tilde{\varphi}: B \longrightarrow X$ extending φ along m, i.e., satisfying $\tilde{\varphi} \circ m = \varphi$) if and only if the functor $X^{(-)}: \mathbb{E}^{op} \longrightarrow \mathbb{E}$ converts monomorphisms $m: A \rightarrowtail B$ (in $\mathbb{E}$) to split epimorphisms $X^m: X^B \longrightarrow X^A$. In fact, because Ω and its powers are injective in any topos, X is injective iff the singleton map $\{\cdot\}_X: X \longrightarrow \Omega^X$ has a retraction; but for each such retraction $\rho: \Omega^X \longrightarrow X$, the composition

$$X^A \xrightarrow[(\{\cdot\}_X)^A]{} (\Omega^X)^A \cong \Omega^{X \times A} \xrightarrow[\exists_{X \times m}]{} \Omega^{X \times B} \cong (\Omega^X)^B \xrightarrow[\rho^B]{} X^B$$

is easily seen to be a section for X^m when $m: A \rightarrowtail B$ is monic; conversely, if X^m has a section, passage to global elements shows that each extension problem (2.1.1) has a solution.

We say an object X of the topos $\mathbb{E}$ is <u>internally injective</u> if the functor $X^{(-)}: \mathbb{E}^{op} \longrightarrow \mathbb{E}$ merely preserves (that is, converts monomorphisms in $\mathbb{E}$ to) epimorphisms. Observe that the (reversible) "deductions"

$$\begin{array}{ll} I^*B \longrightarrow I^*X & (\text{in } \mathbb{E}|_I) \\ \hline 1_I = I^*1 \longrightarrow (I^*X)^{I^*B} \cong I^*(X^B) & (\text{in } \mathbb{E}|_I) \\ \hline \Sigma_I 1_I = I \longrightarrow X^B & (\text{in } \mathbb{E}) \end{array}$$

set up a 1-to-1 correspondence between maps $\tilde{\varphi}: I^*B \longrightarrow I^*X$ in $\mathbb{E}|_I$ <u>extending</u> (as we shall say) φ <u>along</u> m <u>over</u> I (that is, solving the transform under I^* of the extension problem (2.1.1)) and maps $\overline{\varphi}: I \longrightarrow X^B$ rendering

(2.1.2)
$$\begin{array}{ccc} I & \longrightarrow & 1 \\ {\scriptstyle\overline{\varphi}}\downarrow & & \downarrow{\scriptstyle\ulcorner\varphi\urcorner} \\ X^B & \underset{X^m}{\longrightarrow} & X^A \end{array}$$

commutative. When X^m is epic, there are, given $\varphi: A \longrightarrow X$, diagrams (2.1.2) with $I \twoheadrightarrow 1$ epic (take (2.1.2) to be a pullback, for example); so (2.1.1) has solutions <u>locally</u> (that is, over some I with support $\sigma I = 1$) when X^m is epic. In particular, when X is internally injective, $X^{\{\cdot\}_X}$ is epic, and the singleton map $\{\cdot\}_X$ becomes a split monomorphism in $\mathbb{E}|_I$, for some I with $\sigma I = 1$. But for such I, I^*X is then injective in $\mathbb{E}|_I$, i.e., X is <u>locally injective</u>; and it follows, since $I^*(X^m) = I^*X^{I^*m}$, that, for such X and all maps m monic in $\mathbb{E}$, each X^m is a locally split epimorphism. Thus internal injectives and local injectives coincide, and all the maps X^m, for monic maps m, are epic for X internally injective because they are locally split -- indeed, they all split in any $\mathbb{E}|_I$ in which I^*X is injective. To sum up:

(2.1.3) injectives are internally injective;

(2.1.4) internal injectives are injective locally, and conversely.

It then follows easily that

(2.1.5) an object X for which I^*X is internally injective qua object of $\mathbb{E}|_I$, for some I with $\sigma I = 1$, is internally injective in $\mathbb{E}$;

and, as each J^* preserves injectives (because Σ_J preserves monomorphisms),

(2.1.6) J^*X is internally injective if X is, for all J in $\mathbb{E}$.

The proof of Theorem B uses the following amusing characterization of dually algebraic cotriples on topoi.

(2.2) Proposition. A cotriple $\mathbb{G} = (G, \epsilon, \delta)$ on a topos $\mathbb{E}$ is dually algebraic if and only if the functor G preserves equalizers of coreflexive pairs.

Proof. We adopt the terminology of -- and assume known the results in -- §2 of [PCT]. If G preserves such equalizers, $G \circ G$ does too, whence (by the dual of Prop. 3 of [LCA]) $\mathbb{E}_{\mathbb{G}}$ has them and $V_{\mathbb{G}}$ preserves them. It follows that $(V_{\mathbb{G}})^{op}: (\mathbb{E}_{\mathbb{G}})^{op} \longrightarrow \mathbb{E}^{op}$ satisfies the hypotheses of the RTT (cf. [PCT]). But so does the functor $\Omega^{(-)}: \mathbb{E}^{op} \longrightarrow \mathbb{E}$, hence so does their composite. So, applying the RTT, $\mathbb{G}$ is dually algebraic.

Conversely, if $\mathbb{G}$ is dually algebraic, then, because reflexive pairs in $\mathbb{E}^{op}$ are $\Omega^{(-)}$-split (cf. [PCT] again), the tripleable composite $\Omega^{(-)} \circ (V_{\mathbb{G}})^{op}$ is RTT by Beck's theorem; in particular, it preserves coequalizers of reflexive pairs. But $\Omega^{(-)}$ reflects such coequalizers, so $(V_{\mathbb{G}})^{op}$ preserves them, i.e., $V_{\mathbb{G}}$ preserves equalizers of coreflexive pairs. But so does $V_{\mathbb{G}}$'s right adjoint, whose composition with $V_{\mathbb{G}}$ is G, after all; then so does G.

The proof of the sufficiency clause in Theorem B is now at hand. It is convenient to say a topos is σLZ if it satisfies the condition of Theorem B.

(2.3) Lemma (Theorem B -- sufficiency). Every indexed cotriple $\mathbb{G} = (G, \epsilon, \delta)$ on a σLZ topos is dually algebraic.

Proof. Let $\mathbb{G}$ be an indexed cotriple on the σLZ topos $\mathbb{E}$. By (2.2) it suffices to prove G preserves equalizers of coreflexive pairs. So let

$$(\rightarrow\rightrightarrows) \qquad X_1 \longrightarrow X_2 \rightrightarrows X_3$$

be such an equalizer in $\mathbb{E}$. There are four principal steps to take.

Step I. We find an object I having same support $\sigma I = \sigma X_1$ as X_1 for which I^* carries $(\rightarrow\rightrightarrows)$ to an absolute equalizer diagram $I^*(\rightarrow\rightrightarrows)$ in $\mathbb{E}|_I$. To do so, we apply (2.1.4) in each topos $\mathbb{E}|_{\sigma X_i}$ and choose, for each $i = 1, 2, 3$, an object I_i having support $\sigma I_i = \sigma X_i$, for which $I_i^* X_i$ is injective in $\mathbb{E}|_{I_i}$. Writing $I = I_1 \times I_2 \times I_3$, it is clear that $\sigma I = \sigma X_1$ and (from the line before (2.1.6)) that $I^* X_i$ is injective in $\mathbb{E}|_I$, for each i. Hence, writing $\eta_i : X_i \longrightarrow \Omega^{(\Omega^{X_i})}$ for (the exponential transposes of) the evaluation maps (which are monic because Ω is an internal cogenerator), the monomorphisms $I^*(\eta_i)$ are split in $\mathbb{E}|_I$ for each i. Thus, we have at least verified the counterpart of condition (1.3.3) for an eventual application of the dual of the ABS-Lemma (1.4) within $\mathbb{E}|_I$ to $\Omega^{\Omega^{(-)}}$ and to the evaluation $\eta\colon \mathrm{id} \longrightarrow \Omega^{\Omega^{(-)}}$ there. For the counterpart of (1.3.4), note that, being a logical functor, $I^*\colon \mathbb{E} \longrightarrow \mathbb{E}|_I$ satisfies $I^*(\Omega^{(\rightrightarrows)}) \cong \Omega^{I^*(\rightrightarrows)}$. Then, since $\Omega^{(\rightrightarrows)}$ is an absolute coequalizer diagram in $\mathbb{E}$, it follows that $\Omega^{I^*(\rightrightarrows)}$ is an absolute coequalizer diagram in $\mathbb{E}|_I$, whence $\Omega^{\Omega^{I^*(\rightrightarrows)}}$ is an absolute equalizer diagram there, which is (1.3.4). By (1.4), then, $I^*(\rightrightarrows)$ is an absolute equalizer diagram in $\mathbb{E}|_I$, as desired.

<u>Step II</u>. Capitalizing on the hypothesis that $\mathbb{G}$ is indexed, we show that I^* carries $G(\rightrightarrows)$, the transform of $(\rightrightarrows)$ under G, to an equalizer diagram in $\mathbb{E}|_I$. Recall (from [P&S] or [RAF]) that, to be $\mathbb{E}$-indexed, a cotriple $\mathbb{G}$ on $\mathbb{E}$ must, for all I in $\mathbb{E}$, be so accompanied by cotriples $\mathbb{G}_I$ on $\mathbb{E}|_I$ ($\mathbb{G}_1$ being $\mathbb{G}$) that, regardless what the map $j\colon J \longrightarrow I$, each diagram

(D(j))
$$\begin{array}{ccc} \mathbb{E}|_I & \xrightarrow{G_I} & \mathbb{E}|_I \\ {\scriptstyle j^*}\downarrow & & \downarrow{\scriptstyle j^*} \\ \mathbb{E}|_J & \xrightarrow[G_J]{} & \mathbb{E}|_J \end{array}$$

commutes to within a specified equivalence modulo which j^* carries counit to counit and comultiplication to comultiplication. In particular, from the commutativity of $(D(I \longrightarrow 1))$, we see that $I^*(G(\rightrightarrows)) \cong G_I(I^*(\rightrightarrows))$; hence, recalling Step I, $I^*(G(\rightrightarrows))$ is an (absolute!) equalizer diagram in $\mathbb{E}|_I$.

<u>Step III</u>. Since the unique map $I \longrightarrow \sigma I = \sigma X_1$ is epic, pulling back along it gives a functor $\mathbb{E}|_{\sigma X_1} \longrightarrow \mathbb{E}|_I$ that, being a faithful right adjoint, reflects equalizers. In particular, $(\sigma X_1)^*(G(\rightrightarrows))$ is an equalizer diagram in $\mathbb{E}|_{\sigma X_1}$ (because $I^*(G(\rightrightarrows))$ is, in $\mathbb{E}|_I$, by Step II).

<u>Step IV</u>. We finally show $G(\rightrightarrows)$ is an equalizer diagram in $\mathbb{E}$. Suppose $f\colon Y \longrightarrow GX_2$ is a map equalizing the pair $GX_2 \rightrightarrows GX_3$. It follows that the composition $Y \longrightarrow GX_2 \longrightarrow X_2$ of f with the counit ϵ_{X_2} equalizes the pair $X_2 \rightrightarrows X_3$, hence that there is a map $h\colon Y \longrightarrow X_1$ making the box

$$\begin{array}{ccc} Y & \xrightarrow{f} & GX_2 \\ {\scriptstyle h}\downarrow & & \downarrow{\scriptstyle \epsilon_{X_2}} \\ X_1 & \longrightarrow & X_2 \end{array}$$

commute. But then $\sigma Y \subseteq \sigma X_1$, and as f factors uniquely through $G(X_1 \to X_2)$ over σX_1, by Step III, it must do so in $\mathbb{E}$ as well, which ends the proof.

3. <u>The necessity arguments</u>. That the condition in Theorem A is necessary will follow from the necessity of the condition in Theorem B. For the latter, however, we need a suitable criterion for a map $m\colon A \longrightarrow B$ in a

topos $\mathbb{E}$ to be locally a split monomorphism; that criterion will center on the following construction. Given the map m, form all the cokernel pairs

$$X^B \xrightarrow{X^m} X^A \underset{y_X}{\overset{x_X}{\rightrightarrows}} H_m(X) \quad (= H(X))$$

of the maps X^m (X in $\mathbb{E}$). Letting H_m (= H): $\mathbb{E} \longrightarrow \mathbb{E}$ be the unique functor with these values for which the families $x = \{x_X\}_X$ and $y = \{y_X\}_X$ are natural transformations $(-)^A \longrightarrow H$, it is clear that x and y make H_m the cokernel pair of $(-)^m$ in the category of endofunctors on $\mathbb{E}$. In fact, as pulling back preserves colimits and as $(-)^m$ is an indexed natural transformation, H_m inherits an indexed structure by which it serves (still via x and y) as cokernel pair for $(-)^m$ in the category $\mathcal{E}nd(\mathbb{E})$ of indexed endofunctors on $\mathbb{E}$. [Motivation: the Yoneda Lemma provides a sense in which $(-)^A$ is the indexed endofunctor on $\mathbb{E}$ presented by a single free generator -- id_A -- in the value at A; in the same sense, H_m is the indexed endofunctor presented by two generators in $H_m(A)$ -- the Yoneda correspondents $\bar{x}$ and $\bar{y}$ of x and y -- subject to the defining relation $\{H_m(m)\}(\bar{x}) = \{H_m(m)\}(\bar{y})$ in $H_m(B)$. Thus H_m is the generic indexed endofunctor hoping to convert the map m to a non-monomorphism.]

(3.1) Lemma. <u>The following conditions on a monomorphism</u> $m: A \rightarrowtail B$ <u>in a topos</u> $\mathbb{E}$ <u>are equivalent</u>.

(i) m <u>is locally split</u>.

(ii) <u>For each object</u> X, <u>the map</u> $X^m: X^B \longrightarrow X^A$ <u>is epic</u>.

(iii) <u>Where</u> $(-)^B \longrightarrow (-)^A \underset{y}{\overset{x}{\rightrightarrows}} H_m$ <u>is the cokernel pair in the category</u> $\mathcal{E}nd(\mathbb{E})$ (of indexed endofunctors on $\mathbb{E}$) <u>of</u> $(-)^m$, <u>the transition map</u> $H_m(m): HA \longrightarrow HB$ <u>is monic</u>.

(iv) <u>The map</u> $A^m: A^B \longrightarrow A^A$ <u>is epic</u>.

<u>Proof</u>. (iv) ⇒ (i): argue as in the vicinity of (2.1.2).

(i) ⇒ (ii): if I^*m is a split monomorphism, then $I^*(X^m) = (I^*X)^{(I^*m)}$ is a split epimorphism. By (i), there is an object I with $\sigma I = 1$, hence with I^* faithful, for which I^*m is a split monomorphism. But then, as I^* preserves epimorphisms, it reflects them, and X^m is epic.

(ii) ⇒ (iii): by (ii), x = y and $H_m = (-)^A$; so $H_m(m) = m^A$ is monic.

(iii) ⇒ (iv): let $\bar{x} = x_A \circ \ulcorner 1_A \urcorner$ and $\bar{y} = y_A \circ \ulcorner 1_A \urcorner$ be the global elements $1 \longrightarrow H_m A = HA$ corresponding to x and y via Yoneda. By the definition of $H_m = H$, $Hm(\bar{x}) = Hm(\bar{y})$; then by (iii), $\bar{x} = \bar{y}$; so x = y, by an indexed Yoneda Lemma. In particular, $x_A = y_A$, and (iv) follows, as the cokernel pair $\underset{y}{\overset{x}{\rightrightarrows}} H$ is computed pointwise.

We are now ready to prove (somewhat more than) the rest of Theorem B.

(3.2) *Proposition*. *For any topos* $\mathbb{E}$, *the following are equivalent*.

(i) $\mathbb{E}$ *is* σ*LZ*.

(ii) *Every indexed cotriple on* $\mathbb{E}$ *is dually algebraic*.

(iii) *For every indexed cotriple* $\mathbb{G} = (G, \varepsilon, \delta)$ *on* $\mathbb{E}$, *the functor* G *preserves monomorphisms*.

Proof. (i) $\Rightarrow$ (ii) is Lemma (2.3).

(ii) $\Rightarrow$ (iii): every monomorphism in $\mathbb{E}$ is the equalizer of its own (coreflexive) cokernel pair. Hence any endofunctor on $\mathbb{E}$ preserving equalizers of coreflexive pairs preserves monomorphisms. Now apply (2.2).

(iii) $\Rightarrow$ (i): to prove that an object A of $\mathbb{E}$ is internally injective as an object of $\mathbb{E}|_{\sigma A}$, it suffices, by the considerations of (2.1), to show that each monomorphism $m: A \rightarrowtail B$ in $\mathbb{E}$ becomes locally split when pulled back into $\mathbb{E}|_{\sigma A}$; and for this, in turn, it is enough to prove that $(A \times A)^* m$ is locally a split monomorphism in $\mathbb{E}|_{A\times A}$. To this end, let $H = H_m$ be the indexed endofunctor on $\mathbb{E}$ contemplated in (3.1(iii)). As we shall see in a moment, the endofunctor G given by $GX = X \times X \times HX$ is an indexed cotriple on $\mathbb{E}$, so that, by (iii), $Gm = m \times m \times Hm$ is monic. [Were H itself an indexed cotriple, Hm would be monic and, by (3.1), m would be locally split; but this is, in general, too much to hope for.] Now, because

$$\begin{array}{ccc} A\times A\times HA & \xrightarrow{\ m\times m\times Hm\ } & B\times B\times HB \\ & {\scriptstyle A\times A\times Hm}\searrow \qquad \nearrow{\scriptstyle m\times m\times HB} & \\ & A\times A\times HB & \end{array}$$

commutes, $A \times A \times Hm$ is monic in $\mathbb{E}$, whence $(A \times A)^*(Hm)$ is monic in $\mathbb{E}|_{A\times A}$. Using the indexedness of H, however, it is easy to verify that $(A \times A)^*(Hm) = H^{A\times A}((A \times A)^* m)$, and an application of (3.1) in $\mathbb{E}|_{A\times A}$ shows $(A \times A)^* m$ is locally split, as required.

It remains only to indicate how G is an indexed cotriple. Given any object X of $\mathbb{E}$, freely adjoin a (global) zero element by forming $X + 1$; writing $o: 1 \longrightarrow X + 1$ for the injection, endow $X + 1$ with the trivial (constantly zero) semigroup multiplication $(X + 1)^2 \longrightarrow 1 \xrightarrow{o} X + 1$. Next, adjoining another (global) "unit" element, convert the semigroup $X + 1$ into the monoid $X + 2 = (X + 1) + 1$ it freely generates. It is clear that this procedure is perfectly functorial, so that $m: A \rightarrowtail B$ induces a monic map of monoids $m + 2 : A + 2 \rightarrowtail B + 2$. Notice that, whatever the map of monoids $\omega: M' \longrightarrow M$, the induced functors $(-)^M$ and $(-)^{M'}$ are indexed cotriples, and $(-)^{\omega}: (-)^M \longrightarrow (-)^{M'}$ is a map of indexed cotriples. In particular,

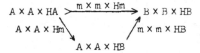

$(-)^{m+2}: (-)^{B+2} \longrightarrow (-)^{A+2}$ is such. To obtain the cotriple $\mathbb{G}$, we use:

(3.3) Lemma. *The forgetful functor $\mathcal{C}otrip(\mathbb{E}) \longrightarrow \mathcal{E}nd(\mathbb{E})$, from the category of indexed cotriples on $\mathbb{E}$ to the category of indexed endofunctors on $\mathbb{E}$, creates colimits.*

Proof. For any monoidal category $\mathcal{V}$, the forgetful functor from the category $\mathcal{M}on(\mathcal{V})$ of monoids in $\mathcal{V}$ to $\mathcal{V}$ itself creates limits. So take $\mathcal{V} = \mathcal{E}nd(\mathbb{E})^{op}$, note that $\mathcal{M}on(\mathcal{V}) = (\mathcal{C}otrip(\mathbb{E}))^{op}$, and dualize for the lemma.

For $\mathbb{G}$, now, take the cokernel pair indicated below:

$$(-)^{B+2} \xrightarrow{(-)^{m+2}} (-)^{A+2} \overset{x'}{\underset{y'}{\rightrightarrows}} G .$$

By the lemma, G "is" an indexed cotriple $\mathbb{G}$; on the other hand, concluding the proof, we have, for every X, a commutative diagram

$$\begin{array}{ccccc} X^{B+2} & \xrightarrow{X^{m+2}} & X^{A+2} & \overset{x'}{\underset{y'}{\rightrightarrows}} & GX \\ \cong\downarrow & & \cong\downarrow & & \cong\downarrow \\ X\times X\times X^{B} & \xrightarrow{X\times X\times X^{m}} & X\times X\times X^{A} & \overset{X\times X\times x_X}{\underset{X\times X\times y_X}{\rightrightarrows}} & X\times X\times HX \end{array} .$$

Finally, we settle the necessity of the condition in Theorem A.

(3.4) Lemma. *If every cotriple on the topos $\mathbb{E}$ is dually algebraic, then every nonzero object $X \neq 0$ of $\mathbb{E}$ has a global section.*

Proof. There is an *idempotent* cotriple $\mathbb{G}$ on $\mathbb{E}$ defined by

$$GX = \begin{cases} X, & \text{if } X \text{ has a global section } 1 \longrightarrow X; \\ 0, & \text{if not.} \end{cases}$$

As $\mathbb{G}$ is dually algebraic and as each X of $\mathbb{E}$ appears as the equalizer

(3.4.1)
$$X \xrightarrow{\text{inj.}} X+1 \overset{X+\text{inj.}_1}{\underset{X+\text{inj.}_2}{\rightrightarrows}} X+1+1 \qquad (X+\text{codiag}: X+1+1 \to X+1)$$

of a coreflexive pair, (2.2) assures that the diagram

$$GX \longrightarrow X+1 \rightrightarrows X+1+1 ,$$

obtained by applying G to (3.4.1), remains an equalizer diagram. But then $GX \cong X$ for all X, and the lemma holds.

(3.5) Lemma. *A σLZ topos $\mathbb{E}$ in which every nonzero object has a global section is coZHD.*

Proof. If I^*X is injective in $\mathbb{E}|_I$ and $\gamma: 1 \longrightarrow I$ is a global section, then $X \cong \gamma^* I^* X$ is injective in $(\mathbb{E}|_I)|_\gamma \cong \mathbb{E}$.

(3.6) Corollary (Theorem A -- necessity). *If every cotriple on* $\mathbb{E}$ *is dually algebraic, then* $\mathbb{E}$ *is coZHD.*

Proof. Apply Lemma (3.5) to Theorem B and Lemma (3.4).

REFERENCES

[IGF] J.R. Isbell. General Functorial Semantics, I. Amer. J. Math. **94** (1972), 535-596.

[LCA] F.E.J. Linton. Coequalizers in categories of algebras. [ZTB], 75-90.

[LDC] F.E.J. Linton. Is the dual of a coalgebra category an algebra category? -- preliminary draft. Multigraph, Dalhousie U., 1970.

[MAT] E.G. Manes. Algebraic Theories (Graduate Texts in Mathematics n° 26). Springer-Verlag, New York-Berlin, 1976.

[McL] S. Mac Lane. Categories for the Working Mathematician (Graduate Texts in Mathematics n° 5). Springer-Verlag, New York-Berlin, 1971.

[MDT] H.-M. Meyer. Injektive Objekte in Topoi. Diss^n., U. Tübingen, 1974.

[PAC] R. Paré. On absolute colimits. J. Alg. **19** (1971), 80-95.

[PCT] R. Paré. Colimits in topoi. Bull. A.M.S. **80** (1974), 556-561.

[P&S] R. Paré & D. Schumacher. Abstract families and the adjoint functor theorems. Indexed Categories and their Applications. Springer Lecture Notes in Mathematics n° 661, 1978, pp. 1-125.

[RAF] R. Rosebrugh. Abstract Families of Algebras. Diss^n., Dalhousie U., 1977.

[ZTB] Seminar on Triples and Categorical Homology Theory. Springer Lecture Notes in Mathematics n° 80, 1969.

F.E.J. Linton
Mathematics Department
Wesleyan University
Middletown, CT 06457 USA

R. Paré
Mathematics Department
Dalhousie University
Halifax, NS B3H 4H8 CANADA

INJECTIVES IN TOPOI, II:

CONNECTIONS WITH THE AXIOM OF CHOICE

P.T. Johnstone (Cambridge), F.E.J. Linton (Wesleyan), and R. Paré (Dalhousie)

The results of Part I (cf. [L&P]) having called attention to the topoidal injectivity conditions (σLZ) and (coZHD), the present sequel begins to chart the positions of these conditions within the sea of topos-theoretic consequences of ZFC by explicitly constructing the conjunctive semilattice $\mathcal{L}$ (with connective &) of topos-theoretic conditions (on nondegenerate elementary topoi) generated by the following nine assertions:

- (SS) supports split (i.e., for each X there is $\sigma X \longrightarrow X$, where σX, the <u>support</u> of X, is the image of the unique map $X \dashrightarrow 1$);
- (SG) subobjects of 1 generate;
- (TV) 2-valuedness (the only subobjects of 1 are 0 and 1);
- (B) Booleanness (every subobject is complemented);
- (IC) internal choice (every object is internally projective);
- (σLZ) every object X is internally injective qua object of $\mathbb{E}|_{\sigma X}$;
- (Kf⇒d) Booleanness of the external lattice $\mathrm{Sub}_{\mathbb{E}}(1)$ of subobjects of 1 (i.e., every subobject of 1 is complemented);
- (DML) De Morgan's Law (i.e., $\neg\neg$-closed subobjects are complemented); and
- (SDML) De Morgan's Law for $\mathrm{Sub}_{\mathbb{E}}(1)$ (i.e., $\neg\neg$-closed subobjects of 1 are complemented).

[The condition (Kf⇒d) is so named because it is equivalent (cf. [JFD], Th. 1.1) to the requirement that all Kuratowski-finite objects be decidable. For a welter of reformulations of (DML), the reader is referred to [JDM$_1$] and [JDM$_2$]. Basic information regarding (SS), (SG), (TV), (B), and (IC) may be found in [JTT], especially in Chapter 5 and Prop. 9.33; for (σLZ), see Part I. The condition (SDML) is included largely in order to retain a certain modicum of draftsmanly rectilinearity in the picture of $\mathcal{L}$.]

Fortunately, the $2^9 = 512$ distinct formal conjunctions of the generators of $\mathcal{L}$ determine only thirtytwo inequivalent members of $\mathcal{L}$, the strongest of which is the strong external axiom of choice (IC)&(SS)&(TV) introduced, in an equivalent formulation, by Lawvere [LCS] (and dubbed (AC1) by Penk [PTF]). Resuming the numbering of results where Part I left off, we give the principal reasons for the reduction from 2^9 to (not more than) 2^5 in Lemmas 4.1 and 4.2; and we exhibit a stock of exemplary topoi $\mathbb{E}_1, \ldots, \mathbb{E}_9$ -- most of them Grothendieck and few of them new -- serving to distinguish among the thirtytwo members of $\mathcal{L}$ -- in the course of proving (4.3). We close with some easy exercises.

But let us begin: we record the relations among the generators of $\mathcal{L}$, writing $(X) \longrightarrow (Y)$ when condition (Y) follows from condition (X).

(4.1) <u>Lemma.</u> <u>The diagram below depicts valid inferences:</u>

$$\begin{array}{ccccccc} (IC) & \longrightarrow & (B) & \longrightarrow & (\sigma LZ) & \longrightarrow & (DML) \\ & & & & \downarrow & & \downarrow \\ & & (TV) & \longrightarrow & (Kf{\Rightarrow}d) & \longrightarrow & (SDML)\,. \end{array}$$

<u>Proof.</u> (i). An object having a global element is injective with respect to inclusions of complemented subobjects (extend a map from such a subobject by sending the complement to the global element). But in $\mathbb{E}|_X$, which is Boolean if $\mathbb{E}$ is, $X^* X$ always has a global element. Thus $(\sigma X)^* X$ is internally injective whenever $\mathbb{E}$ is Boolean, i.e., $(B) \longrightarrow (\sigma LZ)$.

(ii). The relation $(IC) \longrightarrow (B)$ is noted in [JTT] after the proof of Lemma 5.28; the inference $(TV) \longrightarrow (Kf{\Rightarrow}d)$ is trivial, as observed in Remark 1.5 of [JFD], because 0 and 1 are always complemented subobjects of 1; and the relations $(Kf{\Rightarrow}d) \longrightarrow (SDML)$ and $(DML) \longrightarrow (SDML)$ are even more trivial.

(iii). It remains to see that (DML) and $(Kf{\Rightarrow}d)$ both follow from (σLZ). In fact, all we shall use of the condition (σLZ) is that $u+1$ is internally injective whenever u is a subobject $u \rightarrowtail 1$ of 1; but we must recall the observation ((2.4) in [AFD]) that a subobject $m\colon A \rightarrowtail B$ is complemented in $\mathbb{E}$ if and only if there is I in $\mathbb{E}$, with $\sigma I = 1$, for which I^*m makes I^*A a complemented subobject of I^*B in $\mathbb{E}|_I$.

It follows immediately that $\mathbb{E}$ satisfies (DML) if $\mathbb{E}|_I$ does, for some object I with $\sigma I = 1$: for I^* preserves negation. But to say $1+1$ is internally injective is to say there is I of support 1 with $I^*(1+1)$ (which $= I^*1 + I^*1 = 1_I + 1_I$) injective in $\mathbb{E}|_I$, and, by Theorem 1 of [JDM$_1$], that means (DML) holds in $\mathbb{E}|_I$. Thus, $\mathbb{E}$ satisfies (DML) when (σLZ) holds.

For $(Kf{\Rightarrow}d)$, note that a subobject u of 1 is complemented if (and, by the observation heading the proof of (i) above, only if) the extension problem

$$\begin{array}{ccc} u & \rightarrowtail & 1 \\ & {\scriptstyle inj}\searrow & \\ & & u+1 \end{array}$$

has a solution $\tilde{\varphi}\colon 1 \longrightarrow u+1$ -- for then $\tilde{\varphi}^*$ induces a coproduct decomposition $1 \cong \tilde{\varphi}^{-1}u + \tilde{\varphi}^{-1}1$ of 1, and, as the cross-diagonal map $u \rightarrowtail 1$ in the diagram

$$\begin{array}{ccccc} \tilde{\varphi}^{-1}u & \rightarrowtail & 1 & \leftarrowtail & \tilde{\varphi}^{-1}1 \\ \downarrow & \overset{\subseteq}{\nearrow} & \downarrow{\scriptstyle \tilde{\varphi}} & & \downarrow \\ u & \underset{inj}{\rightarrowtail} & u+1 & \underset{inj}{\leftarrowtail} & 1 \end{array}$$

assures, $\tilde{\varphi}^{-1}u \cong u$, so that u has complement $\tilde{\varphi}^{-1}1$. When $u+1$ is internally injective, therefore, there is I with $\sigma I = 1$ and I^*u a complemented subobject of I^*1 in $\mathbb{E}|_I$; but then $u \rightarrowtail 1$ is complemented in $\mathbb{E}$; so $(\sigma LZ) \longrightarrow (Kf{\Rightarrow}d)$.

The following will arise in the next lemma, which identifies certain conjunctions of the generating conditions, and is largely known:

($\exists\gamma$) every nonzero object has a global section;

(LZ) every nonzero object is locally (= internally) injective;

(σZ) every object X is injective qua object of $\mathbb{E}|_{\sigma X}$;

(coZHD) every nonzero object is injective;

(wpt) wellpointedness (1 is a generator);

(AC) external choice (every object is projective); and

(AC1) for each map f with domain $\neq 0$, there is g with $fgf = f$ ([PTF]).

(4.2) Lemma. (i). (SS) & (TV) = ($\exists\gamma$). (ii). (σLZ) & (TV) = (LZ). (iii). (σLZ) & (SS) = (σZ). (iv). (σLZ) & ($\exists\gamma$) = (coZHD). (v). (wpt) = (SS) & (TV) & (B) = (coZHD) & (B). (vi). (AC) = (SS) & (IC). (vii). (AC1) = (AC) & (coZHD) = (AC) & (TV). (viii). (SS) & (B) $\longrightarrow$ (SG). (ix). (SG) & (Kf$\Rightarrow$d) = (SG) & (B). (x). (SG) & (SDML) = (SG) & (DML).

Proof. (i) and (ii): only 0 has zero support, so in 2-valued topoi, ($\exists\gamma$) = (SS) and (LZ) = (σLZ). Conversely, $\sigma X = 1$ if $X \longrightarrow 1$ is (even only locally) a split epimorphism, so both ($\exists\gamma$) and (LZ) entail (TV).

(iii) and (iv): extending $0 \longrightarrow X$ along $0 \rightarrowtail \sigma X$ (resp., along $0 \rightarrowtail 1$, for $X \neq 0$), we see (σZ) $\longrightarrow$ (SS) (resp. (coZHD) $\longrightarrow$ ($\exists\gamma$)). The implications (coZHD) $\longrightarrow$ (σZ) $\longrightarrow$ (σLZ) being trivial, it remains only to note that Lemma 3.5 asserts the converse implication ($\exists\gamma$) & (σLZ) $\longrightarrow$ (coZHD), while an obvious modification of its proof yields (SS) & (σLZ) $\longrightarrow$ (σZ).

(v): the relation (wpt) = (SS) & (TV) & (B) is Freyd's (cf. Prop. 9.33 in [JTT]); the relation (wpt) = (coZHD) & (B) is Meyer's (cf. Satz 6.4 in [MDT]); moreover, either is deducible from the other using (i) and (iv) and (4.1).

(vi) and (viii): -- may be found in [JTT], Chapter 5.

(vii): when f is mono or epi, the relation $fgf = f$ asserts that g is a splitting for f, as is seen by cancelling f from the appropriate side. Epimorphisms with domain zero being isomorphisms, it follows that, in a topos satisfying (AC1), all epimorphisms -- and all monomorphisms with $\neq 0$ domain -- are split, whence (AC1) $\longrightarrow$ (AC) & (coZHD). Conversely, given f with domain $\neq 0$, write $f = me$ with e epi and m mono and, assuming (AC) and (coZHD), let s be a section for e and r a retraction for m: then $g = sr$ satisfies $fgf = mesrme = me = f$, and (AC1) holds, as desired.

(ix) and (x): -- trivial (compare the proof of (1.4) in [JFD]).

We are now in a good position to describe the &-semilattice $\mathcal{L}$ in detail.

(4.3) Proposition. The conjunctive semilattice $\mathcal{L}$ is precisely as depicted in the diagram overleaf, () representing the vacuous condition.

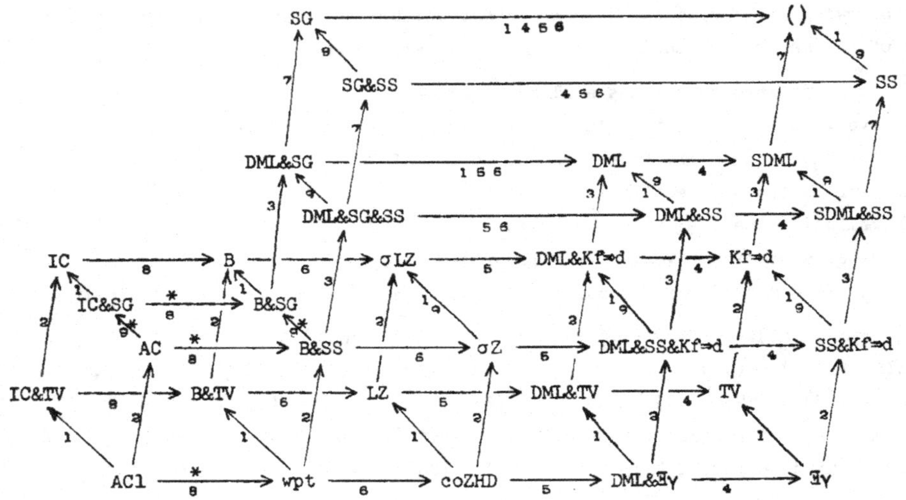

Proof. In view of the preceding lemmas, it suffices for us to exhibit topoi $\mathbb{E}_j$ ($j = 1, \dots, 9$) illustrating the irreversibility of the inferences labeled j. This we now do, choosing Grothendieck topoi where possible.

(i) If G is a nontrivial group, the topos $\mathbb{E}_1 = \mathcal{S}^G$ of G-sets (cf. [JTT], Example 5.26) satisfies (IC)&(TV), but (SG) and (SS) fail, as the regular representation of G itself has no global elements.

(ii) If I is a set with more than one element, the topos $\mathbb{E}_2 = \mathcal{S}^I \simeq \mathcal{S}|_I$ of I-indexed sets satisfies (AC) but not (TV).

(iii) That nearly universal counterexample, the Sierpiński topos $\mathbb{E}_3 = \mathcal{S}^{\mathbf{2}}$, where (SS)&(SG) clearly holds, has three subobjects of 1, hence satisfies (SDML) but not (Kf⇒d), hence also (DML) but not (B). [(DML), in the guise "1+1 is injective", also follows from the curious observation: an object $f: A \longrightarrow B$ of $\mathcal{S}^{\mathbf{2}}$ is injective if and only if $B \neq \emptyset$ and f is surjective.]

(iv) For $\mathbb{E}_4$ take the topos $\mathcal{S}^M$ of (right) M-sets singled out in Corollary 2.4 of [JDM$_1$] as not satisfying (DML). Here M is the three-element monoid $M = \{1, a, b : aa = ab = a,\ bb = ba = b\}$, and so ($\exists\gamma$) holds because all non-empty M-sets have fixed points (and it follows that (SG) fails).

(v) For $\mathbb{E}_5$ take the topos whose objects are pairs of sets (A, B) equipped with pairs of functions (f, g) disposed as follows

(4.3.1)

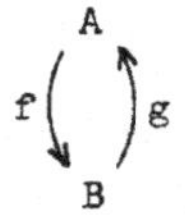

and satisfying $gfg = g$. Certainly each $b \in B$ provides a global section

$$\begin{array}{ccc} 1 & \xrightarrow{g(b)} & A \\ \downarrow\uparrow & & f\downarrow\uparrow g \\ 1 & \xrightarrow[f(g(b))]{} & B \end{array}$$

for (4.3.1), and $\mathbb{E}_5$ satisfies ($\exists\gamma$) since $B \neq \emptyset$ if (4.3.1) is not zero. In fact, (DML) holds in $\mathbb{E}_5$ and (σLZ) fails. To see (σLZ) fail, it suffices, by (4.2(iv)), to exhibit a nonzero object that is not injective. So, choosing an element $x\colon 1 \longrightarrow 2$ in $\mathbf{S}$, consider the map

$$\text{(4.3.2)} \qquad \begin{array}{ccc} 1 & \xrightarrow{x} & 2 \\ x\downarrow\uparrow\text{const} & & \text{id}\downarrow\uparrow x\circ\text{const} \\ 2 & \xrightarrow[\text{id}]{} & 2 \end{array} \;;$$

it is obviously monic, but not split monic, so its domain is not injective. To see (DML) hold, we characterize the injectives of $\mathbb{E}_5$ as those objects (4.3.1) for which $B \neq \emptyset$ and $fg = \mathrm{id}_B$: then nonzero constant objects (like $1+1$) are certainly injective. So, arguing in one direction, let a member $b \in B$ of the vertex B in (4.3.1) be given, and let $h_b\colon 2 \longrightarrow B$ be the function assigning to x the value $f(g(b))$ and taking on, as well, the value b. Any extension along (4.3.2) of the map

$$\begin{array}{ccc} 1 & \xrightarrow{g(b)} & A \\ x\downarrow\uparrow\text{const} & & f\downarrow\uparrow g \\ 2 & \xrightarrow[h_b]{} & B \end{array}$$

may be viewed as a proof that $b = f(g(b))$, and so $fg = \mathrm{id}_B$ when (4.3.1) is injective. Conversely, suppose $B \neq \emptyset$ in (4.3.1), and choose $\hat{b} \in B$.

Given a monomorphism $\begin{array}{ccc} A & \overset{i}{\rightarrowtail} & \bar{A} \\ f\downarrow\uparrow g & & \bar{f}\downarrow\uparrow \bar{g} \\ B & \underset{j}{\rightarrowtail} & \bar{B} \end{array}$, define $p\colon \bar{A} \longrightarrow A$ and $q\colon \bar{B} \longrightarrow B$ by the formulae

$$p(x) = \begin{cases} a\,, & \text{if } i(a) = x \in \mathrm{im}(i)\,; \\ c\,, & \text{if } x \notin \mathrm{im}(i) \text{ but } i(c) = \overline{gf}x \in \mathrm{im}(i)\,; \\ g\hat{b}\,, & \text{if } x \notin \mathrm{im}(i) \text{ and } \overline{gf}x \notin \mathrm{im}(i)\,; \end{cases}$$

$$q(y) = \begin{cases} b\,, & \text{if } j(b) = y \in \mathrm{im}(j); \\ f(p(\bar{g}y))\,, & \text{if } y \notin \mathrm{im}(j)\,. \end{cases}$$

The proof that when $fg = \mathrm{id}_B$ the pair $(p\,,\,q)$ splits the monomorphism $(i\,,\,j)$ in $\mathbb{E}_5$ is a tedious but straightforward case by case argument best left to the diligent reader. Finally, (SG) fails too -- on account of (TV) and the failure of (B).

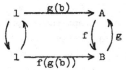

(vi) For a topos satisfying (coZHD) but not (B) (and, hence, not (SG)), take the topos $\mathbb{E}_6$ of sets-with-an-idempotent-endomorphism. Viewed as the topos $\mathcal{S}^M$ of M-sets for the two-element multiplicative monoid $M = \{0, 1\}$, $\mathbb{E}_6$ is clearly not Boolean, M being no group. If we view $\mathbb{E}_6$ instead as a full subcategory of $\mathbb{E}_5$, via the fully faithful functor assigning to the $\mathbb{E}_6$-object $A^{\circlearrowleft e}$ the $\mathbb{E}_5$-object (4.3.1) with $A = A$ and $A \xrightarrow{f} B \rightarrowtail^{g} A$ the factorization of the idempotent e through its image B, every nonzero object of $\mathbb{E}_6$ clearly becomes injective in $\mathbb{E}_5$, hence was injective to begin with, whence (coZHD).

(vii) All spatial topoi Shv(X) satisfy (SG); and, according to [JTT], Exer cises 5.3 and 5.4, Shv(X) will satisfy (SS) if the space X is separable and ze dimensional, but (SDML) fails (as does (DML)) unless X is extremally disconnect So take $\mathbb{E}_7 = \mathrm{Shv}(X)$ with X, say, the Cantor space 2^{ω}.

[Before proceeding to the last two examples, we note that Theorem 5.39 of [JTT asserting the equivalence in Grothendieck topoi of (AC) and (B)&(SG), guarantees no Grothendieck topos can ever testify to the irreversibility of the five inference labeled 8 and 9, adorned with asterisks (*). For a Grothendieck witness to th irreversibility of the two asterisk-free inferences 8, however, one may call the topos $\mathcal{E}$ of [JTT], Exercise 5.5; and, for 9, the topos of presheaves on the pose $(\mathbb{N}, \leq)$ -- (SG)&(DML) holds here, by Propositions 5.34 of [JTT] and 1.1 of [JDM_1], but (SS) fails, the presheaf $\mathbb{N} \xleftarrow{s} \mathbb{N} \xleftarrow{s} \ldots$ having support 1 but no global eleme

(viii) For a topos $\mathbb{E}_8$ in which (wpt) holds and (IC) fails, rely on (4.2(v and take any classical model of ZF witnessing the failure of the Axiom of Choice; o use Freyd's construction (the topos $\mathcal{F}$ of [JTT], Exercise 9.10).

(ix) Heeding the remarks before (viii), we confine our search for a topos $\mathbb{E}_9$ satisfying (IC)&(SG) but not (SS) to candidates of the form $\mathbb{E}_{dKf}$, the topos of decidable Kuratowski-finite objects of a given topos $\mathbb{E}$. For, according to [AFD], such topoi -- rarely Grothendieck -- always satisfy (IC). Moreover, when $\mathbb{E} = \mathrm{Shv}(V)$ is a spatial topos, it is easy to verify that $\mathbb{E}_{dKf}$, which then coincides with the topos $\mathbb{E}_{lcf}$ of locally constant finite objects of $\mathbb{E}$, satisfies (SG) as soon as the space V is zero-dimensional (indeed, if V has a base of clopen sets, then complemented subobjects of 1 already generate in $\mathbb{E}$, but these are precisely the subobjects of 1 in $\mathbb{E}_{dKf}$). We therefore seek a zero-dimensional space V for which (SS) fails in $(\mathrm{Shv}(V))_{dKf}$, and take that as $\mathbb{E}_9$ It is a pleasure, consequently, to record our considerable indebtedness to Frank Tall for his acute suggestion, made during the 1978 NATO Summer School in Set Theory at Peterhouse, Cambridge, in response to a request for a "zero-dimensional analogue of the familiar connected double covering of the circle", that we try splitting Heath's "vee-space" variant (Example 1 in [HCJ]) of Niemytzki's exotic upper half-plane ([A&H], Kap. I, §1, Nr. 6, 2° (p. 31)) along the crack opened (in [RCP], Example 4) by Reed. This we now do.

The underlying point set of Heath's vee-space is the closed upper half-plane $\mathbb{R} \times \mathbb{R}^+ = \{(x, y) / x \in \mathbb{R}, y \in \mathbb{R}, y \geq 0\}$. As basic open neighborhoods of points $(x, 0)$ on the x-axis, take all the "basic open vee's"

$$V_\varepsilon(x) = \{(z, t) / 0 \leq |x - z| = t < \varepsilon\} \qquad (\varepsilon > 0).$$

These constitute a subbase for a topology in which, as is easily verified, each point (x, y) $(y > 0)$ off the x-axis, the complement of each such point, the complement of each basic open vee, and (finally) the complement of each point $(x, 0)$ on the x-axis, is open. The resulting space V is therefore T_1 and zero-dimensional (and T_2 and completely regular).

In what is to come, it will be convenient to have disjointed the left and right fingers of the basic open vee's of V: accordingly, we set

$$L_\varepsilon(x) = \{(z, t) / 0 \leq x - z = t < \varepsilon\}, \quad R_\varepsilon(x) = \{(z, t) / 0 < z - x = t < \varepsilon\},$$

and observe $V_\varepsilon(x)$ is the disjoint union $L_\varepsilon(x) \cup R_\varepsilon(x)$.

To split V, topologize the union $\mathbb{R} \times \mathbb{R}^+ \times \{0, 1\}$ of two disjoint closed half-planes by taking as a subbase (of basic open neighborhoods of points $(x, 0, i)$ on the x-axes) all the "basic broken vee's"

$$V_\varepsilon(x, i) = L_\varepsilon(x) \times \{i\} \cup R_\varepsilon(x) \times \{1 - i\} \qquad (x \in \mathbb{R};\ i = 0, 1;\ \varepsilon > 0).$$

As for V, it is easily verified that the resulting space X, which may be visualized as two nearly discrete half-planes laced, corset-like, together along their x-axes by the basic broken vee's, is zero-dimensional (the basic broken vee's being clopen) and T_1.

Now consider the projection $p\colon X \longrightarrow V$ defined by $p(x, y, i) = (x, y)$. Because the inverse image $p^{-1}(V_\varepsilon(x))$ of a basic open vee is simply the union

$$p^{-1}(V_\varepsilon(x)) = V_\varepsilon(x, 0) \cup V_\varepsilon(x, 1)$$

of two basic broken vee's, p is continuous. Moreover, as the two displayed basic broken vee's are disjoint and clopen, and as the restriction of p to either of them provides a homeomorphism $p|V_\varepsilon(x, i)\colon V_\varepsilon(x, i) \xrightarrow{\cong} V_\varepsilon(x)$, it follows, upon inverting these homeomorphisms, that we obtain a homeomorphism $f\colon V_\varepsilon(x) \times \{0, 1\} \xrightarrow{\cong} p^{-1}(V_\varepsilon(x))$ between the coproduct of two copies of $V_\varepsilon(x)$ and the inverse image $p^{-1}(V_\varepsilon(x))$ satisfying $p(f((z, t), i)) = (z, t)$. But this means that $p\colon X \longrightarrow V$ is (the sheaf space of) a sheaf over V locally isomorphic to $[2] = 1 + 1$ (for basic open vee's cover V), hence belonging to $\mathrm{Shv}(V)_{dKf}$.

It remains to see that p has no global section. Suppose then, arguing by contradiction, that $s\colon V \longrightarrow X$ were one, and let $I\colon V \longrightarrow \{0, 1\}$ be the unique function satisfying $s(x, y) = (x, y, I(x, y))$. Using categorical topology as practiced by Baire, we produce a point of V at which I must take both values. To this end, put $S_i = \{x \in \mathbb{R} / I(x, 0) = i\}$ $(i = 0, 1)$ and note that, since $S_0 \cup S_1 = \mathbb{R}$, at least one S_i is of second Baire category -- say S_0 is.

Now for each $x \in S_0$, s is continuous at $(x, 0)$, and so we can find $\varepsilon(x) > 0$ such that $s(V_{\varepsilon(x)}(x)) = V_{\varepsilon(x)}(x, 0)$. Letting $T_n = \{x \in S_0 \,/\, \varepsilon(x) > 1/n\}$, we have $S_0 = T_1 \cup T_2 \cup \dots \cup T_n \cup \dots$, so some T_n is dense in some interval. Let x and y be points of such a T_n, satisfying $0 < y - x < 2/n$, and put $\varepsilon = 1/n$. Then I is constantly 0 on $L_\varepsilon(y)$ and constantly 1 on $R_\varepsilon(x)$. But $L_\varepsilon(y)$ and $R_\varepsilon(x)$ have a point in common, namely $(\frac{1}{2}(x+y), \frac{1}{2}(y-x))$, at which I thus behaves paradoxically enough that we deduce no such section s exists; and the proof of (4.3) is complete.

Moving on to the promised exercises, observe that Theorem B of Part I assures that some indexed cotriple on the Sierpiński topos ($\mathbb{E}_3$ in the proof of (4.3)) is not dually algebraic -- for, (Kf⇒d) failing, (σLZ) fails too. To see one, do:

(4.4) <u>Exercise.</u> Given $f: A \to B$ in the Sierpiński topos $\mathcal{S}^2$, write $G(f) = (\operatorname{cok}(f) \times A \xrightarrow[\text{proj}]{} A \xrightarrow[f]{} B)$, where $\operatorname{cok}(f)$ is the pushout in $\mathcal{S}$ of f along itself. Verify that the projection helping define $G(f)$ is a split epimorphism, that therefore $\operatorname{cok}(G(f)) = \operatorname{cok}(f)$, and hence that $G(G(f))$ is given by $(\operatorname{cok}(f) \times \operatorname{cok}(f) \times A \xrightarrow[\text{proj}]{} A \xrightarrow[f]{} B)$; then show that, with those projections as counit $G \to \mathrm{id}$, and with the maps $G(f) \to G(G(f))$ induced by the diagonal map of $\operatorname{cok}(f)$ as comultiplication $G \to GG$, G becomes a cotriple $\mathbf{G}$. ($\mathbf{G}$ is an indexed cotriple by Rosebrugh's criterion ([RDD], Ch. I, Ex. 3.9) -- the codomain of $G(f)$ depends only upon the codomain of f.) Choosing an element $x: 1 \to 2$ in $\mathcal{S}$, verify that G fails to preserve the monomorphism

$$\begin{array}{ccc} 1 & \xrightarrow{x} & 2 \\ {\scriptstyle x}\downarrow & \rightarrowtail & \downarrow{\scriptstyle \mathrm{id}} \\ 2 & \xrightarrow[\mathrm{id}]{} & 2 \end{array} .$$

We now exhibit an idempotent cotriple on the topos $\mathbb{E}_5$ of (4.3(v)), preserving monomorphisms but not the equalizer diagram consisting of the monomorphism (4.3.2) and its cokernel pair. This refutes the conjecture, suggested by the proof of Lemma 3.4 in Part I, that an idempotent cotriple on a topos satisfying ($\exists\gamma$) should have to be dually algebraic; likewise, it refutes one of the more extreme conjectures suggested by the equivalence of conditions (ii) and (iii) in Proposition 3.2.

(4.5) <u>Exercise.</u> Writing simply $\mathbb{E}$ for $\mathbb{E}_5$, define a functor G on $\mathbb{E}$ by assigning to the object (4.3.1) the value

$$G\left(f \left(\begin{array}{c} A \\ \downarrow \uparrow \\ B \end{array} \right) g \right) = e \left(\begin{array}{c} A \\ \downarrow \uparrow \\ I \end{array} \right) gm ,$$

where $A \xrightarrow{e} I \xrightarrow{m} B$ is the factorization of f through its image. (That G is well defined on objects is shown by $(gm)e(gm) = gfgm = gm$.) Use the diagrams

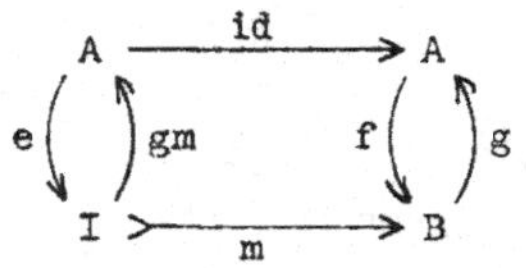

to obtain a counit making G an idempotent cotriple $\mathbb{G}$ on $\mathbb{E}$, leaving fixed precisely those objects (4.3.1) with f surjective. Calculate the cokernel pair of the monomorphism (4.3.2) and show that, while G leaves their domain and codomain fixed, it carries their equalizer to the terminal object 1. Finish by showing that G preserves monomorphisms.

The following addendum to (4.1) is reminiscent of (3.2 (i)⇒(iii)) in Part I.

(4.6) Exercise. Let $\mathbb{T}$ be an internal algebraic theory in (= indexed triple on) a topos $\mathbb{E}$. Improve Proposition 7.7 of [J&W] by proving: $T^1 : \mathbb{E} \longrightarrow \mathbb{E}$ preserves monomorphisms if $\mathbb{E}$ satisfies (σLZ). (Hint: let $m: X \rightarrowtail Y$ be a monomorphism in $\mathbb{E}$. Setting $U = \sigma(T^1(X))$ and choosing I with support $\sigma I = U$ for which $I^*(T^1(X))$ is injective in $\mathbb{E}|_I$, use the argument for case (b), on page 233 of [J&W], to see that $I^*(T^1(m))$ is monic in $\mathbb{E}|_I$. Deduce next that $U^*(T^1(m))$ is monic in $\mathbb{E}|_U$, and then that $T^1(m)$ is monic in $\mathbb{E}$.) Show, likewise, that Corollary 7.10 of [J&W] remains valid after the word "Boolean" is replaced by "(σLZ)".

Peter Freyd has amiably contributed the following (simplest?) example of a Grothendieck topos satisfying (SG) and (DML) but not (SS). The same irreducible sober space X as appears here plays a prominent -- albeit rather different -- role, as well, in recent work of M. Turgi (see §3 of "A sheaf-theoretic interpretation of the Kuroš theorem" in Springer Lecture Notes in Mathematics 616 (1977), pp. 173-196).

(4.7) Exercise (Freyd). Let $X = \mathbf{3} = \{0, 1, 2\}$ be the three-point space with open sets $U = \{0, 1\}$, $V = \{1, 2\}$, $U \cap V = \{1\}$, and X and $\emptyset$. Show X is both connected and extremally disconnected. Deduce that (SG) & (DML) holds in $\mathrm{Shv}(X)$, and, by considering the epimorphism $U + V \twoheadrightarrow U \cup V = X$, that (SS) fails.

REFERENCES

[AFD] O. Acuña-Ortega & F.E.J. Linton. Finiteness and decidability: I. Multigraph, Wesleyan U., 1977 (to appear in [SLN]).

[A&H] P.S. Alexandroff & H. Hopf. Topologie I (Grundlehren der Mathem. Wissenschaften 45). Springer, Berlin, 1935.

[HCJ] R.W. Heath. Screenability, pointwise paracompactness, and metrization of Moore spaces. Canad. J. Mathem. 16 (1964), 763-770.

[JDM_1] P.T. Johnstone. Conditions related to De Morgan's law. Multigraph, U. Cambridge, 1977 (to appear in [SLN]).

[JDM_2] P.T. Johnstone. Another condition equivalent to De Morgan's law. Multigraph, U. Cambridge, 1978 (to appear in Communications in Algebra).

[JFD] P.T. Johnstone & F.E.J. Linton. Finiteness and decidability: II. Math. Proc. Cambr. Phil. Soc. 84 (1978), 207-218.

[JTT] P.T. Johnstone. Topos Theory (L.M.S. Mathematical Monographs 10). Academic Press, London-New York, 1977.

[J&W] P.T. Johnstone & G.C. Wraith. Algebraic theories in toposes. Indexed Categories and their Applications (Springer Lecture Notes in Mathematics 661, pp. 141-242). Springer, Berlin, 1978.

[LCS] F.W. Lawvere. An elementary theory of the category of sets. Proc. Nat. Acad. of Sciences 52 (1964), 1506-1511.

[L&P] F.E.J. Linton & R. Paré. Injectives in topoi, I: Representing coalgebras as algebras. In this volume.

[MDT] H.-M. Meyer. Injektive Objekte in Topoi. Diss^n., U. Tübingen, 1974.

[PTF] A.M. Penk. Two forms of the axiom of choice for an elementary topos. J. Symbolic Logic 40 (1975), 197-212.

[RCP] G.M. Reed. On normality and countable paracompactness. Fund. Math. (in press).

[RDD] R. Rosebrugh. Abstract Families of Algebras. Diss^n., Dalhousie U., 1977.

[SLN] Proc. L.M.S. Durham Symposium on Applications of Sheaf Theory. Springer Lecture Notes in Mathematics (to appear).

P.T. Johnstone
Pure Mathematics
16 Mill Lane
Cambridge
CB2 1SB ENGLAND

F. E. J. Linton
Dept. of Math.
Wesleyan Univ.
Middletown, CT
06457 USA

R. Paré
Mathematics Dept.
Dalhousie Univ.
Halifax, NS
B3H 4H8 CANADA

CATEGORIES OF STATISTIC-METRIC SPACES

(A Co-Universal Construction)

Rainer Bodo Lüschow

0. Introduction

The recent publications on Statistic-Metric Spaces use the concept of distribution-functions always with the assumption of left-continuity (see[1]). There is a natural procedure to receive a monoton increasing, left-continuous function from a given monoton function. This assignment is in a certain sense extreme, such that one obtains -using the concept of morphisms of[2]- a co-universal construction.

1. Preliminaries

From now on let Δ and Γ be sets (non empty) in the following sense:

1.1 Definition

(i) $\Delta := \{\varphi \in [0,1]^{\mathbb{R}} / \varphi$ monoton increasing, φ left-continuous$\}$,

(ii) Γ satisfy the following conditions:

(α) $\Delta \subset \Gamma \subset [0,1]^{\mathbb{R}}$,

(β) $\varphi \in \Gamma \Rightarrow \varphi$ monoton increasing.

1.2 Lemma

Let Δ and Γ be given, than there exist functions $i: \Delta \longrightarrow \Gamma$ and $j: \Gamma \longrightarrow \Delta$ with the following properties:

(i) $j \circ i = 1_{\Delta}$,

(ii) (1) $\varphi \in \Gamma \Rightarrow j(\varphi) \leq \varphi$ (i.e. $(j(\varphi))(x) \leq \varphi(x)\ \forall\ x \in \mathbb{R}$),

(2) $\varphi \in \Gamma,\ x \in R \Rightarrow$

$((j(\varphi))(x) = \varphi(x) \Leftrightarrow \varphi$ is left-continuous in $x)$,

(3) $\varphi_1, \varphi_2 \in \Gamma, \varphi_1 \leq \varphi_2 \Rightarrow j(\varphi_1) \leq j(\varphi_2)$,

(4) $\varphi \in \Gamma \Rightarrow (\varphi \in \Delta \Leftrightarrow j(\varphi) = \varphi)$,

(5) $\varphi \in \Gamma,\ \psi \in \Delta,\ \psi \leq \varphi \Rightarrow \psi \leq j(\varphi)$,

Proof:

Take $i:\Delta \longrightarrow \Gamma$ the injection and define
$j:\Gamma \longrightarrow \Delta$ by $(j(\varphi))(x) := (x^-)$, where $\varphi(x^-) := \sup\{\varphi(x')/x'<x\}$ for all $x \in \mathbb{R}$.

1.3 Agreement

Let i and j be always the maps used in the proof of 1.2.

1.4 Definition

The pair (S,F) is called a SM-Space (Statistic-Metric Space) over Γ, if

(i) S is a set,

(ii) $F : S\times S \longrightarrow \Gamma$ is a map, such that

(I) $F_{pq} = H \Leftrightarrow p=q$ (let always be $F_{pq} := F(p,q)$ for all $(p,q) \in S\times S$ and let H be the Heavisidefunction),

(II) $F_{pq}(0)=0$,

(III) $F_{pq}=F_{qp}$,

(IV) $F_{pq}(x)=F_{qr}(y)=1 \Rightarrow F_{pr}(x+y)=1$, for all $p,q,r \in S$; $x,y \in \mathbb{R}$.

2. The Co-Universal Construction

2.1 Definitions

(i) Let (S,F) be a SM-space over Γ, the points of derivation of (S,F) are $D(S,F) := \{p \in S/E(p)\}$, where the property E(p) is defined to be

$$E(p): \left[\begin{array}{l} \exists\ q,r \in S\ \exists\ x,y>0\ \exists \sigma \in \mathrm{Perm}(\{p,q,r\}) \\ \text{such that } 1=F_{\sigma(p)\sigma(q)}(x)=F_{\sigma(q)\sigma(r)}(y) \text{ and} \\ F_{\sigma(p)\sigma(r)}((x+y)^-)<1, \end{array}\right.$$

(ii) let (S,F) and (S',F') SM-spaces over Γ, let $f:S\longrightarrow S'$ be a map, f is called a morphism (see [2]), if we have

(Mo) $F_{pq} \leq F'_{f(p)f(q)} \quad \forall\ p,p \in S$,

(iii) a morphism $f:(S,F)\longrightarrow(S',F')$ respects the derivation if we have

(D) $f[S\setminus D(S,F)] \subset S'\setminus D(S',F')$.

2.2 Lemma

Let SM (Γ) denote all SM-spaces over Γ with morphisms (Mo), let SM $(\Gamma)/D$ denote all SM-spaces over Γ with morphisms (D), then $SM(\Gamma)$ and $SM(\Gamma)/D$ are categories with the usual compositions of its morphisms.

The proof is obvious, in particular we have:

2.3 Lemma

$SM(\Delta) = SM(\Delta)/D$.

Proof:

For spaces (S,F) in $SM(\Delta)$ the condition (D) is always satisfied because $D(S,F)=\phi$. To see this, let us suppose there is a $p\in D(S,F)$. Then there exist $q,r\in S$; $x,y>0$; $\sigma\in Perm(\{p,q,r\})$ such that $F_{\sigma(p)\sigma(q)}(x)=F_{\sigma(q)\sigma(r)}(y)=1$ and $F_{\sigma(p)\sigma(r)}((x+y)^{-}) < 1$.

From $F_{\sigma(p)\sigma(q)}(x)=F_{\sigma(q)\sigma(r)}(y)=1$ we conclude $F_{\sigma(p)\sigma(r)}(x+y)=1$ (because (S,F) satisfies (ii)(IV) from 1.4). Since $F_{\sigma(p)\sigma(r)}$ is in Δ we know $F_{\sigma(p)\sigma(r)}((x+y)^{-})=1$ and $F_{\sigma(p)\sigma(r)}(x+y)=1$, which is a contradiction.

2.4 Proposition

(i) $L:SM(\Gamma)/D \longrightarrow SM(\Delta)$ defines a functor if we set

(1) $L(S,F):=(jS,jF)$, where

$jS:=S\setminus D(S,F)$,

$jF:\ jS\times jS \longrightarrow \Delta$

$(p,q) \longmapsto (jF)_{pq}:=j(F_{pq})$,

(2) $((S,F)\xrightarrow{f}(S',F')) \longmapsto (L(S,F)\xrightarrow{L(f)}L(S',F'))$,

where $L(f)(p):=f(p)\ \forall\ p\in S$,

(ii) let $E:SM(\Delta)\longrightarrow SM(\Gamma)/D$ be the embedding,

(1) $E(S,F):=(S,F)$, where

$F:S\times S\longrightarrow\Delta$ is identified with

$i\circ F:S\times S\longrightarrow \Gamma$ (use $\Delta\xrightarrow{i}\Gamma$).

(2) $E(f):=f\ \ \forall$ f morphism in $SM(\Delta)$, then E defines a functor,

(iii) $\eta : 1_{SM(\Delta)} \longrightarrow L \circ E$ defines a natural transformation by

$\eta_{(S,F)} : (S,F) \longrightarrow L(E(S,F)) = (jS, jF)$

$p \longmapsto \eta_{(S,F)}(p) = p \ \forall\ p \in S,$

(iv) $\varepsilon : E \circ L \longrightarrow 1_{SM(\Gamma)/D}$ defines a natural transformation by

$\varepsilon_{(S,F)} : E(L(S,F)) \longrightarrow (S,F)$

$p \longmapsto \varepsilon_{(S,F)}(p) = p \ \forall\ p \in jS,$

$(E(L(S,F)) = E(jS, jF) = (jS, jF)).$

Proof:

(i) (1) For $(S,F) \in Ob(SM(\Gamma)/D)$ we have $(jS, jF) \in Ob(SM(\Delta))$, because

(I) $(jF)_{pp} = j(F_{pp}) = jH = H \ \forall\ p \in S$,
let now $p, q \in jS$, $p \neq q$, then there exists an $\varepsilon > 0$ with $F_{pq}(\varepsilon) < 1$. We have (see 1.2(ii)(1))
$(jF_{pq})(\varepsilon) \leq F_{pq}(\varepsilon) < 1$,

(II) Again apply 1.2(ii)(1) to conclude for $p, q \in jS$
$0 \leq (jF_{pq})(0) \leq F_{pq}(0) = 0$,

(III) $F_{pq} = F_{qp} \ \forall\ p, q \in S \Rightarrow jF_{pq} = jF_{qp} \ \forall\ p, q \in jS$,

(IV) Let $p, q, r \in jS$ with $(jF_{pq})(x) = (jF_{qr})(y) = 1$ with certain $x, y > 0$. Then we have $F_{pq}(x) = F_{qr}(y) = 1$ (see 1.2(ii)(1)). Let σ = unit in Perm($\{p,q,r\}$), hence we have $F_{\sigma(p)\sigma(q)}(x) = F_{\sigma(q)\sigma(r)}(y) = 1$. We know $p \in jS$, therefore $p \notin D(S,F)$, then we have $F_{\sigma(p)\sigma(r)}((x+y)^-) = 1$. Hence we know
$jF_{pr}(x+y) = jF_{\sigma(p)\sigma(r)}(x+y) =$
$jF_{\sigma(p)\sigma(r)}((x+y)^-) = 1$.

(2) Let $f : (S,F) \longrightarrow (S',F')$ be a morphism in $SM(\Gamma)/D$. Hence we know $f[jS] \subset jS'$. Therefore we see that $L(f) : (jS, jF) \longrightarrow (jS', jF')$ is a well defined map $(L(f)(p) := f(p) \ \forall p \in jS)$. But $L(f)$ is also a morphism in $SM(\Delta)$, because $p, q \in jS \subset S \Rightarrow F_{pq} \leq F'_{f(p)f(q)} \Rightarrow$ (because 1.2(ii)(3)) $jF_{pq} \leq jF'_{f(p)f(q)}$. The other properties of the functor L are trivial with respect of $\leq$ in Γ.

(ii) and (iii) are trivial, we prove

(iV) (1) By virtue of 1.2(ii)(1) $\varepsilon_{(S,F)}$ is a morphism in $SM(\Gamma)/D$.

(2) Let $f:(S,F) \longrightarrow (S',F')$ a $SM(\Gamma)/D$ morphism. Then the following diagram commutes

$$\begin{array}{ccc} (E\circ L)(S,F)=(jS,jF) & \xrightarrow{\quad L(f) \quad} & (jS',jF')=(E\circ L)(S',F') \\ \Big\downarrow \varepsilon_{(S,F)} & & \Big\downarrow \varepsilon_{(S',F')} \\ (S,F) & \xrightarrow{\quad f \quad} & (S',F'), \end{array}$$

<u>2.5 Lemma</u>

Let $E:SM(\Delta) \longrightarrow SM(\Gamma)/D$ the embedding-functor of 2.4(ii), let $(S,F)\in Ob(SM(\Gamma)/D)$, the following conditions are equivalent:

(i) $(\varepsilon_{(S,F)},(jS,jF))$ is a co-universal pair of (S,F) with respect to E,

(ii) $jS=S$.

<u>Proof:</u>

(i)⇒(ii): Let $jS \neq S$, hence we have $D(S,F) \neq \phi$, let $p \in D(S,F)$, $S':=\{p\}$ and $F'_{pp}:=H$. We have $(S',F') \in Ob(SM(\Delta)) \subset Ob(SM(\Gamma)/D)$. Let $f:E(S',F')=(S',F') \longrightarrow (S,F)$ be the $SM(\Gamma)/D$-morphism with $f(p)=p$. Then there is no $SM(\Delta)$-morphism $\bar{f}:(S',F') \longrightarrow (jS,jF)$ with $f=\varepsilon_{(S,F)} \circ E(\bar{f})$. Since $p \notin jS$, there is no map $\bar{f}:S' \longrightarrow jS$ with $f=\varepsilon_{(S,F)} \circ \bar{f}$ (remember $E(\bar{f})=f$).

Hence in
$$\begin{array}{ccc} (S,F) & \xleftarrow{\quad f \quad} & E(S',F') \\ \Big\uparrow \varepsilon_{(S,F)} & & \\ E(jS,jF) & & \end{array}$$

is no $\bar{f}:(S',F') \longrightarrow (jS,jF)$ ($SM(\Delta)$-morphism), such that
$$\begin{array}{ccc} (S,F) & \xleftarrow{\quad f \quad} & E(S',F') \\ \Big\uparrow \varepsilon_{(S,F)} & \swarrow E(\bar{f}) & \\ E(jS,jF) & & \end{array}$$
commutes.

(ii)$\Rightarrow$(i): Let $jS=S$, let $(S',F')\in Ob(SM(\Delta))$ and let $f:E(S',F')\longrightarrow(S,F)$ a $SM(\Gamma)/D$-morphism. Because of the definition of $\varepsilon_{(S,F)}$ we know that $jS=S$ implies that $\varepsilon_{(S,F)}$ is the identical-map (remember that the identical-map must not be the identity-morphism). Hence in the triangle

$$\begin{array}{ccc} (S,F) & \xleftarrow{\;f\;} & E(S',F') \\ \varepsilon_{(S,F)} \Big\uparrow & & \\ E(S,jF) & & \end{array}$$

there exists at most one morphism $\bar{f}:(S',F')\longrightarrow(S,jF)$ in $SM(\Delta)$ such that the triangle

$$\begin{array}{ccc} (S,F) & \longleftarrow & E(S',F') \\ \varepsilon_{(S,F)} \Big\uparrow & \swarrow{\scriptstyle E(\bar{f})} & \\ E(S,jF) & & \end{array}$$

commutes.

But from set-theory we know further that $\bar{f}$ regarded as a map must be f. It remains to show that $\bar{f}$ is a morphism in $SM(\Delta)$. Let therefore be $p',q'\in S'$. Since f is a morphism in $SM(\Gamma)/D$ we have $F'_{p'q'}\leq F_{f(p')f(q')}$, by virtue of 1.2(ii)(5) we have $F'_{p'q'}\leq jF_{f(p')f(q')}$. Remembering that $f(p')=\bar{f}(p')$ and $f(q')=\bar{f}(q')$, we have $F'_{p'q'}\leq jF_{\bar{f}(p')\bar{f}(q')}$.

2.6 Definition

Let $CSM(\Gamma)/D$ be the full subcategory of $SM(\Gamma)/D$ induced by the class: $\{(S,F)\in Ob(SM(\Gamma)/D)/jS=S\}$.
Let $\hat{L}$ be the restriction of the functor L to $CSM(\Gamma)/D$. We can give the following

2.7 Theorem

$\hat{L}:CSM(\Gamma)/D\longrightarrow SM(\Delta)$ is a bicoreflector.

Proof:

See the foregoing Lemma and the definition of a bicoreflector.

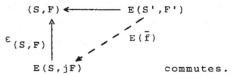

2.8 Proposition

(i) The embedding $E:SM(\Delta)\longrightarrow SM(\Gamma)/D$ has a decomposition over $CSM(\Gamma)/D$ where each factor is an embedding,

(ii) let $E=E_2 \circ E_1$ the decomposition of (i), then E, E_1 and E_2 need not be dense.

Proof:

(i) is trivial (take $E_1:SM(\Delta)\longrightarrow CSM(\Gamma)/D$, $E_2:CSM(\Gamma)/D\longrightarrow SM(\Gamma)/D$ the natural embedding, then $E=E_2 o E_1$),

(ii) let Γ_0 such that there is a $\varphi\in\Gamma_0$ with

(α) $\varphi(x) < 1$ for all $x\in\mathbb{R}$,

(β) $\varphi \notin \Delta$

For such Γ_0 there it is easy to see that E, E_1 and E_2 are not dense (take spaces of two points and define its structures with respect to φ). Therefore we have

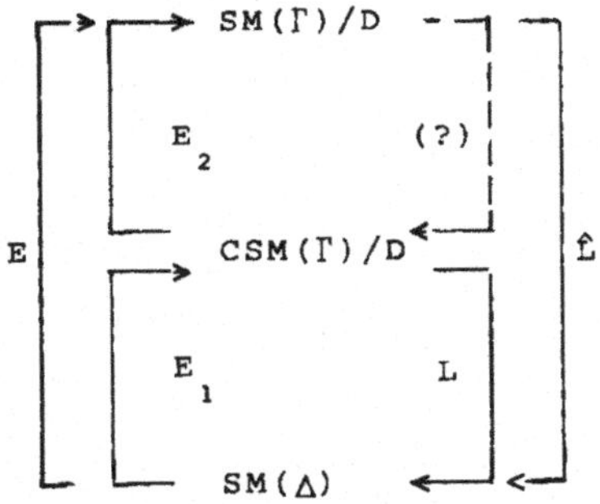

(i) where E, E_1 and E_2 are not (necessary) dense embeddings,

(ii) $\hat{L}$ is bicoreflector.

Open question:

is there a "nice" compatible functor (?) from $SM(\Gamma)/D$ in $CSM(\Gamma)/D$?

Bibliography:

[1] B.Schweizer: Multipl. on the Space of Dis.Fu.,
aequ. math., 1975,

[2] R.B. Lüschow: Which Prod. of SM-Sp. are Cat.Prod.?,
Proc. Yug. Math., 1977

R.B. Lüschow, Santiago, Universidad Tecnica del Estado
August 1978

A Categorical Approach to Primary and Secondary Operations in Topology

by John L. MacDonald

Primary and secondary operations can be introduced in a category having certain (weak) completeness or co-completeness properties. The first section indicates how this is done using the adjoint tower, the shape category associated with a functor K and notions of structure and semantics related to those of Lawvere [4]. This theory may then be applied to the case when $K : \mathcal{P} \to \mathcal{T}$ is the inclusion of the discrete subcategory $\mathcal{P}$ determined by a spectrum into the homotopy category $\mathcal{T}$. In particular for the Eilenberg-MacLane spectrum we show how the theory leads to the classical description of secondary cohomology operations and their associated universal examples as described by Adams for example, in his Hopf invariant paper [1] and later by Mosher and Tangora in [8].

The topological application can be compared with certain algebraic ones appearing in the earlier work of the author [6]. The chief difference appears to be that indeterminancy arises in the topological higher order operations because of the use of weak kernels (i.e. fibrations) in the homotopy category instead of kernels. There is no indeterminacy for the dual case of algebraic higher order operations described in [6].

1. Primary and Secondary Operations

In this section we show how for each collection $\{A_i\}_{i\in I}$ of objects in a category $\mathcal{A}$ there are associated families of primary and secondary operations provided that $\mathcal{A}$ has enough completeness properties for the definitions to make sense

Let

$$\begin{array}{ccccc} \{A_i\}_I & \xrightarrow{K} & \mathcal{A} & \xrightarrow{S_K} & S_K \\ & & \downarrow R & \swarrow J & \\ & & \mathrm{Sets}^I & & \end{array} \tag{1.1}$$

be a diagram of categories and functors where K is the inclusion of a discrete subcategory and R is the contravariant functor with value $RX = \{[X,A_i] \mid i\in I\}$ in the category Sets^I of I-graded sets. The objects of the <u>shape</u> category S_K are the same as those of A and the morphism set $S_K(X,Y)$ is just the set of morphisms $RY \to RX$ in Sets^I and $R = JS_K$ is just the obvious factorization (cf. [3], [5]). We write A° for the opposite (or dual) category of A .

<u>Proposition</u> 1.2: The functor $R: A^{\circ} \to \mathrm{Sets}^I$ has a left adjoint L if A has products.

<u>Proof</u>: Let $L(\{X_i\}_I) = \prod_I[\prod_{X_i} A_i]$ considered as a product in A and let $L(\{f_i\}_I : \{X_i\}_I \longrightarrow \{Y_i\}_I) : \prod_I[\prod_{Y_i} A_i] \to \prod_I[\prod_{X_i} A_i]$ be the unique A morphism such that

(1.3)

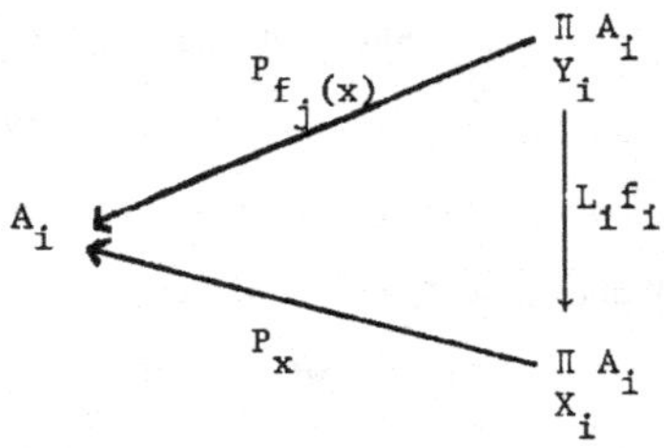

commutes for each $x \in X_i$ and each $i \in I$ where P_x is the product projection corresponding to x and $L_i f_i$ is itself the unique map such that (1.3) commutes for i fixed and all $x \in X_i$. The functor L is the required adjoint. The unit $\eta : 1 \to RL$ of the adjunction is defined on $\{X_i\}_I$ by $\eta(x)$ = the projection $\prod_I \prod_{X_i} A_i \to A_j$ to the x-th component for $x \in X_j$.

The <u>algebraic theory</u> T of $\{A_i\}_{i \in I}$ or <u>algebraic structure</u> of R is by definition just the full subcategory of A° generated by the image of L , that is, by $\{\prod_I \prod_{X_i} A_i] \mid \{X_i\}_I \in \mathrm{Sets}^I\}$.

Let $C_1 \subset \mathrm{Sets}^{T^{\circ}}$ be the full subcategory of all product preserving functors. A functor $F: A^{\circ} \to C_1$ is defined by $FB = [B,-]\,|_{T^{\circ} \subset A}$.

An object G of C_1 may be regarded as a graded set $\{GA_i\}_{i \in I} \in \mathrm{Sets}^I$ with operations corresponding to the morphisms $\alpha_k : \prod_I \prod_{X_i} A_i \to A_k$ defined. That is, to each such morphism there is an operation $G\alpha_k : \prod_I \prod_{X_i} GA_i \to GA_k$ defined.

A <u>primary operation</u> (K,α_k) of type Y, for $Y \in \mathrm{Sets}^I$, on the category A is the unique natural transformation

$$(1.4) \qquad \prod_I [\coprod_{Y_i} -,A_i] \xrightarrow[\approx]{\alpha} \prod_I \prod_{Y_i} [-,A_i] \xrightarrow[\approx]{\psi} [-,\prod_I \prod_{Y_i} A_i] \xrightarrow{[-,\alpha_k]} [-,A_k]$$

determined by $K : \{A_i\}_I \to A$ and $\alpha_k : \prod_I \prod_{Y_i} A_i \to A_k$. Thus theory morphisms $\prod_I [\prod_{X_i} A_i] \to \prod[\prod_{I\, Y_i} A_i]$ correspond to sequences of primary operations. From the preceding paragraph we thus see that the object FB of C_1 may be regarded as the set $\{(B,A_i] \mid i \in I\}$ with all primary operations applied.

<u>Remark</u> 1.5: If $Y_i = \phi$ for $i \neq k$, then (K,α_k) determines a shape morphism $B \to \coprod_{Y_i} B$ in $S_{(K|A_k)}$ for each B in A where $(K|A_k)$ is the restrictioı of K to A_k. Furthermore, a sequence of such primary operations, one for each $k \in I$, determines a shape morphism $B \to \coprod_Y B$ in S_K for each $B \in A$.

<u>Remark</u> 1.6: If $Y_i = \phi$ for $i \neq j$ and Y_j is a 1 element set then $(K,\alpha_k) : [-,A_j] \to [-,A_k]$.

We call C_1 the category of primary operation algebras and $F : A \to C_1$ the primary functor. We record this information in the diagram

(1.7)

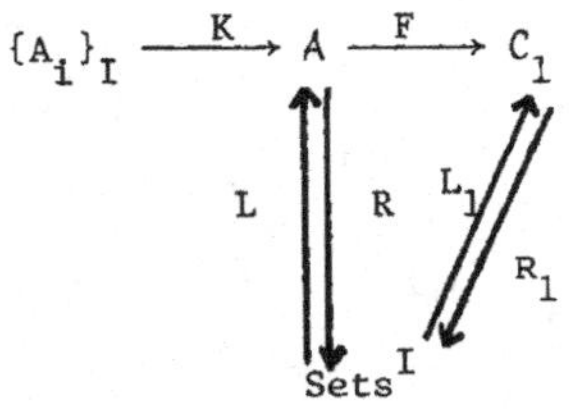
{A_i}_I
K
A
F
C_1
L
R
L_1
R_1
Sets^I

where $L_1 = FL$ and R_1 is defined by $R_1G = \{GA_i\}_I$, hence $R_1F = R$. Furthermore L_1 is left adjoint to $R_1 : C_1 \to \mathrm{Sets}^I$. The functor R_1 is called the <u>semantics</u> of the theory T.

<u>Theorem</u> 1.8: The category C_1 of primary operation algebras is equivalent to the category of algebras over the monad generated by the adjoint pair $L \dashv R : A^{\circ} \to Sets^I$.

<u>Proof</u>: It is routine to verify that $L \dashv R : A^{\circ} \to Sets^I$ and $L_1 \dashv R_1 : C_1 \to Sets^I$ generate the same monad $(RL, \eta, R\epsilon L)$ in $Sets^I$. Then by a result in Mac Lane [7], page 151, 6(ii), it is sufficient to show that if $f,g : G \to H$ in C_1 is any parallel pair for which $R_1 f$, $R_1 g$ has an absolute coequalizer, then C_1 has a coequalizer for f,g, and R_1 preserves and reflects coequalizers for these pairs. Thus suppose that $e = \{e_{A_i}\}_I : \{HA_i\}_I \to \{Z_i\}_I$ is the absolute coequalizer of $R_1 f = \{f_{A_i}\}_I$ and $R_1 g = \{g_{A_i}\}_I : \{GA_i\}_I \to \{HA_i\}_I$. Then each e_{A_i} is a coequali since an absolute coequalizer is preserved by any functor hence, in particular by projections. Let $N : T^{\circ} \to Sets$ be defined on objects by letting $e'_X : HX \to NX$ be a coequalizer of f_X and g_X for $X \in |T^{\circ}|$, requiring that $e'_{A_i} = e_{A_i}$ for $i \in I$. The morphism $N(\alpha : X \to Y)$ is by definition the induced map on coequalizer Upon applying the product functor $\prod_I \prod_{X_i} : Sets^I \to Sets$ to the absolute coequalizers diagram we obtain the top part of the diagram

$$
(1.9)\qquad
\begin{array}{ccccc}
\prod_I \prod_{X_i} GA_i & \overset{\prod_I \prod_{X_i} f_{A_i}}{\underset{\prod_I \prod_{X_i} g_{A_i}}{\rightrightarrows}} & \prod_I \prod_{X_i} HA_i & \xrightarrow{\prod_I \prod_{X_i} e_{A_i}} & \prod_I \prod_{X_i} Z_i \\
\downarrow & & \downarrow & & \downarrow \phi \\
G(\prod_I \prod_{X_i} A_i) & \overset{f_{\prod_I \prod_{X_i} A_i}}{\underset{g_{\prod_I \prod_{X_i} A_i}}{\rightrightarrows}} & H(\prod_I \prod_{X_i} A_i) & \xrightarrow{e'_{\prod_I \prod_{X_i} A_i}} & N(\prod_I \prod_{X_i} A_i)
\end{array}
$$

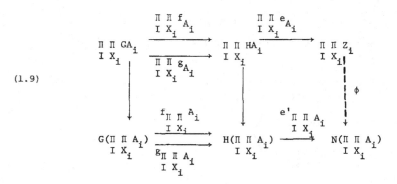

where the vertical maps exist and are isomorphisms and the left squares in f, g, respectively, commute since G and H are product preserving. However, the crucial fact is that the top row is a coequalizer diagram since an absolute coequalizer is preserved by any functor. Thus there is a unique induced map ϕ of coequalizers and furthermore ϕ is an isomorphism. Thus N commutes with products hence is in C_1. Then it is routine to show $e' : H \to N$ coequalizes $f,g : G \to H$ in C and furthermore, by definition of e', $R_1 e' : R_1 H \to R_1 N$ is $\{e'_{A_i}\} = \{e_{A_i}\} : \{HA_i\} \to \{NA_i\} = \{Z_i\}$, thus R_1 preserves coequalizers for such pairs. It is also easy to show that R_1 reflects coequalizers for such pairs f,g since if $\overline{e} : H \to \overline{N}$ is a C_1 morphism with $\overline{e}f = \overline{e}g$ such that $R_1\overline{e}$ coequalizes $R_1 f$, $R_1 g$, then there is a diagram (1.9)', whose vertical arrows are isomorphisms, which is obtained from (1.9) by replacing e' by $\overline{e}$ and e_{A_i} by $\overline{e}_{A_i}$, since $\overline{e} \in C_1$. Furthermore the coequalizer $R_1\overline{e} = \{\overline{e}_{A_i}\}$ is absolute since it is equivalent to the absolute coequalizer e by an isomorphism $R_1\overline{N} = \{\overline{N}A_i\} \to \{Z_i\}$. Hence $\prod_I \prod_{X_i} \overline{e}_{A_i}$ is a coequalizer and by (1.9)' so is $\overline{e}_{\prod_I \prod_{X_i} A_i}$

<u>Remark</u> 1.10: If we identify functors in C_1 which take identical values on the objects A_i, then C_1 is isomorphic to the category of algebras over the monad generated by $L \dashv R$ by [7], page 147, because then the preceding proof shows that R_1 creates coequalizers for those parallel pairs f, g for which the pair $R_1 f, R_1 g$ has an absolute coequalizer.

<u>Corollary</u> 1.11: The primary functor $F : A \to C_1$ is just the comparison functor to the category of algebras over the monad generated by $L \dashv R : A^\circ \to Sets^I$, up to a natural equivalence.

<u>Proof</u>: Let $A^\circ \xrightarrow{C} (Sets^I)^M \xleftarrow{K} C_1$ be the comparison functors determined by the adjoint pairs $L \dashv R : A^\circ \to Sets^I$, and $L_1 \dashv R_1 : C_1 \to Sets^I$ which generate the same monad M over $Sets^I$. It is straightforward to verify that $KF = C$. Then by the theorem K is an equivalence.

The left adjoint C_1 to F at $G \in C_1$, when it exists, is given as the object $C_1 G = C_L \circ KG$ in A for C_L the left adjoint of C. But

$$KG = \langle R_1 G,\ R_1\varepsilon_1 G\colon R_1 L_1 R_1 G \to R_1 G\rangle = \langle \{GA_i\}_I,\ \{(\varepsilon_1 G)A_i\}_I : \{L_1 R_1 GA_i\}_I \to \{GA_i\}_I\rangle$$

for ε_1 the co-unit associated with $L_1 \dashv R_1$. Thus using the description of the adjoint C_L to the comparison functor C as given by Beck's result appearing in [7] we find that C_1G appears in the following equalizer diagram in A ,namely,

(1.12)
$$C_1G \xrightarrow{\sigma} LR_1G = \prod_I\{\prod_{GA_i} A_i\} \underset{\varepsilon_{LR_1G}}{\overset{LR_1\varepsilon_1G}{\rightrightarrows}} LRLR_1G \ .$$

Under these conditions we may define secondary operations in a dual fashion to [6] by constructing a diagram

(1.13)

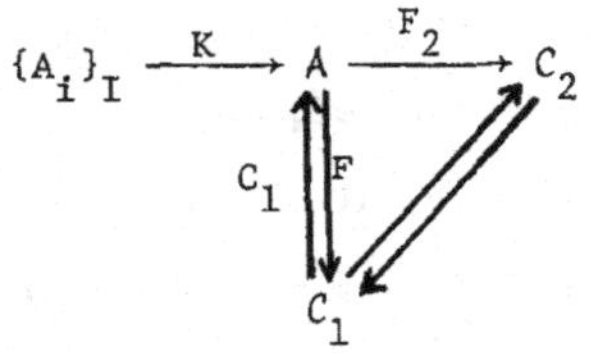

and letting a <u>secondary algebraic theory</u> T_1 be defined as the full image of C_1 in A° . We may then proceed inductively to a description of higher order operations.

We say that a morphism κ is a <u>weak equalizer</u> of a pair of maps $\alpha,\beta\colon A\to B$ if it has the universal property of equalizers but not necessarily the uniqueness.

If the category A has weak equalizers but not equalizers then a weak equalizer diagram (1.12) does not in general define an adjoint to F but only a "weak" adjoint. This does not deter us from defining secondary operations in the same fashion as in [6], however, and has only the effect of introducing indeterminacy

Specifically to each diagram

(1.14)
$$A_k \xleftarrow{\alpha'} C_1G \xrightarrow{\sigma} \prod_I\{\prod_{GA_i} A_i\}$$

with σ a weak equalizer of maps appearing in (1.12) there corresponds a secondary operation (K,α') of type G in C_1 , which by definition is the following sequence of natural transformations

(1.15)
$$\begin{array}{ccccc} \prod_I[\coprod_{GA_i} -, A_i] & \xrightarrow[\approx]{\phi} & \prod_I\prod_{GA_i}[-,A_i] & \xrightarrow[\approx]{\psi} & [-,\prod_I\{\prod_{GA_i} A_i\}] \\ & & & & \uparrow{\scriptstyle [-,\sigma]} \\ & & & & [-,C_1G] \xrightarrow[{[-,\alpha']}]{} [-,A_k] \end{array}$$

The value of the secondary operation (K,α') at a space X is thus the relation $[X,\alpha'] \circ [X,\sigma]^{-1} \circ \psi \circ \sigma$, by analogy to the primary case (1.4), sometimes simply written $[X,\alpha'] \circ [X,\sigma]^{-1}$. These secondary operations may be considered as a subclass of a larger family of secondary operations defined as above from (1.14) be allowing σ to vary over weak equalizers of arbitrary parallel pairs of maps between products of the A_is .

2. Application to Spectra

We now let A be the homotopy category and $\{A_i\}_I$ any discrete subcategory For example, $\{A_i\}_I$ could be the collection of spaces belonging to a spectrum. In this section we simply interpret some of the results from the first section in terms of classical results about the $K(\pi,n)$ spectrum. Thus in the following paragraphs we let $\{A_i\}_I = \{K(\pi,n)\}_{n\in Z}$ for Z the set of positive integers.

A primary operation in the sense of Adams [1] is a natural transformation $\phi\colon \prod_J H^{n_j}(-;\pi) \to H^m(-;\pi)$. Up to isomorphism then ϕ is just a natural transformation $[-, \prod_J K(\pi,n_j)] \to [-,K(\pi,m)]$ which, by the Yoneda lemma, is induced by $\alpha_k : \prod_Z \prod_{J_k} K(\pi,k) = \prod_{n_j\in J} K(\pi,n_j) \to K(\pi,m)$. Thus ϕ is a primary operation (K,α_k) of type $Y = \{J_k\}$ where J_k is the subset of J for which $n_j = k$. Conversely, starting with an operation of type (1.4) we get one of Adams type by letting $J = \bigcup_Z Y_i$ and $n_j = i$ for $j \in Y_i$.

In the homotopy category the fibration induced by a map is a weak equalizer for that map. In particular for (1.12) we let $\sigma : E_\alpha = C_1G \to LR_1G$ be the induced fibration of the difference map $\alpha = (LR_1\varepsilon_1G - \varepsilon_{LR_1G}) : LR_1G \to LRLR_1G$ where $LR_1G = \prod_I \prod_{(R_1G)_i} A_i = \prod_I (\prod_{GA_i} A_i) = \prod_Z \prod_{G(K(\pi,i))} K(\pi,i)$ and $LRLR_1G$ has a similar interpretation.

Thus as in (1.14) to each diagram

$$K(\pi,k) \xleftarrow{\alpha'} C_1G = E_\alpha \xrightarrow{\sigma} \prod_Z \Big(\prod_{GK(\pi,i)} K(\pi,i)\Big) = B \tag{2.1}$$

there is an associated secondary operation Φ which, applied to a given space X , is the relation $[X,\alpha'] \circ [X,\sigma]^{-1}$ illustrated by

(2.2)
$$\begin{array}{ccc} [X,B] & & \\ \Big\uparrow{\scriptstyle [X,\sigma]} & & \\ [X,E_\alpha] & \xrightarrow{[X,\alpha']} & [X,K(\pi,k] \ . \end{array}$$

In the language of Adams [1] the pair (E_α,σ) is a universal example for SX = domain of definition of (2.2) = the subset of those elements of $[X,B]$ which are in the domain of definition of $[X,\sigma]^{-1}$. Furthermore, in the same language, $(E_\alpha,\sigma,\alpha')$ is a universal example for the secondary operation Φ illustrated by (2.2)

As mentioned at the end of section 1 we may allow secondary operations to include relations of the form (2.2) derived from

(2.3)
$$K(\pi,k) \xleftarrow{\ \alpha\ } E_\alpha \xrightarrow{\ \sigma\ } B$$

where σ is allowed to vary over all induced fibrations, that is, over weak equalizers of maps between products of $K(\pi,n)$s.

Finally we consider the approach of Mosher-Tangora to secondary operations in [8]. We let

(2.4)
$$\begin{array}{ccccc} & & E_\alpha & \xrightarrow{\ \alpha'\ } & K(\pi,k) \\ & & \Big\downarrow{\scriptstyle \sigma} & & \\ X & \xrightarrow{\ \delta\ } & K(\pi_1,n_1) & \xrightarrow{\ \alpha\ } & K(\pi_2,n_2) \end{array}$$

be a diagram of topological spaces where α is a primary operation of type $(\pi_1,n_1;\pi_2,n_2)$, σ is the fibration incuded by α and $\alpha' \in [E_\alpha,K(\pi,k)]\approx H^k(E_\alpha;\pi)$ is a given cohomology class. Then in the notation of [8] a secondary operation Φ_1 is defined on $[\delta] \in [X,K(\pi_1,n_1)]$ if $\alpha\delta$ is nullhomotopic and furthermore, using a stability assumption, Φ_1 takes its values in a quotient group of $H^k(X,\pi)\approx[X,K(\pi,k)]$. The explicit value of Φ_1 at δ is $[\alpha'\tilde{\delta}]$, which is the coset of $\alpha'\tilde{\delta}$ in $[X,K(\pi,k)]$ determined by various liftings $\tilde{\delta}\colon X \to E_\alpha$ of δ through σ .

Of course some mystery is taken out of the preceding by observing that in the language of (2.2) and (2.3) we are saying that to the relation

(2.5)
$$K(\pi,k) \xleftarrow{\ \alpha'\ } E_\alpha \xrightarrow{\ \sigma\ } K(\pi_1,n_1) = B$$

there corresponds the relation $\Phi_2 = [X,\alpha']\circ[X,\sigma]^{-1}$ illustrated by (2.2). As noted in [6] the morphism $\delta\colon X \to K(\pi_1,n_1) = B$ is in the domain of definition of Φ_2 if and only if $\alpha\delta$ is nullhomotpic, hence $\Phi_1 = \Phi_2$. As a final remark we note that one can consider relations associated with a tower of fibrations as well.

References

1. J.F. Adams, On the non-existence of elements of Hopf invariant one, Ann. Math. (2) 72(1960), 20-104.
2. G.M. Bergman, Some category-theoretic ideas in algebra, Proc. Int. Congress, Vancouver (1974).
3. A. Deleanu and P. Hilton, Borsuk shape and Grothendieck categories of pro-objects, Math. Proc. Cam. Phil. Soc. 79(1976).
4. F.W. Lawvere, Functorial semantics of algebraic theories, Thesis, Columbia (1963).
5. J. MacDonald, Categorical shape theory and the back and forth property, J. Pure Appl. Alg. 12(1978), 79-92.
6. J. MacDonald, Cohomology operations in a category, to appear.
7. S. Mac Lane, Categories for the working mathematician, Springer (1973).
8. R. Mosher and M. Tangora, Cohomology operations and applications in homotopy theory, Harper and Row (1968).
9. C. Maunder, Cohomology operations of the nth kind, Proc. London Math. Soc. 13(1963), 125-154.

Department of Mathematics
University of British Columbia
Vancouver, B.C., Canada

LIMIT-METRIZABILITY OF LIMIT SPACES AND UNIFORM LIMIT SPACES

Thomas Marny

Limit spaces and uniform limit spaces, introduced by H.R. Fischer [3] and C.H. Cook and H.R. Fischer [2], are generalizations of topological spaces and uniform spaces. H.J. Biesterfeldt [1], H.H. Keller [6], J.F. Ramaley and O.Wyler [8], and O. Wyler [11] have established limit-uniformization theorems for limit spaces. In this paper a metrization theory for limit spaces and uniform limit spaces is developed. For this the metric limit spaces are introduced which are a generalization of metric spaces. We obtain limit-metrization theorems for limit spaces and uniform limit spaces. In particular, the limit-metrizable topological spaces are the first countable Hausdorff spaces. Furthermore, a limit-uniformization theorem for pseudometric limit spaces is proved which has no analogue in classical metrization theory. The reader is supposed to be familiar with the concept of topological categories by H. Herrlich [4], [5] and O.Wyler [9], [10].

Let $\mathbb{R}$ denote the set of real numbers, $[0,\infty]=\{r\in\mathbb{R}\ /\ r\geq 0\}\dot{\cup}\{\infty\}$ and define $r\leq\infty$ and $r+\infty=\infty+r=\infty$ for each $r\in[0,\infty]$. For any set X the set $X^*=[0,\infty]^{X\times X}$ of all functions from XxX to $[0,\infty]$ is a complete lattice with respect to the partial ordering: $d_1\leq d_2$ iff $d_2(x,y)\leq d_1(x,y)$ for each $(x,y)\in X\times X$. The supremum $\bigvee_{i\in I} d_i$ of a family $(d_i)_{i\in I}$ is given by $(\bigvee_{i\in I} d_i)(x,y)=\inf_{i\in I} d_i(x,y)$. A subset D of X^* is called a $\vee$-ideal iff $d_1\vee d_2\in D$ is equivalent to $d_1\in D$ and $d_2\in D$. We define a binary operation on X^* by $d_1\circ d_2(x,y)=\inf\{d_2(x,z)+d_1(z,y)\ /\ z\in X\}$. Then X^* is a monoid with unit element δ for which $\delta(x,y)=0$ if $x=y$ and $\delta(x,y)=\infty$ otherwise. For $d\in X^*$ let $d^{-1}\in X^*$ be determined by $d^{-1}(x,y)=d(y,x)$. Then the following compatibility relations hold:

$d_1 \leq d_2$ implies $d_1^{-1} \leq d_2^{-1}$, $d_1 \leq e_1$ and $d_2 \leq e_2$ implies $d_1 \circ d_2 \leq e_1 \circ e_2$, $(d_1 \circ d_2)^{-1} = d_2^{-1} \circ d_1^{-1}$, $(\bigvee_{i \in I} d_i)^{-1} = \bigvee_{i \in I} d_i^{-1}$, $(\bigvee_{i \in I} d_i) \circ (\bigvee_{j \in J} e_j) = \bigvee \{d_i \circ e_j / (i,j) \in I \times J\}$ for finite sets I and J. If $f: X \longrightarrow Y$ is a map and $e \in Y^*$, then we define $f^* e \in X^*$ by $f^* e = e \circ (f \times f)$. The following relations hold: (0) $f^* e_1 \vee f^* e_2 = f^*(e_1 \vee e_2)$, (1) $\delta_X \leq f^* \delta_Y$, (2) $(f^* e)^{-1} = f^*(e^{-1})$, (3) $f^* e_1 \circ f^* e_2 \leq f^*(e_1 \circ e_2)$.

Definition 1. A limit pseudometric -shortly l-pseudometric- or a pseudometric limit structure on X is a $\vee$-ideal D in X^* which satisfies the conditions:

(ML1) $\delta \in D$.

(ML2) $d \in D$ implies $d^{-1} \in D$.

(ML3) $d_1 \in D$ and $d_2 \in D$ implies $d_1 \circ d_2 \in D$.

A limit metric -shortly l-metric- or a metric limit structure on X is a l-pseudometric D on X such that:

(ML4) $d(x,y)=0$ for some $d \in D$ implies $x=y$.

The pair (X,D) is called a pseudometric limit space (resp. metric limit space) iff D is a l-pseudometric (resp. l-metric) on X. If (X,D) and (Y,E) are pseudometric limit spaces, then a function $f: X \longrightarrow Y$ is called a m-continuous map from (X,D) to (Y,E) iff for each $d \in D$ there exists an $e \in E$ such that $d \leq f^* e$. The category of pseudometric limit spaces and m-continuous maps is denoted by PMetrL. Its full subcategory whose objects are the metric limit spaces is denoted by MetrL.

Proposition 1. PMetrL is a topological category such that T_0PMetrL=MetrL.

For the definition of T_0-objects of a topological category see the paper [7] of the author. The connection to the classical concepts of pseudometric and metric is given by:

Proposition 2. Let $d\in X^*$ such that $d(x,y)\neq\infty$ for each $(x,y)\in X\times X$ and let $D(d)=\{e\in X^*/\ e\leq d\}$. Then d is a pseudometric (resp. metric) on X if and only if $D(d)$ is a l-pseudometric (resp. l-metric) on X.

Let $\underline{F}(X)$ denote the complete lattice of all filters on X including the null filter $0=\underline{P}(X)$. The supremum $\bigvee_{i\in I}\underline{F}_i$ of a family $(\underline{F}_i)_{i\in I}$ of filters is given by $\bigvee_{i\in I}\underline{F}_i=\bigcap_{i\in I}\underline{F}_i$. If $d\in X^*$, then let $\underline{U}(d)\in\underline{F}(X\times X)$ be the filter which is generated by the filter base $\{U(d,2^{-n})\ /\ n\in\mathbb{N}\}$ where $U(d,2^{-n})=\{(x,y)\in X\times X\ /\ d(x,y)<2^{-n}\}$. Then the following relations hold:

(0) $\underline{U}(d_1)\vee\underline{U}(d_2)=\underline{U}(d_1\vee d_2)$.
(1) $[\Delta]=\underline{U}(\delta)$, where $[\Delta]$ is the diagonal filter.
(2) $\underline{U}(d)^{-1}=\underline{U}(d^{-1})$.
(3) $\underline{U}(d_1)\circ\underline{U}(d_2)\leq\underline{U}(d_1\circ d_2)$.
(4) $\dot{x}\times\dot{y}\leq\underline{U}(d)$ iff $d(x,y)=0$.
(5) $(f\times f)(\underline{U}(f^*e))\leq\underline{U}(e)$.

Thus for each l-pseudometric D on X we can define an induced uniform limit structure $\psi_X D$ on X by

$$\psi_X D=\{\ \underline{V}\in\underline{F}(X\times X)\ /\ \underline{V}\leq\underline{U}(d)\text{ for some } d\in D\}\ .$$

Let UnifL denote the topological category of uniform limit spaces and uniformly continuous maps. By the assignments $\Psi(X,D)=(X,\psi_X D)$ and $\Psi(f)=f$ we obtain a functor Ψ: PMetrl $\longrightarrow$ UnifL such that $\Psi(X,D)$ is separated iff D is a l-metric.

Definition 2. A uniform limit space (X,J) is called l-pseudometrizable (resp. l-metrizable) iff J is induced by a l-pseudometric (resp. l-metric) on X.

A necessary and, as wo shall see, sufficient condition for the l-pseudometrizability of a uniform limit space (X,J) is that the $\vee$-ideal J has a $\vee$-ideal base such that each of its filters has a countable filter base. Those uniform limit spaces which satisfy this condition form the objects of a bicoreflective subcategory PMUnifL of UnifL. The PMUnifL-coreflection of a uniform limit space (X,J) is the map $1_X\colon (X,J_{pm}) \longrightarrow (X,J)$ where $J_{pm} = \{\underline{W} \in \underline{F}(X\times X) \mathbin{/} \underline{W} \leq \underline{V}$ for some $\underline{V} \in J$ which has a countable filter base$\}$.

For a set X let $\underline{B}(X) = \{(F_n)_{n\in\mathbb{N}} \mathbin{/} (F_n)_{n\in\mathbb{N}}$ is a sequence in $\underline{P}(X)$ such that $F_{n+1} \subset F_n$ for each $n \in \mathbb{N}\}$. $\underline{B}(X)$ is a complete lattice with respect to the partial ordering: $(F_n) \leq (G_n)$ iff $F_n \subset G_n$ for each $n \in \mathbb{N}$. Then $(F_n) \vee (G_n) = (F_n \cup G_n)$. If (V_n), $(W_n) \in \underline{B}(X\times X)$, then let $(V_n)^{-1} = (V_n^{-1})$ and $(V_n) \circ (W_n) = (V_n \circ W_n)$. Furthermore we define $d((V_n)) \in X^*$ by $d((V_n))(x,y) = \inf\{2^{-n} \mathbin{/} (x,y) \in V_n\}$. If $(F_n) \in \underline{B}(X)$ and $x \in X$, then let $d((F_n)) = d((F_n \times F_n))$ and $d((F_n),x) = d((F_n \times \{x\}))$. The following relations hold:

(0) $d((V_n)) \vee d((W_n)) = d((V_n) \vee (W_n))$.
(1) $\delta = d((\Delta))$.
(2) $d((V_n))^{-1} = d((V_n)^{-1})$.
(3) $d((V_n)) \circ d((W_n)) \leq d((V_n) \circ (W_n))$.
(4) $d((V_n))(x,y) = 0$ iff $(x,y) \in V_n$ for each $n \in \mathbb{N}$.
(5) $d((V_n)) \leq f^* d(((f\times f)(V_n)))$.

Thus for each uniform limit structure J on X we can define an <u>induced l-pseudometric</u> $\varphi_X J$ on X by

$$\varphi_X J = \{e \in X^* \mathbin{/} e \leq d((V_n)) \text{ for some } (V_n) \in \underline{B}(X\times X) \text{ such that } [(V_n)] \in J\}$$

where $[(V_n)]$ denotes the filter based on $\{V_n \mathbin{/} n \in \mathbb{N}\}$. By the assignments $\Phi(X,J) = (X, \varphi_X J)$ and $\Phi(f) = f$ we obtain a functor $\Phi\colon$ UnifL $\longrightarrow$ PMetrL such that $\Phi(X,J)$ is a metric limit space iff (X,J) is separated.

<u>Proposition 3</u>. The functor $\Phi\colon$ UnifL $\longrightarrow$ PMetrL is a right adjoint of $\Psi\colon$ PMetrL $\longrightarrow$ UnifL. In particular, Φ preserves initial sources and Ψ preserves final sinks.

<u>Proof</u>. We must show (a) $D \subset \varphi_X \psi_X D$ and (b) $\psi_X \varphi_X J \subset J$.
(a) If $d \in D$, we define $(V_n) \in \underline{B}(X\times X)$ by $V_1 = X\times X$ and $V_n = U(d, 2^{-n+1})$ for $n \geq 2$. Then $[(V_n)] = \underline{U}(d)$ is in $\psi_X D$ such that $d \leq d((V_n))$, hence $d \in \varphi_X \psi_X D$.

(b) Since $U(d((V_n)),2^{-n}) \subset V_n$, we have $\underline{U}(d((V_n))) \leq [(V_n)]$, whence (b).

<u>Theorem 1</u>. A uniform limit space (X,J) is 1-pseudometrizable (resp. 1-metrizable) if and only if J has a $\vee$-ideal base such that each of its filters has a countable filter base (resp. and in addition is separated).

<u>Proof</u>. We show that $J \subset \psi_X \varphi_X J$ provided that (X,J) is a <u>PMUnifL</u>-space. Given $\underline{W} \in J$ there exists a $(V_n) \in \underline{B}(X\times X)$ such that $[(V_n)] \in J$ and $\underline{W} \leq [(V_n)]$. Since $U(d((V_n)),2^{-n}) = V_{n+1}$, we have $\underline{U}(d((V_n))) = [(V_n)]$ where $d((V_n)) \in \varphi_X J$. Hence $\underline{W} \in \psi_X \varphi_X J$.

Let $\underline{A}$ be the set of all monotone unbounded sequences a in $\mathbb{N}$ such that $a(n) \leq n$ for each $n \in \mathbb{N}$. If $a \in \underline{A}$ and $(F_n) \in \underline{B}(X)$, then we define $(F_n)_a = (G_n) \in \underline{B}(X)$ by $G_1 = X$ and $G_n = F_{a(n-1)}$ for $n \geq 2$. Then $[(F_n)_a] = [(F_n)]$. For $a \in \underline{A}$ and $d \in X^*$ we define $d_a \in X^*$ by $d_a(x,y) = \inf\{4^{-1}, \inf\{2^{-(n+1)} / d(x,y) < 2^{-a(n)}\}\}$. Then $d \leq d_a$. If we let $s: \mathbb{N} \longrightarrow \underline{A}$ be the map for which $s(m)(n) = 1$ if $1 \leq n \leq m$ and $s(m)(n) = n$ if $m < n$, then we have $\overline{2^{-m}} \leq (d_{s(1)})_{s(m)}$ for each $d \in X^*$. $\overline{2^{-m}}$ denotes the constant function on 2^{-m}.

<u>Definition 3</u>. A pseudometric limit space (X,D) is called <u>1-uniformizable</u> iff D is induced by a uniform limit structure on X.

<u>Theorem 2</u>. A pseudometric limit space (X,D) is 1-uniformizable if and only if it has the property that $d \in D$ and $a \in \underline{A}$ implies $d_a \in D$.

<u>Proof</u>. The condition is necessary since $d((V_n))_a \leq d(((V_n)_{s(1)})_a)$. To show that the condition is sufficient we need only prove that $\varphi_X \psi_X D \subset D$. For this we verify that for $d \in D$ and $(V_n) \in \underline{B}(X\times X)$ such that $\overline{4^{-1}} \leq d$ and $[(V_n)] \leq \underline{U}(d)$ there exists an $a \in \underline{A}$ such that $d((V_n)) \leq d_a$. There exists a strictly monotone sequence b in $\mathbb{N}$ such that

$b(1)=1$, $m \leq b(m)$, and $V_{b(m)} \subset U(d,2^{-m})$ for each $m \in \mathbb{N}$. The unique sequence a in $\mathbb{N}$ for which $b(a(n)) \leq n < b(a(n)+1)$ is in $\underline{A}$. Given $(x,y) \in X \times X$ the relation $d_a(x,y) \leq d((V_n))(x,y)$ is trivial if $d((V_n))(x,y)=\infty$. In case $d((V_n))(x,y)=0$, $[(V_n)] \leq \underline{U}(d)$ implies $d(x,y)=0$, hence $d_a(x,y)=0$. Now suppose $d((V_n))(x,y)=2^{-m}$. Then it follows from $(x,y) \in V_m \subset V_{b(a(m))} \subset U(d,2^{-a(m)})$ that $d_a(x,y) \leq \inf \{ 2^{-(n+1)} / d(x,y) < 2^{-a(n)} \} \leq 2^{-m} = d((V_n))(x,y)$.

<u>Corollary 1</u>. If (X,D) is a l-uniformizable pseudometric limit space, then $\bar{\varepsilon} \in D$ for each positive real number ε. In particular, the trivial pseudometric $\bar{0}$ is the only l-uniformizable pseudometric on X.

<u>Proposition 4</u>. The full subcategory <u>UPMetrL</u> of <u>PMetrL</u> whose objects are the l-uniformizable pseudometric limit spaces is bireflective in <u>PMetrL</u>. The <u>UPMetrL</u>-bireflection of a pseudometric limit space (X,D) is the map $1_X : (X,D) \longrightarrow (X,D_u)$ where $D_u = \{ e \in X^* / e \leq (\overrightarrow{d \vee 4^{-1}})_a \text{ for some } (d,a) \in D \times \underline{A} \}$.

<u>Proof</u>. It follows from the theory of topological categories (see [9], [10]) that $1_X : (X,D) \longrightarrow (X,\varphi_X \psi_X D)$ is a <u>UPMetrL</u>-bireflection for (X,D). Thus we need only verify that $D_u = \varphi_X \psi_X D$. Since $D \subset \varphi_X \psi_X D$ and since $\varphi_X \psi_X D$ is l-uniformizable, we have by corollary 1 and theorem 2 that $D_u \subset \varphi_X \psi_X D$. Given $e \in \varphi_X \psi_X D$ there exists a $d \in D$ and a $(V_n) \in \underline{B}(X \times X)$ such that $e \leq d((V_n))$ and $[(V_n)] \leq \underline{U}(d)$. The proof of theorem 2 shows that there exists an $a \in \underline{A}$ such that $d((V_n)) \leq (d \vee 4^{-1})_a$. Thus $e \in D_u$.

<u>Proposition 5</u>. The categories <u>PMUnifL</u> and <u>UPMetrL</u> are isomorphic.

<u>Definition 4</u>. A limit space is called <u>l-pseudometrizable</u> (resp. <u>l-metrizable</u>) iff it is induced by a l-pseudometrizable (resp. l-metrizable) uniform limit space.

<u>Definition 5</u>. A limit space (X,q) is called <u>first countable</u> iff for each point $x \in X$ the $\vee$-ideal $q(x)$ has a $\vee$-ideal base such that each of its filters has a countable filter base.

Since $\underline{F} \times \dot{x} \leq \underline{V}$ iff $\underline{F} \leq [\{y \in X \,/\, (y,x) \in V\} \,/\, V \in \underline{V}]$, it can easily be seen that each l-pseudometrizable limit space is first countable. By Keller [6] or Ramaley and Wyler [8] a limit space (X,q) is l-uniformizable if and only if it satisfies the axiom:

(S_1) $\underline{F}qx$ and $\underline{G}qx$ and $\underline{G}qy$ and $\underline{F} \wedge \underline{G} \neq 0$ implies $\underline{F}qy$.

The fine uniform limit structure $\tau_X q$ of a S_1-space (X,q) is $\tau_X q = \{ \underline{V} \in \underline{F}(X \times X) \,/\, \underline{V} \leq [\Delta] \vee \underline{F}_1 \times \underline{F}_1 \vee \ldots \vee \underline{F}_k \times \underline{F}_k$ for some finite set of filters $\underline{F}_1, \ldots, \underline{F}_k \in \bigcup \{q(x) \,/\, x \in X\}\}$. Obviously, $\tau_X q$ has a $\vee$-ideal base such that each of its filters has a countable filter base if and only if (X,q) is first countable. By theorem 1 we get:

<u>Theorem 3</u>. A limit space is l-pseudometrizable (resp. l-metrizable) if and only if it is a first countable S_1-space (resp. a first countable T_2-space).

<u>Theorem 4</u>. The induced l-pseudometric of the fine uniform limit structure of a first countable S_1-space (X,q) is $D = \{ d \in X^* / \, d \leq \delta \vee d((F_{1n})) \vee \ldots \vee d((F_{kn}))$ for some finite set of elements $(F_{1n}), \ldots, (F_{kn}) \in \underline{B}(X)$ such that $[(F_{in})] \in \bigcup_{x \in X} q(x)$ for $i = 1, \ldots, k\}$.

<u>Proof</u>. We must show $D = \varphi_X \tau_X q$. Clearly, D is a $\vee$-ideal which has properties (ML1) and (ML2). If $m \in \mathbb{N}$, then let $\emptyset_n = X$ for $1 \leq n \leq m$ and $\emptyset_n = \emptyset$ for $m < n$. Since $\overline{2^{-m}} \leq d((\emptyset_n))$ and $[(\emptyset_n)] = 0$ is in $q(x)$ for each $x \in X$, we obtain $\overline{2^{-m}} \in D$ if $X \neq \emptyset$. In case $X = \emptyset$ we have $\overline{2^{-m}} = \emptyset = \delta$. Thus $\overline{2^{-m}} \in D$ for each $m \in \mathbb{N}$. If (F_n), $(G_n) \in \underline{B}(X)$ such that $F_m \cap G_m = \emptyset$ for some $m \in \mathbb{N}$, then $d((F_n)) \circ d((G_n)) \leq \overline{2^{-m}}$. Hence $d((F_n)) \circ d((G_n)) \in D$ in this case. If $[(F_n)] \in q(x)$ and $[(G_n)] \in q(y)$ such that $F_m \cap G_m \neq \emptyset$ for each $m \in \mathbb{N}$, then $[(F_n \cup G_n)] \in q(x)$ since (X,q) is a S_1-space. Thus

it follows from $d((F_n))\circ d((G_n))\leq d((F_n\cup G_n))$ that $d((F_n))\circ d((G_n))\in D$. This proves (ML3).

The relations $(d_1 \vee d_2)_a = d_{1a} \vee d_{2a}$, $\delta_a = \delta \vee \overline{4^{-1}}$, and $d((F_n))_a \leq d(((F_n)_{s(1)})_a)$ imply that D has the property stated in theorem 2. Thus, by this theorem we see that D is l-uniformizable. Recall that $\underline{U}$: $X^* \longrightarrow \underline{F}(XxX)$ is a monotone map such that $\underline{U}(d_1 \vee d_2)=\underline{U}(d_1)\vee \underline{U}(d_2)$, $\underline{U}(\delta)=[\Delta]$, $\underline{U}(d((V_n))=[(V_n)]$, especially $\underline{U}(d((F_n))=[(F_n)]x[(F_n)]$. This together with the first countability of (X,q) implies $\psi_X D=\tau_X q$. Since D is l-uniformizable, we finally conclude that $D=\varphi_X\tau_X q$.

(X,D) is called a <u>quasipseudometric limit space</u> iff the $\vee$-ideal D satisfies (ML1) and (ML3), but not necessarily (ML2). Likewise one has the <u>quasiuniform limit spaces</u>. Theorem 1 and 2 also hold for quasipseudometric limit spaces with respect to quasiuniform limit spaces. A limit space (X,q) is l-quasiuniformizable if and only if it satisfies the axiom: (QU) $\underline{F}qx$ and $\dot{x}qy$ implies $\underline{F}qy$.

The fine quasiuniform limit structure of a QU-space (X,q) is $\nu_X q=\{\underline{V}\in\underline{F}(XxX) \;/\; \underline{V}\leq[\Delta]\vee \underline{F}_1 x\dot{x}_1 \vee \ldots \vee \underline{F}_k x\dot{x}_k$ for some finite set of elements $(\underline{F}_1,x_1),\ldots,(\underline{F}_k,x_k)\in q$ (cf. Wyler [11] whose definition of a quasiuniform limit space is slightly different since he does not require $[\Delta]\in J$ but only $\dot{x}x\dot{x}\in J$ for each $x\in X$). The following theorems may be proved similarly to their corresponding theorems 3 and 4.

<u>Theorem 5</u>. A limit space is l-quasipseudometrizable (resp. l-quasimetrizable) if and only if it is a first countable QU-space (resp. a first countable T_1-space).

<u>Theorem 6</u>. The induced l-quasipseudometric of the fine quasiuniform limit structure of a first countable QU-space (X,q) is $D=\{d\in X^* / \; d\leq \delta \vee d((F_{1n}),x_1) \vee \ldots \vee d((F_{kn}),x_k)$ for some finite set of elements $((F_{1n}),x_1),\ldots,((F_{kn}),x_k)\in\underline{B}(X)xX$ such that $([(F_{in})],x_i)\in q$ for $i=1,\ldots,k\}$.

Applying theorem 3 and 5 to topological spaces we get:

Theorem 7. A topological space is l-quasipseudometrizable (resp. l-pseudometrizable, l-quasimetrizable, l-metrizable) if and only if it is a first countable space (resp. a first countable R_1-space, a first countable T_1-space, a first countable T_2-space).

References

[1] H.J. Biesterfeldt, Uniformization of convergence spaces, Part I. Definitions and fundamental constructions. Math. Annalen 177 (1968), 31-42.

[2] C.H. Cook and H.R. Fischer, Uniform convergence structures. Math. Annalen 173 (1967), 290-306.

[3] H.R. Fischer, Limesräume. Math. Annalen 137 (1959), 269-303.

[4] H. Herrlich, Cartesian closed topological categories. Math. Coll. Univ. Cape Town 9 (1974), 1-16.

[5] H. Herrlich, Topological structures. Math. Centre Tracts 52 (1974), 59-112.

[6] H.H. Keller, Die Limes-Uniformisierbarkeit der Limesräume. Math. Annalen 176 (1968), 334-341.

[7] T. Marny, On epireflective subcategories of topological categories. to appear in General Topology and its Applications.

[8] J.F. Ramaley and O.Wyler, Cauchy spaces I. Structure and uniformization theorems. Math. Annalen 187 (1970), 175-186.

[9] O. Wyler, On the categories of general topology and topological algebra. Archiv der Math. 22 (1971), 7-17.

[10] O.Wyler, Top categories and categorical topology. Gen. Top. Appl. 1 (1971), 17-28.

[11] O. Wyler, Filter space monads, regularity, completions, TOPO 72-General Topology and its Applications, Lecture Notes in Mathematics 378 (1974), 591-637.

Institut für Mathematik I
Freie Universität Berlin
Hüttenweg 9
1000 Berlin 33
Federal Republic of Germany

BANACH SPACES OVER A COMPACT SPACE

Christopher J. Mulvey, University of Sussex

For any topological space X , there is an equivalence between the categories of Banach sheaves on X , of Banach spaces over X and of Banach spaces in the category Sh(X) of sheaves on X . This equivalence, together with the definitions of Banach sheaf on X and of Banach space over X , may be found in [11] . The existence of this equivalence allows one to introduce sheaves directly in working with these other concepts. One particular possibility lies in applying the internal mathematics of the category Sh(X) to problems concerning Banach sheaves and Banach spaces over X : the Hahn-Banach theorem established in [3] provides an example of this kind of technique. The intuitionistic context in which one is operating [8] is reflected in the unfamiliar form of the axiomatisation of Banach spaces in Sh(X) . In particular, the requirement that a Banach space be complete with respect to the norm must be modified in the absence of any axiom of (even countable) choice to one stated in terms of Cauchy *approximations* rather than Cauchy *sequences* [5] . Although these complications allow the development of functional analytic techniques in the category Sh(X) to proceed more smoothly [3,4,13] , the task of verifying that a particular sheaf is a Banach space may be difficult.

In this paper it is proved that the Cauchy completeness required of a Banach space in Sh(X) may be expressed in a particularly simple form in the case that the topological space X is compact. This provides the means for applying the methods of [9] to obtain various results concerning categories of Banach modules [12] extending those established earlier by Hofmann and Keimel [6] using different techniques. The result is also closely related to the characterisation established by Banaschewski [1] of Banach sheaves on a paracompact space: other results concerning Banach sheaves on compact spaces may be found in [2] . Throughout any compact space X is assumed to be Hausdorff. The notation used in dealing with sheaves on a topological space X is that of the language of the category Sh(X) , of which descriptions may be found in [7,8] . Although the Banach spaces considered will be over the reals, an analogous result exists for those over the complexes.

Consider a linear space B in the category Sh(X) of sheaves on a topological space X : that is, a module over the sheaf $\mathbb{R}_X$ of continuous real functions on X . Then B will be said to be normed provided that there is given a map

$$\mathbb{Q}^+_X \xrightarrow{N} \Omega^B \quad ,$$

from the sheaf of strictly positive, locally constant rational functions on X to the sheaf of subsheaves of B , satisfying the following conditions in the category Sh(X) :

a) $a \in N(q) \leftrightarrow \exists_{q' < q} \; a \in N(q')$;

b) $\exists_q \; a \in N(q)$;

c) $a \in N(q) \wedge a' \in N(q') \rightarrow a + a' \in N(q + q')$;

d) $\lambda \in N_{\mathbb{R}}(r) \wedge a \in N(q) \rightarrow \lambda a \in N(rq)$;

e) $a = 0 \leftrightarrow \forall_q \; a \in N(q)$.

The bounding universal quantifiers of a,a' over B , r,q,q' over $\mathbb{Q}^+_X$ and λ over $\mathbb{R}_X$ have been omitted, and the formula expressing that $-r < \lambda < r$ in the sheaf of continuous real functions has been written $\lambda \in N_{\mathbb{R}}(r)$. These conditions describe in the language of the category of sheaves on X the properties of the open balls about the origin in a normed space B [5] , the predicate $a \in N(q)$ expressing intuitively that the norm of $a \in B$ is less than the positive rational q .

Then a Banach space in the category Sh(X) is defined to be a normed linear space which is complete in the following sense: a Cauchy approximation on a normed linear space B in Sh(X) is defined to be a map

$$\mathbb{N}_X \xrightarrow{C} \Omega^B \quad ,$$

from the sheaf of locally constant, natural number functions on X to the sheaf of subsheaves of B , satisfying the following axioms internally in Sh(X) :

f) $\forall_n \exists_a \; a \in C_n$;

g) $\forall_k \exists_m \forall_{n,n' \geq m} \; a \in C_n \wedge a' \in C_{n'} \rightarrow a - a' \in N(1/k)$.

The normed linear space B is said to be complete provided that the internally quantified formula expressing that every Cauchy approximation on B converges is satisfied in the category of sheaves on X .

For any normed linear space B in the category Sh(X) , the bounded sections, by which are meant those $a \in B(X)$ for which there exists some rational number $q > 0$ such that $a \in N(q)$ for the *constant* rational function determined by q , form a linear space which may be normed by defining

$$\|a\| = \inf \{ q \mid a \in N(q) \} \quad .$$

Moreover, if B is a Banach space in Sh(X) then this yields a Banach space: any Cauchy sequence of bounded sections determines a Cauchy approximation in B consisting of singleton subsheaves of B . In the case that X is compact, any $a \in B(X)$ is necessarily bounded: the condition expressed in the internal language that every section is locally bounded yields external boundedness on choosing a finite subcovering on which the section is bounded.

One therefore has immediately one implication in the following theorem:

THEOREM *Let* X *be a compact space. Then a normed linear space* B *in the category* Sh(X) *of sheaves on* X *is a Banach space if and only if the normed space* B(X) *is a Banach space.*

Proof. It remains to establish that if every Cauchy sequence in the space B(X) converges then the condition that every Cauchy approximation in B converges is satisfied internally in Sh(X) . Before proving this, we recall the interpretation in Sh(X) of the formula expressing that every Cauchy approximation converges.

An approximation in B over an open subset $U \subset X$ is given by taking for each natural number n a subsheaf C_n of B over U for which there exists an open covering (U_α) of U over each open subset of which there is an $a_\alpha \in C_n(U_\alpha)$. The approximation is Cauchy provided that for any natural number k there exists an open covering (U_α) of U together with for each α an m_α such that for any open subset $U' \subset U_\alpha$, any $n,n' \geq m_\alpha$, and any $a \in C_n(U')$ and $a' \in C_{n'}(U')$, it is the case that $a - a'$ lies in the subsheaf $N(1/k)$ over U' .

Then the completeness of B means that given any Cauchy approximation C over an open subset $U \subset X$ there exists, locally but then by uniqueness actually over U, an element $b \in B(U)$ such that for each k there exists an open covering (U_α) of U together with for each α an m_α such that for any $U' \subset U$, any $n \geq m_\alpha$, and any $a \in C_n(U')$, it is the case that $a - b$ lies in the subsheaf $N(1/k)$ over U'. It will now be proved that this is satisfied in B provided that $B(X)$ is complete, by showing that any Cauchy approximation in B can be lifted locally to a Cauchy sequence in $B(X)$.

Consider then a Cauchy approximation C in the normed space B over an open subset $U \subset X$. By the compactness of X there exists a family (f_α) of continuous real functions on X such that:

i) for each α, $0 \leq f_\alpha \leq 1$;

ii) the support F_α of each f_α is contained in the open subset U;

and iii) the open sets U_α on which each f_α equals the identity in the sheaf $\mathbb{R}_X$ form an open covering of U.

Since C is an approximation, there exists for each n an open covering $(V_i^n)_{i \in I_n}$ of U together with for each $i \in I_n$ an element of $C_n(V_i^n)$. The compactness of X allows this covering to be chosen in such a way that there exists a partition of unity $(p_i^n)_{i \in I_n}$ by non-negative continuous real functions subordinate to the open covering $(V_i^n)_{i \in I_n}$ together with for each $i \in I_n$ an element $a_i^n \in B(X)$ for which

$$a_i^n \mid V_i^n \in C_n(V_i^n) .$$

For each α, choose a finite subfamily $(p_i^n)_{i \in I_n^\alpha}$ giving a partition of unity over the compact subset F_α and let

$$a_n^\alpha = \Sigma_{i \in I_n^\alpha} \; p_i^n \, a_i^n .$$

It is then asserted that (a_n^α) is a Cauchy sequence over the closed subset F_α.

Since C is Cauchy, there exists for each k an m_α beyond which members of C are within $1/3k$ over the compact subset F_α. For each $n, n' \geq m_\alpha$ there is an open covering of F_α over each open subset of which the expression

$$a_n^\alpha - a_{n'}^\alpha = \Sigma_i \, p_i^n \, a_i^n - \Sigma_{i'} \, p_{i'}^{n'} \, a_{i'}^{n'}$$

may be taken with summations extending over only those $i \in I_n$, $i' \in I_{n'}$ for which the restriction of a_i^n , $a_{i'}^{n'}$ lie respectively in C_n , $C_{n'}$. Choosing particular i_o , i'_o with this property, the expression may be rewritten in the form

$$a_n - a_{n'} = \Sigma_i\, p_i^n(a_i^n - a_{i_o}^n) - \Sigma_{i'}\, p_{i'}^{n'}(a_{i'}^{n'} - a_{i'_o}^{n'}) + (a_{i_o}^n - a_{i'_o}^{n'})$$

over each open subset of the covering of F_α : this then lies in $N(1/k)$ since $n,n' \geq m_\alpha$. The sequence is therefore Cauchy over the compact subset F_α .

Defining for each α

$$b_n^\alpha = f_\alpha a_n^\alpha ,$$

it follows that (b_n^α) is a Cauchy sequence in the Banach space $B(X)$. Moreover, that over the open subset U_α the sequence converges to a section b_α which provides a limit for the Cauchy approximation C over U . The uniqueness of this limit establishes that these sections patch over the open covering (U_α) to give a limit b for the approximation C over the open subset U .

The Banach space B is therefore complete with respect to Cauchy approximations, which completes the proof of the theorem.

It may be added that the norm

$$\mathbb{Q}_X^+ \xrightarrow{N} \Omega^B$$

is also determined uniquely by the norm on the Banach space $B(X)$ of bounded sections: the subsheaf $N(q)$ over an open subset $U \subset X$ assigned to a locally constant rational function $q > 0$ over U is exactly that of which the sections over any $U' \subset U$ are those $a \in B(U')$ for which there exists an open covering (U_α) of U' together with for each α an $a_\alpha \in B(X)$ such that

i) $a_\alpha | U_\alpha = a | U_\alpha$,

and ii) $\| a_\alpha \| < q | U_\alpha$.

Taken together with the observation [9] that any linear space B over a compact space X is uniquely determined to within isomorphism by the

$\mathbb{R}(X)$-module $B(X)$, it follows that one has the following corollary:

COROLLARY *Let* X *be a compact space. Then a Banach space* B *in the category* $Sh(X)$ *of sheaves on* X *is uniquely determined to within isometric isomorphism by the Banach space* $B(X)$.

Evidently the corollary remains true for a normed space in the category of sheaves on X . Moreover, the space X need only be paracompact.

There follow similar characterisations of Banach sheaves and of Banach spaces over a compact space, of which the details will be omitted. It may also be remarked that the property of the sheaf $\mathbb{R}_X$ on which the proofs ultimately rest is that the compactness of $(X,\mathbb{R}_X)$ in the sense of [10] may be established using continuous real functions lying in the closed unit ball of the sheaf $\mathbb{R}_X$. The extension of these results to which this observation gives rise will be examined elsewhere.

REFERENCES

1. Banaschewski, B.: Sheaves of Banach spaces. Quaest. Math. 2 , 1-22 (1977)

2. Banaschewski, B.: Recovering a space from its Banach sheaves. *This volume*

3. Burden, C.W.: The Hahn-Banach theorem in a category of sheaves. J. Pure and Applied Algebra. *To appear*

4. Burden, C.W.: Normed and Banach spaces in categories of sheaves. D. Phil. thesis : University of Sussex 1978

5. Burden, C.W., Mulvey, C.J.: Normed spaces in a category of sheaves. *In* Applications of sheaves. Lecture Notes in Mathematics. Berlin and New York: Springer. *To appear*

6. Hofmann, K.H., Keimel, K.: Sheaf theoretical concepts in analysis. *In* Applications of sheaves. Lecture Notes in Mathematics. Berlin and New York: Springer. *To appear*

7. Johnstone, P.T.: Topos Theory. L.M.S. Monographs 10 . London and New York: Academic Press 1977

8. Mulvey, C.J.: Intuitionistic algebra and representations of rings. *In* Recent advances in the representation theory of rings and C*-algebras

by continuous sections. Mem. Amer. Math. Soc. 148 , 3-57 (1974)

9. Mulvey, C.J.: A categorical characterisation of compactness. J. London Math. Soc. (2) , 17 , 356-362 (1978)

10. Mulvey, C.J.: Compact ringed spaces. J. Algebra 52 , 411-436 (1978)

11. Mulvey, C.J.: Banach sheaves. J. Pure and Applied Algebra. *To appear*

12. Mulvey, C.J.: Categories of Banach modules. *To appear*

13. Pelletier, J.W., Rosebrugh, R.: The category of Banach spaces in sheaves. Preprint : York University 1978

Mathematics Division,
University of Sussex,
Falmer, BRIGHTON, BN1 9QH,
England.

A NOTE ON (E,M)-FUNCTORS

by

Ryosuke Nakagawa (Ibaraki, Japan)

Introduction In a recent paper [5] Herrlich and Strecker defined (E,M)-functors and showed that various nice functors can be considered as (E,M)-functors by taking suitable E and M. The notion of (E,M)-functors is also a generalization of that of (E,M)-categories. In this paper we show that some basic properties of (E,M)-categories do not hold for (E,M)-functors, for example M is not uniquely determined by E, and show that they hold for (E,M)-functors with some additional conditions. We shall also give some sufficient conditions in order that a functor is an (E,M)-functor.

The author is indebted to G. E. Strecker and A. Melton for many useful discussions during the year he stayed at Kansas State University.

§1.

Let $\underline{A}$ and $\underline{X}$ be categories, $T:\underline{A} \to \underline{X}$ be a functor, E be a class of T-morphisms and M be a collection of $\underline{A}$-sources. Throughout this section we assume that E, E', M and M' are closed under composition with isomorphisms.

An (E,M)-factorization of a T-source $(X,(f_i:X \to TA_i)_I)$ is a pair which consists of a T-morphism $e:X \to TA$ in E and an $\underline{A}$-source $(A,(m_i:A \to A_i)_I)$ in M such that $f_i = Tm_i \cdot e$ for each $i \in I$.

If every T-source has an (E,M)-factorization, the functor T is called (E,M)-factorizable.

A functor T is called to have a diagonalization property if for any

T-morphism $e:X \to TA$ in E, any $\underline{A}$-source $(B,(m_i:B \to C_i)_I)$ in M, any T-morphism $f:X \to TB$ and any $\underline{A}$-source $(A,(g_i:A \to C_i)_I)$ with $Tg_i \cdot e = Tm_i \cdot f$ for each $i \in I$, there exists a unique[1)] $\underline{A}$-morphism $d:A \to B$ such that $Td \cdot e = f$ and $m_i d = g_i$ for each $i \in I$.

An (E,M)-factorizable functor T with a diagonalization property is called an (E,M)-<u>functor</u>.

As examples of (E,M)-functors we note the following.

(1) regular functor $\Rightarrow$ (Extremal generating T-mor, initial source)-functor [1],

(2) (E,M)-topological functor $\Rightarrow (E,M_T)$-functor,

(3) Topologically-algebraic functor $\Leftrightarrow$ (Semi-universal, initial)-functor [5],

(4) right adjoint functor $\Longleftrightarrow$ (Universal, source)-functor [5].

A category $\underline{X}$ is called an (E,M)-<u>category</u> if the identify functor $I:\underline{X} \to \underline{X}$ is an (E,M)-functor.

<u>Proposition 1.1</u> If T is an (E,M)-functor, each member in E is a generating T-morphism (= T epi-morphism).

For the proof see [5]. This is also shown in [7].

<u>Proposition 1.2</u> Suppose that T is an (E,M)-functor. Then $f:X \to TA$ belongs to E if and only if it satisfies the following condition (*).

For any $\underline{A}$-source $(B,(m_i:B \to C_i)_I)$ in M, any $g:X \to TB$ and any
(*) $(A,(h_i:A \to C_i)_I)$ with $Tm_i \cdot g = Th_i \cdot f$ for each $i \in I$, there exists a unique $\underline{A}$-morphism $d:A \to B$ with $Td \cdot f = g$ and $m_i d = h_i$ for each $i \in I$.

1) G. E. Strecker pointed out that Prop. 1.1 below is proved without the uniqueness of diagonal morphisms in this definition and the uniqueness is deduced from Prop. 1.1. Thus we may define (E,M)-functors without this uniqueness assumption. However it is needed in order to define morphism-(E,M)-functors in §3.

Corollary 1.3 If T is an (E,M)-functor and an (E',M)-functor, then $E = E'$.

For a collection M of $\underline{A}$-sources, a class E consisting of all T-morphisms satisfying the condition (*) will be denoted by $\gamma(M)$.

Now we shall consider the following conditions.

(EI) For any $f:TA \to TB$ in E there exists an $\underline{A}$-morphism $f':A \to B$ with $Tf' = f$.

(EII) E is closed under the formation of compositions.

(EIII) For any $\underline{A}$-object A, the identity 1_{TA} belongs to E.

(MI) Any $\underline{A}$-source in M is T-initial.

(MII) An $\underline{A}$-source $(B,(m_i:B \to C_i)_I)$ belongs to M if and only if it satisfies the following condition (**).

For any $f:X \to TA$ in E, any $g:X \to TB$ and any $\underline{A}$-source

(**) $(A,(h_i:A \to C_i)_I)$ with $Tm_i \cdot g = Th_i \cdot f$ for each $i \in I$, there exists a unique $\underline{A}$-morphism $d:A \to B$ with $Td \cdot f = g$ and $m_i d = h_i$ for each $i \in I$.

Theorem 1.4 Suppose that T is an (E,M)-functor. Then we have the following implications.

$(EI) \Longrightarrow (EII) \Longrightarrow (EIII) \Longleftrightarrow (MI) \Longrightarrow (MII)$

Proof $(EIII) \Longleftrightarrow (MI)$ has been proved by Tholen [7]. We shall show $(EII) \Longrightarrow (EIII)$. The other implications can be proved by the same technique.

For an $\underline{A}$-object A, let $(e:TA \to TB, m:B \to A)$ be an (E,M)-factorization of 1_{TA}. By considering an (E,M)-factorization of e, we can show that 1_B belongs to M. Let $(e':TB \to TC, m':C \to B)$ be an (E,M)-factorization of 1_{TB}. Then there exists an $\underline{A}$-morphism $d':B \to C$ with $Td' \cdot e = e'e$ and $m'd' = 1_B$. We have that $T(d'm')e'e = e'e$ and $m'(d'm') = m'$. Since $e'e$ belongs to E, $d'm' = 1_C$. Therefore we have that 1_{TB} belongs to E. Since $Tm(e \cdot Tm) = Tm \cdot 1_{TB}$, there exists an $\underline{A}$-morphism $d'':B \to B$ with $Td'' = e \cdot Tm$ and $md'' = m$. Again by the uniqueness of diagonal morphisms we have $d'' = 1_B$, that is, $e \cdot Tm = 1_{TB}$. This shows that Tm is an isomorphism and that 1_{TA} belongs to E.

It is noted that the equivalence of the following is given by Herrlich and Strecker [5]. Topologically-algebraic functor = (Semi-universal, initial)-functor = (E,M)-functor with (EIII) = (E,M)-functor with (MI).

Corollary 1.5 Suppose that T is an (E,M)-functor and an (E,M')-functor. Then if M and M' satisfy (MII), $M = M'$ (of course, if E satisfies (EIII), $M = M'$).

Example 1.6 Let $\underline{X}$ be a category consisting of two objects A, B and four morphisms 1_A, 1_B, $f:A \to B$, and $g:B \to A$ with $gf = 1_A$ and $fg = 1_B$, $\underline{A}$ be a subcategory of $\underline{X}$ consisting of A, B, 1_A, 1_B and g, $E = \text{Mor}\,\underline{X}$ and M be a collection of all $\underline{A}$-sources containing an identity. Then the embedding functor $T:\underline{A} \to \underline{X}$ is an (E,M)-functor satisfying (EII) but not (EI).

Example 1.7 Let $\underline{X}$ be a category consisting of sets $A_n = \{1,2,\ldots,n\}$, $n = 1,2,\ldots$ and all monotone maps $f:A_m \to A_n$, m, $n = 1,2,\ldots$, i.e. maps satisfying $f(i) \leq f(j)$ for $i < j$, $\underline{A}$ be a subcategory of $\underline{X}$ consisting of all objects A_n and all morphisms $f:A_m \to A_n$ with $f(m) = n$, $E = \{e_n:A_n \to A_{n+1} \mid e_n(i) = i,\ i \leq n,\ n = 1,2,\ldots\}$ and M be a collection of all $\underline{A}$-sources. Then E and M are closed under composition with isomorphisms and the embedding functor $T:\underline{A} \to \underline{X}$ is an (E,M)-functor. M satisfies (MII) but E does not satisfy (EIII).

Let $M' = M \setminus \{1_{A_1}\}$. Then T is an (E,M')-functor, too. M' does not have (MII) and some properties in Prop. 2.2.

For $m \geq n$ define a map $e_{m,n}:A_m \to A_n$ by $e_{m,n}(i) = i$ for $i \leq n$

and $e_{m,n}(i) = n$ for $i > n$. Let $E" = E \cup \{e_{m,n} \mid m = 1,2,\ldots,m \geq n\}$ and $M"$ be a collection of all $\underline{A}$-sources $(A_m, (f_k : A_m \to A_{n(k)})_K)$ such that $f_k(m-1) < n(k)$ for some $k \in K$. Then T is an $(E",M")$-functor and hence it is topologically-algebraic.

Example 1.8 Let $\underline{X}$ be a category consisting of a set N of all positive integers and all monotone maps $f:N \to N$, $\underline{A}$ be a subcategory of $\underline{X}$ consisting of N and $f:N \to N$ with $f(1) = 1$, $E = \{e:N \to N \mid e(i) = i + 1$ for $i \in N\}$ and M = (all $\underline{A}$-sources). Then $T:\underline{A} \to \underline{X}$ is an (E,M)-functor. If T is an (E,M')-functor, then we have that $M' = M$. (EIII) is not satisfied. T is also topologically algebraic.

Example 1.9 Let $\underline{A}$ be a category of compact T_2-spaces, $\underline{X}$ a category of sets, $T:\underline{A} \to \underline{X}$ the forgetful functor, $E = \{e: X \to TA \mid e(X)$ is dense in $A\}$ and M = (all closed embedding sources). Then T is an (E,M)-functor and E satisfies (EIII) but not (EII) (the author is indebted to H. Herrlich for this example).

§2. For two sources $(A, (h_i)_I)$ and $(B, (k_i)_I)$, let (L,f,g) be a limit of a diagram

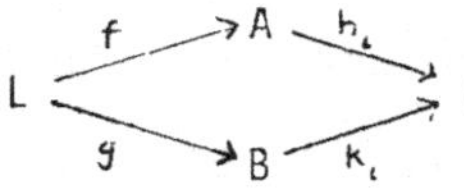

C_i. f is called a pullback of $(B, (k_i)_I)$. Multiple pullbacks of sources are similarly defined (see [6]).

For an $\underline{A}$-source $(A, (f_i:A \to B_i)_I)$ and a collection of $\underline{A}$-sources $(B_i, (g_{ij}:B_i \to C_j)_{J(i)})_I$, an $\underline{A}$-source $(A, (g_{ij}f_i:A \to C_j)_K)$ with $K = \bigcup_I J(i)$ is called their composition. If $(A, (f_i)_J)$ is a subsource of $(A, (f_i)_I)$ with a subclass J of I, $(A, (f_i)_I)$ is called an extension of $(A, (f_i)_J)$. If for each $i \in I$ there exist $j(i) \in J$ and an isomorphism $v_i:B_{j(i)} \to B_i$ with $f_i = v_i f_{j(i)}$, $(A, (f_i)_J)$ is called a trivial restriction of $(A, (f_i)_I)$.

Proposition 2.1 (E,M)-functors preserve limits. (E,M)-functors with (EIII) are faithful.

Proposition 2.2 Suppose that $T:\underline{A} \to \underline{X}$ is an (E,M)-functor and that M satisfies (MII). Then we have the following.

$[M1_0]$ M contains every $\underline{A}$-source consisting of an isomorphism.

[M2] M is closed under the formation of compositions.

[M3] M is closed under the formation of extensions.

[M4] M is closed under the formation of trivial restrictions.

[M5] M is closed under the formation of pullbacks.

[M6] M is closed under the formation of multiple pullbacks.

[M7] M is closed under the formation of joint pullbacks (for the definition see [6]).

[M8] M is left cancellative.

Moreover, if $\underline{A}$ has pullbacks and pushouts, or if E satisfies (EIII),

[M1] M contains every extremal mono-source.

The proofs are routine and so omitted.

Now let $T:\underline{A} \to \underline{X}$ be a functor and M a collection of $\underline{A}$-sources. We will give a sufficient condition in order that T is an (E,M)-functor for some E.

Definition 2.3 $\underline{X}$ is called T-co-wellpowered provided that for any $\underline{X}$-object X a T-source $(X, (f_i:X \to TA_i)I)$ consisting of all generating T-morphisms with domain X has a trivial T-restriction $(X,(f_i)_J)$ with a set J, i.e. J is a subclass of I and for each $i \in I$ there exist $j(i) \in J$ and an $\underline{A}$-isomorphism $v_i:A_{j(i)} \to A_i$ with $f_i = Tv_i \cdot f_{j(i)}$

If $T:\underline{A} \to \underline{X}$ is an (E,M)-functor and $\underline{A}$ is co-wellpowered, $\underline{X}$ is T-co-wellpowered.

Theorem 2.4 Suppose that $\underline{A}$ is wellpowered, co-wellpowered and complete, $\underline{X}$ is T-co-wellpowered, $T:\underline{A} \to \underline{X}$ preserves limits and M is a collection of $\underline{A}$-sources satisfying [M1] ∿ [M6]. Then T is an (E,M)-functor with $E = \gamma(M)$.

It is noted that [M8] implies [M4]. The proof is accomplished by the standard technique for constructing factorizations and so omitted.

§3.

Let $T:\underline{A} \to \underline{X}$ be a functor, E a class of T-morphisms and M a class of $\underline{A}$-morphisms. T is called a morphism-(E,M)-functor provided that every T-morphism has an (E,M)-factorization and the diagonalization property holds.

If T is an (E,M)-functor, it is a morphism-(E,M_*)-functor with a class M_* consisting of all morphisms in M as singleton sources.

Suppose that $\underline{A}$ has products and let M be a class of $\underline{A}$-morphisms. Define a collection M^* of $\underline{A}$-sources as follows. An $\underline{A}$-source $(A, (f_i:A \to A_i)_I)$ belongs to M^* if and only if there exists an $\underline{A}$-source $(A, (g_j:A \to B_j)_J)$ with a set J such that for each $i \in I$ there exist $j(i) \in J$ and a morphism $v_i:B_{j(i)} \to A_i$ in M with $f_i = v_i g_{j(i)}$, $J = \bigcup_I \{j(i)\}$ and a morphism $g:A \to \pi_J B_j$ defined by $p_j g = g_j$, $j \in J$ belongs to M.

Theorem 3.1 Suppose that $\underline{A}$ is complete, $\underline{X}$ is complete and T-co-wellpowered and $T:\underline{A} \to \underline{X}$ is a morphism-(E,M)-functor preserving products. Then T is an (E,M^*)-functor if and only if E consists of generating T-morphisms.

Proof Prop. 1.1 implies 'only if' part. The following diagram indicates the proof of (E,M^*)-factorizability, where (g_i,h_i) and (g,h) are (E,M)-factorizations. The diagonalization property can be verified.

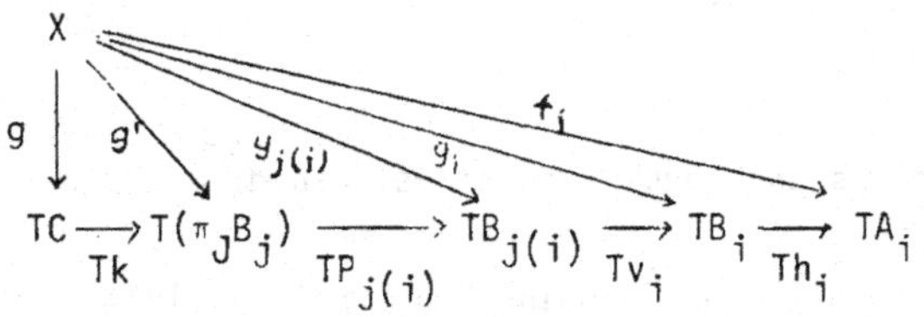

<u>Corollary 3.2</u> A complete, co-wellpowered morphism-(E,M)-category with E consisting of epimorphisms is an (E,M^*)-category.

It is known that a wellpowered and complete category is an (epi, extremal mono)-category [4]. Then if it is co-wellpowered, (extremal mono) = $(\text{extremal monomorphism})^*$. It is noted that Top is a morphism-(dense, closed embedding)-category, but it is not a (dense, $(\text{closed embedding})^*$)-category.

For a morphism-(E,M)-functors, Prop. 1.1 does not hold but the other results in §1 hold.

REFERENCES

1. H. Herrlich, Regular categories and regular functors, Canad. J. Math 26(1974) 709-720.

2. H. Herrlich, Topological functors, Gen. Topology and Appl. 4 (1974) 125-142.

3. H. Herrlich, G. Salicrup and R. Vázquez, Dispersed factorization structures, Preprint.

4. H. Herrlich and G. E Strecker, Category Theory, Boston Allyn and Bacon 1973.

5. H. Herrlich and G. E. Strecker, Semi-universal maps and universal initial completions, Preprint.

6. G. E. Strecker, Perfect sources, Proc. Categorical Topology, Lecture Notes in Math. 540 (Springer, 1976) 605-624.

7. W. Tholen, Semi-topological functors I, J. Pure Appl. Algebra (to appear).

Department of Mathematics
University of Tsukuba
Ibaraki
Japan

CONVENIENT TOPOLOGICAL ALGEBRA AND REFLEXIVE OBJECTS

L.D. Nel, Ottawa, Ontario

Reflexiveness for Banach spaces is analogous to reflexiveness for locally compact abelian groups. The analogy becomes even more striking when these classical categories are embedded into better endowed categories such as convergence vector spaces and convergence abelian groups respectively, which allow a more comprehensive theory. Such categories are axiomatized in this paper in order to study reflexiveness and related matters in abstract context. Results obtained enable us to recover several known facts in a unified manner and also to find new reflexive objects in several special situations, including extensions of the abovementioned ones. Known Pontryagin-type dualities thus become automatically extended to larger categories.

The abstract setting is chosen so as to form a general foundation for topological algebra. Topological behaviour is encoded as a convenient category $\underline{X}$ (i.e. a cartesian closed topological category). For the algebraic aspects we start off with an algebraic category $\underline{A}$ over $\underline{X}$, equip it further with a compatible function algebra functor $\Phi:\underline{X}^{*}\times\underline{A} \to \underline{A}$ (to make $\underline{A}$ <u>strongly algebraic</u> over $\underline{X}$) and also with a compatible internal hom-functor $\Omega:\underline{A}^{*}\times\underline{A} \to \underline{A}$ (to make $\underline{A}$ <u>autonomously algebraic</u> over $\underline{X}$). This Ω turns out to relate in the desired way to the "natural" tensor product that we construct for algebraic categories.

1 Motivating examples

1.1 The category Ban of Banach spaces with continuous linear (= bounded linear) maps embeds into several categories which allow all usual operations to be performed in them. To describe some of these categories, recall first the "base" categories Cv (convergence spaces [Fi59] with continuous maps), Cg (compactly generated spaces i.e. the coreflective hull of compact T_2-spaces in the usual category Top of topological spaces), Bo (bornological spaces [Hg71] with boundedness preserving maps). One forms over Cv, Top, Cg, Bo respectively, in the manner of universal algebra, the categories CvVec, TopVec, CgVec, BoVec of vector spaces with scalar field K . Each of these contains Ban as a subcategory (= full isomorphism-closed subcategory in this paper). For a given A in Ban we can form its "dual" space ΔA in each of Ban, TopVec, CvVec, CgVec, and BoVec by putting some "topological" structure on the vector space of morphisms $A \to K$. Since Cv, Cg and Bo are cartesian closed there is a canonical structure for ΔA , different in each category, with only the BoVec-dual coinciding with the classical normed dual. In TopVec there is no canonical structure and every topological structure for ΔA has some pathology, so we exclude TopVec from further consideration: the objects of TopVec are important, not TopVec itself. The CvVec-duals, while not coinciding with normed duals, have nevertheless a very pleasing property : every complete locally convex topological vector space is reflexive in the CvVec sense. For a theory to apply to a specific situation the analyst should make the most expeditious choice from among CvVec, CgVec, BoVec or other conveniently based categories.

1.2 The category LAb of locally compact Hausdorff abelian groups is a subcategory of CvAb (convergence groups [Bi75]) and also of CgAb (compactly generated groups [Lm77]). For each of CvAb, CgAb there is a canonical "topological" structure for the dual group ΔE (of morphisms $E \to K$ = circle group). The CvAb structure (continuous convergence) coincides with the compact open topology just when $E \in$ LAb; that of CgAb is in general finer than the compact-open topology, even when E is in LAb.

1.3 The reflexiveness of compact 0-dimensional semilattices (ZSL) studied in [HMS74] is quite analogous to that of compact abelian groups. This duality has been extended in the conveniently-based category CvSl of convergence semi-lattices [HN7-].

The examples 1.1, 1.2, 1.3 illustrate a common phenomenon. A theory is developed first for technically simple objects like those of Ban, LAb or ZSL which do not form a good category. Embedding into a conveniently based category which permits all usual constructions, such as CvVec, CvAb or CvSl paves the way for a more highly developed theory.

2 Strongly and autonomously algebraic categories

2.1 S will denote our chosen category of sets. X will

be a convenient category i.e. it is equipped with functors

$$!\,!:\underline{X} \longrightarrow \underline{S} \quad \text{and} \quad \nabla:\underline{X}^*\times\underline{X} \to \underline{X}$$

as follows. The "underlying set" functor $!\,!$ makes $\underline{X}$ initially complete (see e.g. [HN77]); its fibres on any set S has a representative set which moreover reduces to a single object whenever S has cardinality 0 or 1. The "function space" functor ∇ gives for every T in $\underline{X}$ a functor $T\nabla-$ right adjoint to $T\times-$. Moreover, $!\,!\circ\nabla = \#$, where $\#$ denotes, for any category $\underline{C}$, the hom set functor $\#:C^*\times\underline{C} \to \underline{S}$. The counit of the above adjunction will be denoted by $ev_{XY}:X\times(X\nabla Y) \to Y$; at set level it carries (x,f) to $f(x)$ and for each t in $!X!$ it induces an $\underline{X}$-morphism $@t:X\nabla Y \to Y$ such that $!@t!(f) = f(t)$.

For each point object P (i.e. $!P! = \{p\}$) and every X in $\underline{X}$ there is a canonical isomorphism $eq_X:X \to P\nabla X$. For further basic properties and several examples of convenient categories see [Ne76], [HN77] and [Ne77].

2.2 Recall that a category $\underline{A}$ is algebraic over $\underline{X}$ if it is equipped with an "underlying space" functor $|\,|:\underline{A} \to \underline{X}$ which is algebraic (i.e. regular in the sense of [He74a]). We will call such $\underline{A}$ strongly algebraic over $\underline{X}$ if it is further equipped with a "function algebra" functor $\Phi:X^*\times\underline{A} \to \underline{A}$ such that

$$\begin{array}{ccc} \underline{X}^*\times\underline{A} & \xrightarrow{\Phi} & \underline{A} \\ {\scriptstyle Id\times|\,|}\downarrow & & \downarrow{\scriptstyle |\,|} \\ \underline{X}^*\times\underline{X} & \xrightarrow[\nabla]{} & \underline{X} \end{array}$$

commutes. Then $eq:|A| \to P\nabla|A|$ (2.1) lifts to an isomorphism $A \to P\Phi A$. Hence each $@x:X\nabla|A| \to |A|$ ($x\in !X!$) lifts to $X\Phi A \to A$, thus inducing a natural monomorphism $X\Phi A \to \Pi_{!X!}A$.

2.3 Examples of strongly algebraic categories include all the categories constructed over $\underline{X}$ by means of finitary operations as in universal algebra or by means of actions (such as the continuous scalar multiplication of CvVec which cannot be expressed by unary operations as in $\underline{S}$-based universal algebra). Every (regular epi)-reflective subcategory of a strongly algebraic category is strongly algebraic. Thus subcategories of the above defined by equations or implications between equations [BH76] must be strongly algebraic. Thus CvVec is strongly algebraic over Cv, as likewise its (regular epi)-reflective subcategroy CvHVec of Hausdorff objects; similarly for CgVec, CgHVec, CgAb, CgHAb etc.

When $\underline{X} = \underline{S}$, every algebraic category $\underline{A}$ over $\underline{X}$ is already strongly algebraic : $X\Phi A = \Pi_X A$ in this case. In general an algebraic category over X need not be strongly algebraic. It is so when defined by a $\underline{X}$-monad ([Du70] p.104).

2.4 The faithful functor $||$ induces a natural $\underline{S}$-monomorphism $||_{AB}: A\#B \to |A|\#|B| = !|A|\nabla|B|!$ which lifts to a natural $\underline{X}$-embedding (= initial monomorphism) $ne_{AB}: A\Psi B \longrightarrow |A|\nabla|B|$. Thus $||$ induces a "spectral space" functor

$$\Psi: \underline{A}^* \times \underline{A} \to \underline{X}$$

such that $!!\cdot\Psi = \#$. Moreover, composition as a function $(A\#B)\times(B\#C) \to A\#C$ lifts to an $\underline{X}$-morphism $\circ: (A\Psi B)\times(B\Psi C) \to A\Psi C$. Thus $||$ induces on $\underline{A}$ the structure of an $\underline{X}$-category such that $||$ is a $\underline{X}$-functor. We mention in passing that when $\underline{A}$ <u>is strongly algebraic, there exists a natural isomorphism</u>

$ni_{XAB}: (X\Phi A)\Psi * B = B\Psi(X\Phi A) \to X\nabla(B\Psi A)$ <u>such that</u>, at set level, $ni(h)(x)(b) = h(b)(x)$. (Facts 2.3, 2.4 will be proved in a forthcoming paper in which enriched categorical aspects of $\underline{A}$ will be studied). As a corollary, every functor $-\Phi A: \underline{X}^* \to \underline{A}$ has $-\Psi A$ as $\underline{X}$-left adjoint. The back adjunction is the morphism $\tau_X: X \to (X\Phi A)\Psi A$ such that, at set level $\tau(x)(f) = f(x)$.

<u>2.5</u> We will call $\underline{A}$ <u>autonomously algebraic</u> over $\underline{X}$ if it is strongly algebraic and further equipped with an "internal hom" functor

$$\Omega: \underline{A}^* \times \underline{A} \to \underline{A}$$

such that $|\,|\cdot\Omega = \Psi$ and with a natural monomorphism $\tilde{\tilde{ne}}_{AB}: A\Omega B \to |A|\Phi B$ such that $|\tilde{\tilde{ne}}| = ne: A\Psi B \to |A|\nabla|B|$ (see 2.4). We thus arrive at the following commutative diagram:

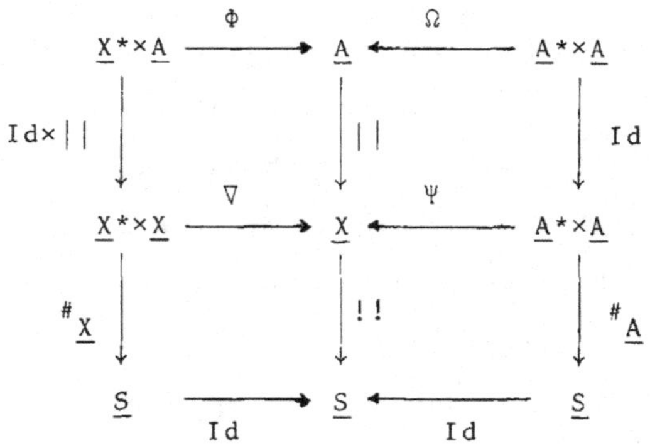

(This usage of "autonomous" is consistent with that in [Li66] even if the technical context is somewhat different). Examples of autonomously algebraic categories include those of abelian groups, semilattices with 1 and vector spaces formed over $\underline{X}$.

A decent internal hom functor should have a left adjoint. We will show this for Ω but to this end we first construct a tensor product for any algebraic category.

2.6 Let A, B, C be objects in the algebraic category $\underline{A}$ over $\underline{X}$. As usual, an $\underline{X}$-morphismm $g:|A\times B| \to |C|$ will be called an ($\underline{A}$-)bimorphism when the composition $g\cdot(id,\hat{b})$: $|A| \to |C|$ and $g\cdot(\hat{a},id)$: $|B| \to |C|$ underly $\underline{A}$-morphisms $A \to C$ and $B \to C$ respectively, where $\hat{k}:X \to Y$ is the constant morphism with value $\hat{k}$ in !Y!)

2.6a Proposition. There exists a functor $\otimes:\underline{A}\times\underline{A} \to \underline{A}$ and a natural transformation θ_{AB}: $|A\times B| \to |A\otimes B|$ such that for every bimorphism g: $|A\times B| \to |C|$ there is precisely one $\underline{A}$-morphism $g^{\dagger}:A\otimes B \to C$ with $|g^{\dagger}|\cdot\theta_{AB} = g$.

Proof. The algebraic functor $|\,|$ has a left adjoint < > with front adunction η. Let g_i: $|A\times B| \to |C_i|$(i in I) be the source in $\underline{X}$ of all bimorphisms with domain $|A\times B|$. This induces a source $\underline{h}_i$:$<|A\times B|> \to C_i$ (i in I) in $\underline{A}$ such that $g_i=|h_i\cdot\eta_{|A\times B|}$. Let

$$<|A\times B|>\xrightarrow{r_{AB}} A\otimes B\xrightarrow{m_i} C_i \quad (i \text{ in } I)$$

be the regular factorization in $\underline{A}$ of the source $(h_i)_I$; this defines the object $A\otimes B$. Every $\underline{A}$-morphism $k:B' \to B$ induces a bimorphism $g = |r|\cdot\eta_{|A\times B|}\cdot|id_A\times k|$ hence an $\underline{A}$-morphism $A\otimes k:A\otimes B' \to A\otimes B$ such that $|A\otimes k|\cdot|r_{AB'}|\cdot\eta_{|A\times B'|} = g$. The definition of $f\otimes B$ is similar, in fact symmetical. Clearly $\theta_{AB}=|r_{AB}|\cdot\eta$ is the required natural transformation./

A systematic discussion of this tensor product must also be delayed. We prove here only the following basic result.

2.6b Theorem. For every object A in an autonomously algebraic category $\underline{A}$, $A\otimes-$ is left adjoint to $A\Omega-$.

Proof. let us denote the composition

$$|A|\times|A\Omega C| \xrightarrow{id\times ne} |A|\times(|A|\nabla|C|) \xrightarrow{ev} |C|$$

by $k = k_{AC}$ and note that it is a bimorphism. Hence k induces the $\underline{A}$-morphism $te_{AC} : A\otimes(A\,\Omega\, C) \to C$ (tensored evaluation) such that $|te|\cdot\theta_{A,A\Omega C} = k$. Moreover, every bimorphism $g: |A\times B| \to |C|$ can be seen to factorize as $g = k\cdot|id_A \times g\dagger|$ for some $\underline{A}$-morphism $g\dagger : B \to A\,\Omega\, C$, by the definition of bimorphism. Thus for every $\underline{A}$-morphism $h: A\otimes B \to C$ we have, via the bimorphism $g = |h|\cdot\theta_{AB}$, the $\underline{A}$-morphism $h^{\theta} = (|h|\cdot\theta)\dagger$ such that the following diagram commutes:

$$\begin{array}{ccccc}
|A|\times|A\Omega C| & \xrightarrow{\theta} & |A\otimes(A\Omega C)| & \xrightarrow{|te|} & |C| \\
\uparrow{\scriptstyle |A\times h^{\theta}|} & & \uparrow{\scriptstyle |A\otimes h^{\theta}|} & & \uparrow{\scriptstyle |h|} \\
|A\times B| & \xrightarrow[\theta]{} & |A\otimes B| & \longrightarrow & \lrcorner
\end{array}$$

By 2.6a, $h = te\cdot(A\otimes h^{\theta})$ as required for the universality of te. Uniquness of h^{θ} follows as for exponential transposes./

2.7 Remark. It will be shown in the planned forthcoming paper that the last adjunction enriches to a natural isomorphism $(A\otimes B)\Omega C \simeq (A\,\Omega(B\,\Omega C)$ and that one also has related natural isomorphisms $\langle X\rangle\,\Omega\, A \simeq X\,\Phi\, A$, $A\otimes B \simeq B\otimes A$, $\langle P\rangle\otimes E \simeq E$, $A\otimes(B\otimes C) \simeq (A\otimes B)\otimes C$, the last three with coherence.

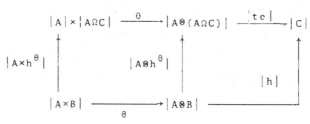

3. Reflexive objects and duality of Pontryagin type

Henceforth K wil be a fixed chosen object in the autonomously algebraic category $\underline{A}$ over $\underline{X}$. For brevity we put

$$\Delta = (-\Omega K)^*:\underline{A} \to \underline{A}^* \text{ and } \Delta^2 = \Delta^* \cdot \Delta.$$

$\underline{A}$-morphisms $A \to K$ will be called <u>characters</u> on $\underline{A}$.

<u>3.1</u> <u>Proposition</u>. The functor Δ is left adjoint to Δ^*, with front adjunction $\delta_A : A \to \Delta^* \cdot \Delta A$ such that, at set level, $\delta(a)(c) = c(a)$.

Proof. Routine verification - one uses standard properties of cartesian closedness and the fact that every monosource in $\underline{A}$ is $|\,|$-initial./

<u>3.2</u> In view of 3.1 we can define an object A to be <u>reflexive</u> (more precisely K-reflexive) if the front adjunction δ_A is an isomorphism.

<u>3.3</u> Remark. Any subcategory $\underline{B}$ of $\underline{A}$ consisting of reflexive objects is automatically dual to its image $\Delta(\underline{B})$ and furnishes a Pontryagin-type duality. Thus the problem of extending a given Pontryagin-type duality to larger categories reduces to the problem of finding more reflexive objects. Of course, the duality is usually more interesting if $\underline{B}$ as well as $\Delta(\underline{B})$ have independent explicit description.

<u>3.4</u> <u>Remark</u>. To be of interest, K itself should at least be K - reflexive. We remark that K = <P> (the "free algebra on one generator") always satisfies $\Delta K = K$ and is thus K-reflexive. This follows from the isomorphisms (2.7) $P \simeq P \Phi K \simeq \langle P \rangle \Omega K$. However, not all interesting K have the form <P> e.g. the circle group in example 1.2.

The next result furnishes a useful criterion for a function algebra $X\Phi K$ to be reflexive. In special situations these function algebras are often objects of considerable interest. The criterion says, roughly, that $X\Phi K$ is reflexive whenever the point evaluations span a "dense" subalgebra of $\Delta(X\Phi K)$. Its statement uses the morphism $\tau_X : X \to (X\Phi K)\Psi K$ (2.4).

3.5 Proposition. Suppose X is such that for every pair of characters $c,d\colon \Delta(X\Phi K) = (X\Phi K)\Omega K \to K$ the equation $|c|\cdot\tau = |d|\cdot\tau$ implies $c = d$. Then $X\Phi K$ is reflexive.

Proof. Using $\tau : X \to (X\Phi K)\Psi K$ and $\tilde{n}\tilde{e} : ((X\Phi K)\Omega K)\Omega K \to (((X\Phi K)\Psi K)\Phi K)$ (2.4) we form $\gamma = (\tau\Phi K)\cdot\tilde{n}\tilde{e}$. The stated assumptions allow us to conclude that $(\delta\cdot\gamma)(c) = c$ (at set level). It is straight forward to show $\gamma\cdot\delta = id$./

We proceed to show that certain subobjects inherit reflexiveness. The category $\underline{L}$ introduced in 3.6 will in applications be one in which a Hahn-Banach-type theorem holds.

3.6 Proposition. An object A in $\underline{A}$ is reflexive whenever there exist $w\colon A \to B$, $u,v\colon B \to C$ with B,C reflexive such that $u\cdot w = v\cdot w$ as an equalizer diagram and K is w-injective. Such w, u, v exist in particular when there exist a faithful functor $G\colon \underline{L} \to \underline{A}$ with left adjoint F such that the front adjunction at A,K reduces to identities $A \to G(F(A)) = A$, $K \to G(F(K))$ and there exist an equalizer diagram $s\cdot r = t\cdot r$ with $r\colon F(A) \to M$, $s,t\colon M \to N$ in $\underline{L}$ such that $G(M)$, $G(N)$ are reflexive and $F(K)$ is r-injective.

Proof. It follows by adjointness that for any $c\colon A \to K$ we have

$c = G(F(c))$. Hence G carries the given equalizer diagram in $\underline{L}$ to an equalizer diagram as described in the first statement. Because K is w-injective, we have Δw epimorphic and $\Delta^2 w$ monomorphic. Now we have, from the commutative diagram

$$\begin{array}{ccccc} A & \xrightarrow{w} & B & \overset{u}{\underset{v}{\rightrightarrows}} & C \\ \downarrow{\scriptstyle \delta_A} & & \downarrow{\scriptstyle \delta_B} & & \downarrow{\scriptstyle \delta_C} \\ \Delta^2 A & \xrightarrow[\Delta^2 w]{} & \Delta^2 B & \overset{\Delta^2 u}{\underset{\Delta^2 v}{\rightrightarrows}} & \Delta^2 C \end{array}$$

that $u\cdot\delta_B^{-1}\cdot\Delta^2 w = \delta_C^{-1}\cdot\delta_C\cdot u\cdot\delta_B^{-1}\cdot\Delta^2 w = \delta_C^{-1}\cdot \Delta^2 u\cdot\delta_B\cdot\delta_B^{-1}\cdot \Delta^2 w$ $=\delta_C^{-1}\cdot \Delta^2(u\cdot w)$ and similarly that $v\cdot\delta_B^{-1}\cdot\Delta^2 w = \delta_C^{-1}\cdot\Delta^2(v\cdot w)$. Equating the left hand sides we conclude that there exists precisely one $q\colon \Delta^2 A \to A$ such that $w\cdot q = \delta_B^{-1}\cdot\Delta^2 w$. Thus $\delta_B\cdot w\cdot q = \Delta^2 w\cdot\delta_A\cdot q = \Delta^2 w$ which implies $\delta_A\cdot q = id$. Moreover $w\cdot q\cdot\delta_A = \delta_B^{-1}\cdot\Delta^2 w\cdot\delta_A = w$ and so $q\cdot\delta_A = id$./

<u>Henceforth we assume the autonomous category $\underline{A}$ to be semiadditive</u> (see [HS73]).

<u>3.7 Proposition.</u> Every finite $\underline{A}$- product $\Pi_J G_j$ of reflexive objects is reflexive.

Proof. In the semiadditive category $\underline{A}$ we have for each j a commutative diagram

$$\begin{array}{ccc} \coprod_J G_j & \xrightarrow{u} & \Pi_J G_j \\ \uparrow{\scriptstyle in_j} & & \downarrow{\scriptstyle pr_j} \\ G_j & \xrightarrow[id]{} & G_j \end{array}$$

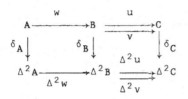

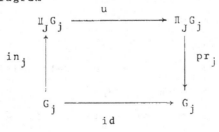

where u is an isomorphism and in_j (resp. pr_j) the canonical injection (resp. projection). The $\Delta(in_j)$ form a coproduct in $\underline{A}^*$ thus a product in $\underline{A}$ and one verifies that the $\Delta(pr_j)$ form a coproduct in $\underline{A}$, thus a product in $\underline{A}^*$. Hence the $\Delta^*\Delta(pr_j):\Delta^2\Pi_J G_j \to \Delta^2 \Pi G_j$ must form a product diagram in $\underline{A}$. The conclusion follows quickly from this./

3.8 Proposition. Suppose $\Pi_I A_i$ is an $\underline{A}$-product of reflexive objects such that for every character $c: \Pi_I A_i \to K$ there exists a finite set $J \subset I$ such that c factors through the projection $pr_J: \Pi_I A_i \to \Pi_J A_j$. Then $\Pi_I A_i$ is reflexive.

Proof. For varying c and corresponding $J = J(c)$, the morphisms $(pr_{J(c)})$ form an epi-sink, hence the $\Delta^2(pr_{J(c)})$ form a mono-source. By using this, the proof can be completed with arguments similar to those already presented./

As the following proposition will show, the requirement stated in 3.8 is satisfied in all the major special situations of interest.

3.9 Proposition. In each of the categories CvVec, CgVec and BoVec (with K the scalar field), CvAb and CgAb (with K the circle group or the discrete group of integers), CvSl and CgSl (with K the discrete two point semilattice) a character of the form $c: \Pi_I A_i \to K$ depends on finitely many coordinates (i.e. 3.8 is applicable).

Proof. For CvVec, CgVec with K = scalar field and for CvAb, CgAb with K = discrete integer group: If the statement were false, there would exist for every finite $J \subset I$ a point x in $\Pi_I A_i$ and h in $I \setminus J$ such that $c_h(x_h) = 1$, where c

is the restriction of c to the canonical subspace of $\Pi_I A_i$ formed by A_h. Hence there exists an infinite subset H of I and z in $\Pi_I A_i$ such that $z_h = 1$ for all h in H and $z_i = 0$ when $i \in I \backslash H$. Since c is continuous, we force the absurd conclusion that $c(z) = \lim_J \Sigma_J c_j (z_j) = \lim_J \text{card } J > n$ for all natural numbers n. For BoVec we form H and z as above and let $z(h)$ denote the canonical image of z_h in $\Pi_I A_i$. Then c carries the bounded set $\{Z(h) | h \in H\}$ to an unbounded set - an absurd conclusion. The proof for CgAb is given in [Lm77] and it applies also to CvAb. For CvSl the proof is given in [Hn7-] and it applies also to CgSl.

3.10 Proposition. If Δ carries products in $\underline{A}$ to products in $\underline{A}^*$ (i.e. coproducts in $\underline{A}$), then every product and every coproduct of reflexive objects is reflexive./

The requirement in 3.10 is again satisfied in several special categories, cf. [FJ72], [Lm77], [Da78a], for reasons that we have not been able to generalize satisfactorily to abstract level.

4. Applications in special situations

For the categories CvVec, CgVec, BoVec we always choose K to be the scalar field. We will show how the machinery of section 3 provides for new categorized proofs of certain known facts as well as for a discovery of many new reflexive objects. Lcs will denote the category of locally convex Hausdorff topological vector spaces and $\underline{CT}_2$ that of compact Hausdorff topological spaces.

4.1 Proposition. (a) [Bu72] Every complete space in Lcs is reflexive in CvVec.

(b) Every product of L_c-spaces is reflexive in CvVec.

Proof. (a) A complete A in Lcs admits a closed embedding $w:A \to B = \Pi_I Q_i \Phi K$ where each Q_i is in CT_2 and therefore (as shown in [Bu72] or [Bi75, th. 88] via Krein-Milman theorem) each $Q_i \Phi K$ is reflexive in CvVec. The Hahn-Banach theorem and its standard corollaries enables one to find, for each x in $B \setminus w(E)$ a character c which separates w(E) from x, hence a pair $u,v:B \to \Pi_{x \in B \setminus w(E)} K$ such that $u \cdot w = v \cdot w$ as an equalizer diagram to which 3.6 applies via 3.8, 3.9. (b) L_c-spaces are the duals of complete Lcs-spaces [BBK72], hence reflexive, so that 3.8, 3.9 applies./

Thus complete Lcs-spaces and L_c-spaces are in Pontryagin-type duality, which extends by 4.1b to a larger duality.

4.2 Proposition [FJ72] Let $G:$Lcs $\to$ CgHVec be the obvious modification functor and F its left adjoint. A space $A = G(F(A))$ is reflexive in CgHVec provided that F(A) is a complete space.

Proof. F(A) admits an embedding $r:F(A) \to M = \Pi_I F(Q_i \Phi K)$ with each Q_i compact and therefore each $Q_i \Phi K$ refelexive in CgHVec, as in the proof of 4.1. Again as in 4.1 we find a pair $s,t:M \to N$ such that $s \cdot r = t \cdot r$ is an equalizer diagram in Lcs. The result follows by applying 3.6, 3.8, 3.9., since G, F have the required properties./

As shown in [FJ72], the spaces described in 4.2 actually account for all reflexive objects in CgHVec. This suggests that the conditions imposed on the abstract results in section 3 were not prohibitively restrictive.

Before discussing applications to BoVec, we should point out that the concept of reflexiveness employed in [Hg71], [Hg77] has the dual of a BoVec object lying in Lcs, equippped with a polar topology. We will call this notion "polar reflexiveness" to distinguish it from reflexiveness as used here.

4.3 Propositon. Let $G:Lcs \to BoVec$ be the obvious underlying functor and F its left adjoint. A space $A = G(F(A))$ is reflexive in Bovec provided that there exist reflexive objects $B_i = G(F(B_i)$ and a closed embedding $r:F(A) \to \Pi_I F(B_i)$ in Lcs.

Proof. Since G, F have the properties required in 3.6, the result follows immediately via 3.8, 3.9. and the Hahn-Banach theorem, as explained in the proof of 4.1./

Known reflexive spaces in BoVec include all reflexive Banach spaces (in classical sense) e.g. all L_p-spaces with $1 < p < \infty$, all spaces of the form G(E) where E is a complete bornological Schwartz space (see [Be7-]) and G as in 4.3, and of course all duals of these spaces, which include all Silva spaces. By choosing the B_i in 4.3 from among these spaces, the class of known reflexive spaces in Bovec becomes considerably extended by 4.3. Also, since reflexiveness implies polar reflexiveness (see [Be7-]), we have thus found many new polar reflexive spaces as well. Let us also point out that 3.6 provides a categorized proof for the well known fact that a closed subspace of a reflexive Banach space is reflexive (cf. [DS58, II.3.23].

For several applications of section 3 which extend considerably the duality studied in [HMS74], we refer the reader to [HN7-].

In connection with CvAb we should again warn that the reflexiveness studied in [Ka48], [Ka50], [Ve75] differs from our "canonical" reflexiveness in that these authors use the compact-open topology for dual groups. Thus 3.8, 3.9 gives a new result for CvAb, namely that a product of reflexive groups is refelexive. The penetrating analysis in [Bi76] showed that $Y \Phi K$ is reflexive in CvAb when Y is a connected and locally path connected topological space with finite fundamental group. The subcategory of such Y in Cv is not closed under coproducts, but since $- \Phi K : \underline{X}^* \to \underline{A}$ carries coproducts in $\underline{X} = \underline{Cv}$ to products in CvAb we have at once the following extension of Binz's result.

4.4 Proposition If Y is a coproduct in Cv of connected and locally path connected topological spaces with finite fundamental group, then $Y \Phi K$ is reflexive in CvAb./

Section 3 provides also a simple alternative proof for the fact [Lm77] that a product of reflexive objects in CgHAb is reflexive.

For related recent results supplementary to this paper, see [Da78a], [Da78b], [Da7-a] [Da7-b].

REFERENCES

[BBK72] E. Binz, H.P. Butzmann and K. Kutzler, Über den c-Dual eines topologische Vektorräumes, Math. Z. 127 (1972) 70-74.

[Be7-1 R. Beattie, On the mackey convergence structure, (preprint).

[BH76] B. Banaschewski and H. Herrlich, Subcategories defined by implications, Houston J. of Math. 2 (1976) 149-171.

[Bi75] E. Binz, Continuous convergence on C(X), Springer Lecture Notes in Math. 469 (1975).

[Bi76] E. Binz, Charaktergruppe von Gruppen von S^1-wertigen stetigen Funktionen, Categorical Topology (Proceedings of Conference at Mannheim, July 1975), Springer Lecture Notes in Math. 540 (1976) 43-91.

[BN76] B. Banaschewski and E. Nelson, Tensor produts and Bimorphisms, Canad. Math. Bull. 19 (1976) 385-401.

[Bu72] H.-P. Butzmann, Über die c-Reflexivität von $C_c(X)$, Comm. Math. Helv. 49 (1972) 92-101.

[Bu77] H.-P. Butzmann, Pontrjagin-Dualität für topologische Vektorräume, Arch. Math. 28 (1977) 632-637.

[Da78b] B. Day, On Pontryagin duality, Glasgow Math. J. (to appear).

[Da78a] B. Day, Note on duality in Kelleyspace products, Bull. Austral. Math. Soc. 19 (1978) 273-275.

[Da7-a] B. Day, On topological modules and duality (preprint)

[Da7-b] B. Day, Note on duality in a closed category (preprint).

[Ds58] N. Dunford and J.T. Schwartz, Linear Operators Part I, Interscience, New York (1958).

[Du70] E. Dubuc, Kan Extensions, in Enriched Category Theory, Springer Lecture notes in Math. 145 (1970).

[Fi59] H.R. Fischer, Limesräume, Math. Ann. 137 (1959) 269-303.

[FJ72] A. Frölicher and H. Jarchow, Zur Dualitäts theorie Kompakt erzeugter und lokalkonvexer Vektorräume, Comm. Math. Helv. 47 (1972) 289-310.

[He74a] H. Herrlich, Regular categories and regular functors, Canad. J. Math. 26 (1974) 709-720.

[He74b] H. Herrlich, Cartesian closed topological categories Math. Colloq. Univ. Cape Town 9 (1974) 1-16.

[Hg71] H. Hogbe-Nlend, Theorie des Bornologie et applications, Springer Lecture Notes Math. 213 (1971).

[Hg77] H. Hogbe-Nlend, Bornologies and Functional Analysis, North Holland, Amsterdam (1977).

[HMS74] K.H. Hofmann, M. Mislove and A. Stralka, The Pontryagin Duality of Compact 0-dimensional Semilattices and its applications, Springer Lecture Notes in Math. 396 (1974).

[HN77] H. Herrlich and L.D. Nel, Cartesian closed topological hulls, Proc. Amer. Math. Soc. 62 (1977) 215-222.

[HN7-] S.S. Hong and L.D. Nel, Reflexive topological semilattices and duality (preprint).

[HS73] H. Herrlich and G.E. Strecker, Category Theory, Allyn and Bacon, Boston (1973).

[Ka48] S. Kaplan, Extensions of the Pontryjagin duality I: Infinite products, Duke Math. J. 15 (1948), 649-658.

[Ka50] S. Kaplan, Extensions of the Pontrjagin dualtiy II: Direct and inverse sequences, Duke Math. J. 17 (1950) 419-435.

[Li66] F.E.J. Linton, Autonomous Equational Categories, Journal of Mathematics and Mechanics 15 (1966) 637-642.

[Lm77] W.F. Lamartin, Pontryagin duality for products and coproducts of abelian k-groups, Rocky Mountain J. Math. 7 (1977) 725-731.

[Ne76] L.D. Nel, Cartesian closed topological categories, Categorical Topology (Proceedings of Conference at Mannheim, July 1975) Springer Lecture Notes in Math. 540 (1976) 439-451.

[Ne77] L.D. Nel, Cartesian closed coreflective hulls, Questiones Math. 2 (1977) 269-283.

[Ve75] R. Venkataraman, Extensions of Pontryagin Duality, Math. Z. 143 (1975) 105-112.

This research was aided by an NSERC grant. I thank Sung Sa Hong for valuable conversations.

EXISTENCE AND APPLICATIONS OF MONOIDALLY CLOSED STRUCTURES IN TOPOLOGICAL CATEGORIES

H.-E. Porst and M.B. Wischnewsky

We construct internal hom-functors and monoidal structures by means of cone-factorization techniques; this is used in particular to lift the corresponding data along semitopological functions. Then we show, how these results may be applied to obtain duality theories, Morita theorems, and general Galois theorems.

Introduction

It is well known that a lot of fundamental mathematical constructions can be done in an arbitrary category if it is equipped with a monoidally closed structure. Hence our aim is to show, that in very general situations we can lift internal hom-functors and monoidal structures, which deliver monoidally closed structures in many concrete cases. Since we can carry out our construction simply by cone-factorization techniques, this paper is also a contribution to the idea that the notion of cone-factorization is a fundamental for category theory. Obviously one can vary the factorizations and modify the cones in our construction in order to get similar results but other applications. We want to mention that following the same idea G. Greve has independently obtained similar results.

1. Extensions and coextensions

1.0 Notations

Let $S: \underline{A} \to \underline{X}$ be a functor. A S-cone is a triple $(X, \psi, D(\underline{A}))$ where X is an $\underline{X}$-object, $D(\underline{A}): \underline{D} \to \underline{A}$ is an $\underline{A}$-diagram ($\underline{D}$ may be void or large) and $\psi: \Delta X \to SD(\underline{A})$ is a functorial morphism (Δ denotes the "constant functor into the functor category). We shall abbreviate often $(X, \psi, D(\underline{A}))$ by ψ .

$\underline{\text{Cone}}(S)$ denotes the class of all S-cones. If $\underline{D} = \underline{1}$

(i.e. the one point category) then Ψ is called a <u>S-morphism</u> denoted by (A,a) where A is an <u>A</u>-object and a: X → SA is an <u>X</u>-morphism. The dual notions are <u>S-cocone</u> and <u>S-comorphism</u>. The corresponding classes of S-morphisms, S-cocones and S-comorphisms are denoted by <u>Mor</u>(S), <u>Co-cone</u>(S), <u>Co-Mor</u>(S). <u>Epi</u>(S) denotes the class of all S-epimorphisms (A,e: X → SA) i.e. the class of all S-morphisms (A,e) with the property: for all <u>A</u>-morphisms p,q: A ⇉ B the equation (Sp)e = (Sq)e implies p = q. <u>Iso</u>(S) denotes the class of all S-<u>iso-morphisms</u> i.e. of all objects (A,a) in <u>Mor</u>(S) with a an isomorphism in <u>X</u>. <u>Init</u>(S) denotes the class of all S-<u>initial cones</u> i.e. of all <u>A</u>-cones $\alpha: \Delta A \to D(\underline{A})$ such that for any <u>A</u>-cone $\beta: \Delta B \to D(\underline{A})$ and any <u>X</u>-morphism $x: SB \to SA$ with $(S\beta) = (S\alpha)\cdot(\Delta x)$ there exists a unique <u>A</u>-morphism $a: B \to A$ with $\beta = \alpha(\Delta a)$ and $Sa = x$.

Let $S: \underline{A} \to \underline{X}$ be a functor and $\underline{Iso}(S) \subset \Phi \subset \underline{Mor}(S)$. Let $SD(\underline{A}) \xleftarrow{\varphi} D(\underline{X}) \xrightarrow{\psi} X$ be a functorial chain with $(D(\underline{A}), \varphi: D(\underline{X}) \to SD(\underline{A}))$ being pointwise in Φ. We call a functorial chain of this type a Φ-functorial double-morphism over the index-category <u>D</u>. If $\underline{D} = \underline{1}$ we speak about a Φ-double-morphism. The class of all Φ-double-morphisms is denoted by $\underline{Mor}^2(S)$. Let $\gamma(\Phi) \subset \underline{Mor}^2(S)$ be a subclass.

<u>1.1 DEFINITION</u>: (<u>Semifinal</u> <u>extensions</u>)

Let (φ, ψ) be a Φ-functorial double-morphism. Let $(A, e: X \to SA) \in \underline{Mor}(S)$. A functorial morphism $\alpha: D(\underline{A}) \to \Delta A$ is called a <u>Φ- extension of</u> (φ, ψ) <u>by</u> (A,e) iff $(S\alpha)\varphi = (\Delta e)\psi$.
Let $\alpha': D(\underline{A}) \to \Delta A'$ be a Φ-extension of (φ, ψ) by (A',e'). A morphism between the given Φ-extensions is an <u>A</u>-morphism $a: A \to A'$ such that $e' = (Sa)e$ and $\alpha' = (\Delta a)\alpha$. This defines the category of all Φ-extensions of (φ, ψ).

An initial object $\alpha: D(\underline{A}) \to \Delta A$ with $(A,e) \in \Phi$ in this category is called a Φ-<u>semifinal</u> <u>extension</u> of (φ, ψ).

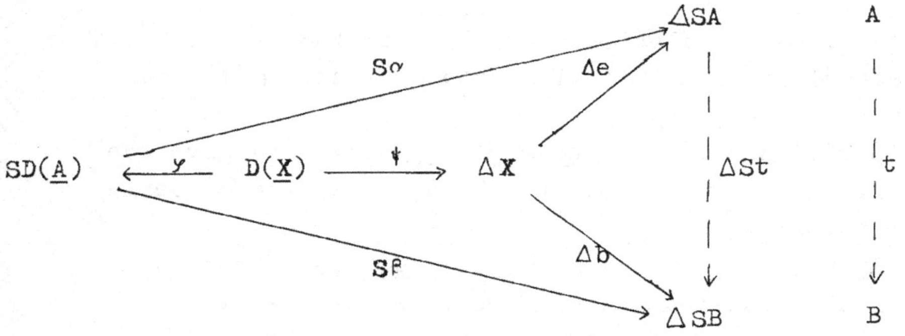

The functor S is called a $(\Upsilon(\Phi),\Phi)$-<u>semifinal</u> <u>functor</u> if every Φ-functorial double morphism which is pointwise in $\Upsilon(\Phi)$ has a Φ-semifinal extension.

<u>1.2 DEFINITION</u>: (<u>Semiinitial</u> <u>coextension</u>)

Let $(D(\underline{A}), \varphi: \Delta X \to SD(\underline{A}))$ be a S-functorial morphism. Let $(A, e: X \to SA) \in \Phi$ and $\alpha: \Delta A \to D(\underline{A})$ with $\varphi = (S\alpha)(\Delta e)$. (A,e,α) is called a $(\Upsilon(\Phi),\Phi)$ <u>semiinitial</u> <u>coextension</u> of $(D(\underline{A}),\varphi)$ iff

SI 1) for all $(B, SB \xleftarrow{f} Y \xrightarrow{x} X) \in \Upsilon(\Phi)$ and all $\beta: \Delta B \to D(\underline{A})$ with $(S\beta)(\Delta f) = \varphi(\Delta x)$ there exists a unique morphism $a: B \to A$ with $\beta = \alpha(\Delta a)$ and $(\Delta e)(\Delta x) = (\Delta Sa)(\Delta f)$

SI 2) for any morphism $t: A \to A$ the equations $e = (St)e$ and $\alpha = \alpha(\Delta t)$ imply $t = id_A$.

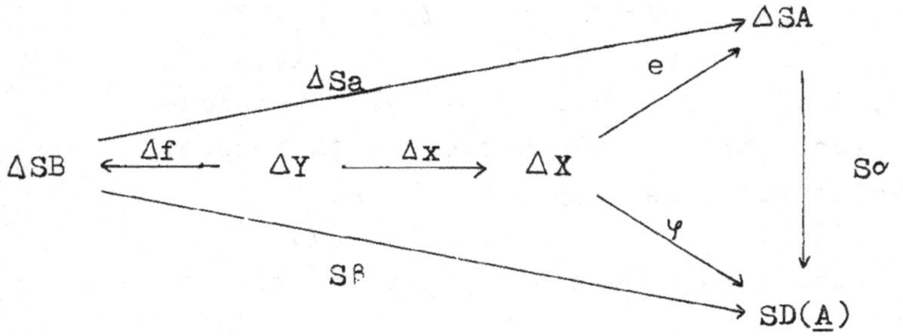

A functor S is called a $(\Upsilon(\Phi),\Phi)$-semiinitial functor if every S-functorial morphism φ: $(D(\underline{A}),\ \varphi:\Delta X \to SD(\underline{A}))$ has a $(\Upsilon(\Phi),\Phi)$-semiinitial coextension.

1.3 Example: Semitopological functors (cp [18],[7],[17],[19]) are semifinal and semiinitial functors. Φ and $\Upsilon(\Phi)$ can be chosen in different ways.

(a) $\Phi = \underline{\mathrm{Mor}}(S)$ and $\Upsilon(\Phi) = \underline{\mathrm{Co\text{-}Mor}}(S)$; these data give the semifinal resp. semiinitial characterization of semitopological functors.

(b) $\underline{\mathrm{Iso}}(S) \subset \Phi \subset \underline{\mathrm{Mor}}(S)$ and $\Upsilon(\Phi) \subset \underline{\mathrm{Mor}}^2(S)$; these data give the locally-orthogonal resp. left-extension characterization of semitopological functors.

1.4 Remarks: While semifinal extensions in the sense of 1.3(a) are essentially unique, semiinitial coextensions are not; they are unique, if we assume $\Phi = \underline{\mathrm{Epi}}(S)$, which can be done in case of semitopological functors. Hence in the following we always refer to this situation, if we use the terms semifinal resp. semiinitial without specifying the data Φ and $\Upsilon(\Phi)$.

2. Evaluation cones and section cocones

It is well known, that in the category Top a monoidally closed structure is given as follows (P: Top $\to$ Set is the underlying functor):

For topological spaces X and Y

$X \otimes Y$ - is $PX \times PY$ with the final topology with respect to all sections $PX \to PX\times PY$, $x \mapsto (x,y)$ and $PY \to PX\times PY$, $y \mapsto (x,y)$

$[X,Y]$ is $\underline{\mathrm{Top}}(X,Y)$ with the topology of pointwise convergence, i.e. the initial topology with respect to all evaluation maps $\underline{\mathrm{Top}}(X,Y) \to PY$, $f \mapsto f(x)$.

The unit of this structure is the one point codiscrete space.

Let now $\underline{V} = (\underline{V}_o, -\otimes-, \underline{V}(-,-), I)$ be a (symmetric) monoidally closed category, $\underline{A}$ a $\underline{V}$-category, and $P: \underline{A} \to \underline{V}$ a $\underline{V}$-functor. (For the different steps of our construction we need only parts of this assumption, but in order not to complicate the presentation, we always refer to this situation).

For given $\underline{A}$-objects A and B we now form the evaluation-cone

$$\eta = (\eta_x^{AB})_{x \in \underline{V}_o(I,PA)}$$

$$\eta_x^{AB} := \underline{A}(A,B) \xrightarrow{P_{AB}} \underline{V}(PA,PB) \xrightarrow{\underline{V}(x,PB)} \underline{V}(I,PB) \simeq PB$$

and the section-cocone

$$\sigma = (\sigma_c^{AB})_{c \in \underline{V}_o(I,PA) \cup \underline{V}_o(I,PB)}$$

where for $c \in \underline{V}_o(I, PA)$

$$\sigma_c^{AB} := PB \simeq I \otimes PB \xrightarrow{c \otimes PB} PA \otimes PB$$

and for $c \in \underline{V}_o(I,PB)$

$$\sigma_c^{AB} := PA \simeq PA \otimes I \xrightarrow{PA \otimes c} PA \otimes PB$$

3. Lifted tensorproducts and inner hom-functors

In the following we only use semifinal extensions of section cocones and semiinitial (resp. cosemifinal) coextensions of evaluation cones in order to lift tensorproducts and inner hom-functors. Obviously one can vary the factorizations (see section 1) and modify the cones in order to get similar results but other applications.

3.1 PROPOSITION: <u>If all section-cocones allow a P-semifinal extension, then these extensions define a functor</u> $-\Box-: \underline{A} \times \underline{A} \to \underline{A}$, <u>which is called the semifinal tensorproduct on</u> $\underline{A}$.

<u>Proof</u>: If $e: PA \otimes PB \to PC$ is the semifinal extension of σ^{AB}, define $A \Box B := C$ and extend this to a functor by means of the universal property of semifinal extensions.

3.2 PROPOSITION: If all evaluation-cones allow a P-cosemifinal (resp. semiinitial) coextension, then these coextensions define a functor

$[-,-]: \underline{A}^{op} \times \underline{A} \to \underline{A}$ (resp. $(-,-): \underline{A}^{op} \times \underline{A} \to \underline{A}$)

which is called the cosemifinal (resp. semiinitial) hom-functor on $\underline{A}$.

Proof: Similar to the proof of 3.1.

Obviously the hom-functors of 3.2. in general fail to be inner hom-functors in the usual sense, since we only have morphisms

$m: P[A,B] \to \underline{A}(A,B)$ (resp. $e: \underline{A}(A,B) \to P(A,B)$)

instead of the desired isomorphisms.

Now it is an immediate consequence from the definitions that if m is a P-semiinitialcoextension of η with m an isomorphism, then m^{-1} is a P-cosemifinal coextension of η and vice versa. Hence if P allows the construction of $[-,-]$ and all the coextensions occuring in this construction are isomorphisms, then the same is true for $(-,-)$ and vice versa; moreover in this situation we have $[-,-] = (-,-)$. Let us call P a functional functor, if it has the property just described. For the compatibility of this notion with the notion of a functional inner hom-functor in the sense of [1] see the forthcoming paper [12].

3.3 PROPOSITION: Every Set-based $(\underline{E},\underline{M})$-topological functor is functional.

Proof: The evaluation cones are in $\underline{M}$.

3.4 THEOREM: Let $P: \underline{A} \to \underline{V}$ be a functional $\underline{V}$-functor, where the underlying functor $U: \underline{V} \to \underline{Set}$ is faithful and the evaluation cones η are U-initial. Then the semifinal tensorproduct, if it exists, is a tensormultiplication for the cosemifinal (= semiinitial) inner hom-functor on $\underline{A}$.

Proof: If P_0 is faithful, this is an immediate consequence

of a $\underline{V}$-enriched version of the results of [1] (cp. [12]). Otherwise we have to construct unit and counit for an adjunction $- \square A \dashv [A,-]$ as follows:
The unit $\xi = (\xi_B)$ we get by the universal cosemifinal property from the following commutative diagram, where (m,λ) is the cosemifinal extension of η and (e,μ) is the semifinal extension of σ, and where r is the morphism with underlying map $b \mapsto \mu_b$ $(b \in \underline{V}_o(I,PB))$ (η is U-initial!)

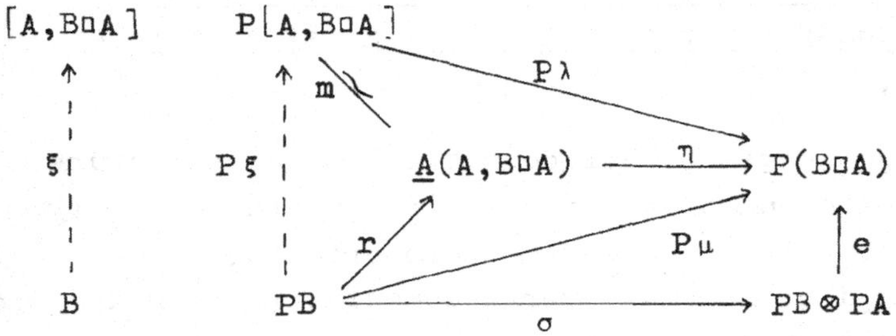

The counit $\zeta = (\zeta_B)$ we get correspondingly from the commutative diagram

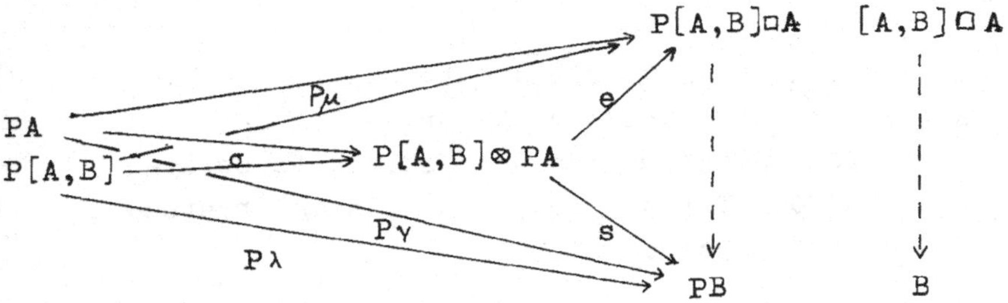

where the notation is the same as in the first diagram, and where $(\gamma_x) = (Um(x))$ $(x \in \underline{V}_o(I,P[A,B]) = UP[A,B])$ and

$$s := P[A,B] \otimes PA \xrightarrow{P_{AB} \circ m \otimes PA} \underline{V}(PA,PB) \otimes PA \xrightarrow{i} PB$$

where i corresponds to the identity on $\underline{V}(PA,PB)$ by the adjunction of the tensormultiplication on $\underline{V}_o$.

<u>3.5 REMARK</u>: The additional assumptions on U in the above theorem are fulfilled in many cases, e.g.:

(i) $\underline{V}$ = <u>Set</u>

(ii) U: $\underline{V} \to$ Set is $(\underline{E},\underline{M})$-topological
(iii) U: $\underline{V} \to$ Set is regular (cp [6], [13]) and P_o is faithful

Further investigations of our construction show, that at least in the Set-based $(\underline{E},\underline{M})$-topological case we get a symmetric monoidally closed structure on $\underline{A}$ [12]. Hence by the general results of [2] we have as a corollary:

3.6 PROPOSITION: Let P: $\underline{A} \to$ Set be $(\underline{E},\underline{M})$-topological. Then with respect to the semifinal tensorproduct and the cosemifinal inner hom-functor on $\underline{A}$ there is an exponential-law: $[A\Box B,C] \simeq [A,[B,C]]$

3.7 Examples: In Top we get by our construction the structure described at the beginning of section 2; in Lim, the category of limitspaces, we get the so-called S-tensorproduct; in the category R-mod of modules over a commutative ring R we also get the ordinary monoidally closed structure.

4. Applications

In this final section we want to indicate how the monoidal closed structure introduced in the preceding paragraph may be applied. Since we can only sketch these applications we have to refer the reader who is interested in details to the references.

While the application to function algebras (4.1) depends heavily on the special structure we just have described, the further results (4.2) and (4.3) are available in arbitrary monoidal closed categories; therefor we conclude this section by an interpretation of the notion of a finite right A-object, which lies at the heart of (4.2) and (4.3), in a concrete topological situation with semifinal momoidal closed structure.

4.o Monoids, actions, and algebras

Given a monoidal closed category $(\underline{C}, -\otimes-, [-,-], I)$, MonC (resp. cMonC, ComonC, Hopf MonC, cHopf MonC) denotes the category of monoids (resp. commutative monoids, Comonoids, Hopfmonoids, Commutative Hopfmonoids) in $\underline{C}$, where a monoid in $\underline{C}$ is to be understood as in [11], and a Hopfmonoid in $\underline{C}$ is defined in the same way as a Hopfalgebra in k-mod is defined in [16]. A left A-object in $\underline{C}$ with respect to some $A \in$ MonC is a pair (C, ∇_C) where ∇_C is a left action in the sense of [11]. Dually one defines left A-coobjects with respect to some $A \in$ ComonC. ${}_A\underline{C}$ (resp. $\underline{C}_A$, ${}_A\underline{C}_B$, ${}^A\underline{C}$, $\underline{C}^A$) denotes the category of left A-objects (resp. right A-objects, A-B-biobjects, left A-coobjects, right A-coobjects). Similarly one has the notion of groups, cogroups, or co-Hopfmonoids in $\underline{C}$.

The categories $\underline{C}$, ${}_A\underline{C}$, ${}_A\underline{C}_B$ etc. are related by the following two functors, provided $\underline{C}$ has equalizers and coequalizers (cp. [15]):
Let $A, B \in$ MonC. Then any $M \in \underline{C}_A$ defines (by means of a coequalizer construction) a functor

$$M \otimes_A - : {}_A\underline{C} \to \underline{C}$$

and any $M \in {}_B\underline{C}$ defines (by means of an equalizer construction) a functor

$${}_B[M,-] : {}_B\underline{C} \to \underline{C}$$

where ${}_B[M,M]$ again becomes a monoid.
If now $M \in {}_B\underline{C}_A$, these functors may be interpreted as functors between categories ${}_A\underline{C}$, $\underline{C}_B$, ${}_A\underline{C}_B$ etc. in various ways, such that they become adjoint.
Using these facts, one gets for example for $B \in$ MonC, $P \in {}_B\underline{C}$ (with $A := {}_B[P,P]$, $Q := {}_B[P,B]$) a morphism
$\tilde{g}: Q \otimes_A P \to B$ corresponding to id_Q, and a morphism
$\tilde{f}: P \otimes_B Q \to A$ corresponding to $id_P \otimes_B \tilde{g}$
By means of these morphisms one gives the following definitions:

4.o.1 DEFINITION: P is called

(a) finite over B, if $\tilde{f}$ is an isomorphism

(b) finitely generated projective over B, if

$$P \otimes Q \xrightarrow{\text{can}} P \otimes_B Q \xrightarrow{f} A$$

is rationally surjective ($h \in \underline{C}(A,B)$ is rationally surjective if $\underline{C}(I,h)$ is surjective)

(c) faithfully projective over B if P is finite and $\tilde{g}$ is an isomorphism

(d) a progenerator over B if P is finitely generated projective over B and

$$Q \otimes P \xrightarrow{\text{can}} Q \otimes_A P \xrightarrow{g} B$$

is rationally surjective.

Furthermore we use the obvious notion of a ring R in $\underline{C}$ and of R-algebras in $\underline{C}$ (cp. [14]); the category of R-algebras in $\underline{C}$ is denoted by $\underline{R\text{-Alg}(C)}$. Underlying functors, if needed, are denoted by $| \ |$.

Befor stating generalizations of classical results to the categories just introduced, we will indicate how these categories arise "in nature" (besides the obvious ones). So assume that there is a Gelfand-type duality, as discussed in (4.1)

$$C: \underline{U} \simeq \underline{B}^{op} \subset \underline{R\text{-Alg}(V)}^{op}$$

In well behaved situations $\underline{U}$ and $\underline{B}$ are momoidal (closed) and C is a monoidal functor. Then obviously C lifts to a duality

$$C^*: \underline{\text{MonU}} \simeq \underline{\text{ComonB}}^{op}$$

or similarly to a duality between groups in $\underline{U}$ and cogroups in $\underline{B}$. As far as commutativity is involved, one is led to the notion of (Co)-Hopfmonoids in this context.

4.1 Function algebras and Gelfand duality (cp [14])

Let $P: \underline{V} \to \underline{\text{Set}}$ be an $(\underline{E},\underline{M})$-topological category. By (3.6) $\underline{V}$ is monoidally closed with respect to the semifinal tensorproduct. It is shown by the authors in [14] that with respect to this structure the category $\underline{\text{R-Alg}(V)}$ becomes a $\underline{V}$-category, which moreover is cotensored. So especially we know

4.1.1 PROPOSITION:

(i) *For any* $A, B \in \underline{\text{R-Alg}(V)}$ *the continuous* R-*algebramorphisms form a* $\underline{V}$-*object* $\underline{A}(A,B)$

(ii) *For any* $V \in \underline{V}$ *there is an object* $C_R(V) \in \underline{\text{R-Alg}(V)}$ *such that*

- *the functor* $\underline{A}(-,R)\colon \underline{\text{R-Alg}(V)}^{op} \to \underline{V}$ *has as a left adjoint the functor* C_R
- $|C_R(V)| = [V,R]$

Hence the semifinal monoidal closed structure canonically gives a suitable notion of function algebras in topological categories. The objects $[A,R]$ are called spectral objects. Having this at hand one gets a general description of dualities of Gelfand (or spectral) type:

4.1.2 THEOREM ([14]): *Let* $\underline{V}_o$ *additionnaly have a proper* $(\underline{E},\underline{M})$-*factorization such that the unit* $\epsilon\colon 1_{\underline{V}} \to \underline{A}(C_R-,R)$ *is pointwise in* $\underline{E}$. *Then the full subcategories* $\underline{B}$ *of all function algebras in* $\underline{\text{R-Alg}(V)}$ *and* $\underline{U}$ *of all spectral objects in* $\underline{V}$ *are the largest subcategories which are dual equivalent under* C_R *and* $\underline{A}(-,R)$:

$$\begin{array}{ccc} \underline{B} & \hookrightarrow & \underline{\text{R-Alg}(V)} \\ \wr\!\wr & & C_R \uparrow \downarrow \underline{A}(-,R) \\ \underline{U}^{op} & \hookrightarrow & \underline{V}^{op} \end{array}$$

As results one gets a lot of old and new dualities for topological categories (cp [8], [14]).

4.2. Morita theorems (cp [3], [10], [15])

With the notation of (4.0) we first give the following definition:

4.2.1 DEFINITION: A *Morita-context* (or a *set of pre-equivalence data*) (A,B,P,Q,f,g) consists of

(a) $A,\ B \in \underline{\mathrm{Mon}\mathbf{C}}$

(b) $P \in {}_A\underline{\mathbf{C}}_B$ and $Q \in {}_B\underline{\mathbf{C}}_A$

(c) $f \in {}_A\underline{\mathbf{C}}_A(P\otimes_B Q, A)$ and $g \in {}_B\underline{\mathbf{C}}_B(Q\otimes_A P, B)$

such that the following diagrams commute

$$\begin{array}{ccc} P\otimes_B Q\otimes_A P & \xrightarrow{f\otimes_A \mathrm{id}_P} & A\otimes_A P \\ {\scriptstyle \mathrm{id}_P\otimes_B g}\downarrow & & \downarrow\wr \\ P\otimes_B B & \xrightarrow{\ \sim\ } & P \end{array} \qquad \begin{array}{ccc} Q\otimes_A P\otimes_B Q & \xrightarrow{g\otimes_B \mathrm{id}_Q} & B\otimes_B Q \\ {\scriptstyle \mathrm{id}_Q\otimes_A f}\downarrow & & \downarrow\wr \\ Q\otimes_A A & \xrightarrow{\ \sim\ } & Q \end{array}$$

Now we can state the following general Morita theorem:

4.2.2 THEOREM ([15]): *Let* $(\underline{\mathbf{C}}_o, -\otimes-, \underline{\mathbf{C}}(-,-), I)$ *be a monoidally closed category and* (A,B,P,Q,f,g) *a Morita-context in* $\mathbf{C}$. *Then the following assertions hold:*

(a) f *and* g *are rationally surjective iff* P *is faithfully projective over* A *and* Q *faithfully projective over* B.

(b) *If* f (resp. g) *is rationally surjective, then* f (resp. g) *is an isomorphism.*

(c) *If* f *and* g *are rationally surjective, then*

(i) *one has canonical isomorphisms*

$$\begin{array}{llll} P & \cong & {}_B[Q,B] & \text{in } {}_A\underline{\mathbf{C}}_B \\ Q & \cong & {}_A[P,A] & \text{in } {}_B\underline{\mathbf{C}}_A \\ A & \cong & {}_B[Q,Q] & \text{in } {}_A\underline{\mathbf{C}}_A \text{ and in } \underline{\mathrm{Mon}\mathbf{C}} \\ B & \cong & {}_A[P,P] & \text{in } {}_B\underline{\mathbf{C}}_B \text{ and in } \underline{\mathrm{Mon}\mathbf{C}} \end{array}$$

(ii) $P\otimes_B - : {}_B\underline{\mathbf{C}} \to {}_A\underline{\mathbf{C}}$ and $Q\otimes_A - : {}_A\underline{\mathbf{C}} \to {}_B\underline{\mathbf{C}}$ *are inverse equivalences of categories.*

4.2.3 COROLLARY: Let $\underline{C}$ be an $(\underline{E},\underline{M})$-topological category (over Set) with the semifinal monoidal structure. Let (A,B,P,Q,f,g) be a Morita-context in $\underline{C}$. If f and g are surjective then $P\otimes_B - : {}_B\underline{C} \to {}_A\underline{C}$ and $Q\otimes_A - : {}_A\underline{C} \to {}_B\underline{C}$ are inverse equivalences of categories.

4.3. Galois theorems in topological categories ([9])

With the notation of (4.o) we first give the following definition:

4.3.1 DEFINITION: Let $S \in \underline{\text{Mon}C}$, $A \in \text{Hopfmon}C$, and $\alpha \in \underline{C}(S,S\otimes A)$.

(a) S is called A-co-object-monoid if

  (i) $(S,\alpha) \in \underline{C}^A$

  (ii) $\alpha \in \underline{\text{Mon}C}\ (S,S\otimes A)$

(b) Let (S,α) be a A-co-object-monoid

  Let $\gamma := (\nabla_S \otimes id_A)(id_S \otimes \alpha) \in \underline{\text{Mon}C}(S\otimes S,S\otimes A)$.

  S is called A-Galois over I if

  (i) S is faithfully projective over I, and

  (ii) γ is an isomorphism in $\underline{C}$.

Example: Let $\underline{C}$ = K-Mod (K is a field) and let S be a finite dimensional field-extension of K. Then S is KG-Galois over K (where KG is the group-algebra over a finite group G) iff S is a separable Galois-extension with Galois-group G in the classical sense.

Let $A \in \underline{\text{Hopfmon}C}$, A finite over I and $(M,\alpha) \in \underline{C}^A$. The fix-object M^{A^*} is defined as the equalizer of the following pair:

$$M^{A^*} \to M \underset{id\otimes\eta}{\overset{\alpha}{\rightrightarrows}} M\otimes A$$

4.3.2 THEOREM ([9]): Let $A,B_1,B_2 \in c\,\text{Hopfmon}C$, A finite over I, $c_i \in \underline{\text{Hopfmon}C}(A,B_i)$ a retraction in $\underline{C}$.

S $\in$ cMonC, and S A-Galois over I. Then the following assertions hold:

(a) $S^{A^*} = I$

(b) With $T := S^{B^*}$ S is $T \otimes B$-Galois over T; the canonical morphism $S^{B^*} \rightarrow S$ is a section in ${}_T\mathrm{C}$.

(c) $B_1^* \subset B_2^*$ $\quad S^{B_2^*} \subset S^{B_1^*}$

$B_1^* = B_2^*$ $\quad S^{B_1^*} = S^{B_2^*}$

By the above theorem we have an injective order reversing mapping from the lattice of those subhopf-monoids of A^* which are sections in C, into the lattice of the submonoids of S. By specializing the data one obtains all known Galois theorems over various categories.

4.4 The category of quasi-complete T_2-barreled vector spaces

Denote by QcHb the subcategory of all quasi-complete Hausdorff barreled vector spaces in the category of all Hausdorff, locally convex topological vector spaces over $\mathbb{K}$ ($\mathbb{K} = \mathbb{R}$ or $K = \mathbb{C}$). Then QcHb is a co-semi-topological category, precisely it is a monocoreflective subcategory of the category of all Hausdorff, locally convex topological vector spaces. Then QcHb is a symmetric, monoidal closed category with respect to the semifinal tensorproduct.

$\underline{QcHb}(E \otimes_s F, G) = \{f \in \mathrm{Bilin}(E \times F, G);\ f \text{ separately continuous}\}$

(cp Bourbaki: Espaces vectoriels Topologiques)

h $\in$ QcHb(A,B) is

(i) a monomorphism $\Leftrightarrow$ h is injective

(ii) an epimorphism $\Leftrightarrow$ Bi(h) is dense in B

(iii) an equalizer $\Leftrightarrow$ h is injective, $\mathrm{Bi}(h) = \overline{\mathrm{Bi}(h)}$ and A has the coarsesttopology for which h is continuous

(iv) h is coequalizer $\Leftrightarrow$ Bi(h) is strictly dense in B and h is open.

The problem in general and in particular in this example is to characterize intrinsically the "finite" objects. S. Ligon [9] gave the following characterization (with notation as in Bourbaki):

<u>4.4.1 PROPOSITION</u>: <u>Let</u> $E \in$ <u>QcHb</u>. <u>Assume</u> <u>that</u> E'_β <u>and</u> $L_\beta(E,E)$ <u>are</u> <u>barreled</u>. <u>Then</u> E <u>is</u> <u>finite</u> <u>iff</u> E <u>is</u> <u>finite</u> <u>dimensional</u>.

Then we obtain for example:
For a C^∞-manifold M there are equivalent:
(i) $\mathcal{E}(M)$ is finite in <u>QcHb</u>
(ii) M is a finite set
(iii) $\mathcal{E}(M)$ is finite dimensional.

<u>References</u>

[1] BANASCHEWSKI, B., NELSON, E.: Tensorproducts and bimorphisms, Canad. Math. Bull. <u>19</u>, 385-4o2 (1976)

[2] EILENBERG, S., KELLY, G.M.: Closed categories, Proc. Conf. Cat. Alg., La Jolla 1965, Springer, Berlin-Heidelberg-New York 1966, 421-562

[3] FISCHER-PALMQUIST, I., PALMQUIST, P.H.: Morita contexts of enriched categories, Proc. Amer. Math. Soc. <u>5o</u>, 55-6o (1975)

[4] GREVE, G.: M-geschlossene Kategorien, Dissertation, Fernuniversität Hagen 1978

[5] HERRLICH, H.: Topological functors, Gen. Top. Appl. <u>4</u>, 125-142 (1974)

[6] HERRLICH, H.: Regular categories and regular functors, Can. J. Math. <u>26</u>, 7o9-72o (1974)

[7] HOFFMANN, R.-E.: Semi-identifying lifts and a generalization of the duality theorem for topological functors, Math. Nachr., <u>74</u>, 297-3o7 (1976)

[8] HONG, S.S., NEL, L.D.: Dualities of algebras in cartesian closed topological categories, Preprint

[9] LIGON, S.: Galoistheorie in monoidalen Kategorien, Algebraberichte <u>35</u>, Uni-Druck München 1978

[1o] LINDNER, H.: Morita-Äquivalenzen von Kategorien über einer geschlossenen Kategorie, Dissertation, Universität Düsseldorf 1973

[11] MAC LANE, S.: Categories for the working mathematician, Springer, Berlin-Heidelberg-New York 1971

[12] PORST, H.-E.: Functional Hom-functors in enriched categories, to appear

[13] PORST, H.-E.: On underlying functors in general and topological algebra, manuscripta math. 2o, 2o9-225 (1977)

[14] PORST, H.-E., WISCHNEWSKY, M.B.: Every topological category is convenient for Gelfand duality, manuscripta math. 25, 169-2o4 (1978)

[15] PAREIGIS, B.: Non-additive ring and module theory III, to appear: Publ. Math. Debrecen

[16] SWEEDLER, M.E.: Hopf Algebras, W.A. Benjamin, New York 1969

[17] THOLEN, W.: Semitopological functors I, to appear in J. Pure and Appl. Algebra

[18] TRNKOVA, V.: Automata and categories, Lecture Notes Computer Sci. 32, 138-152 (1975)

[19] WISCHNEWSKY, M.B.: A lifting theorem for right adjoints, to appear

Hans-E. Porst
Manfred B. Wischnewsky

Fachbereich Mathematik

Universität Bremen

28oo Bremen 33

Federal Republic of Germany

CONNECTION PROPERTIES IN TOPOLOGICAL CATEGORIES AND RELATED TOPICS

by

Gerhard Preuß, Berlin

§ 0 Introduction

In 1966, HERRLICH [4] introduced E-compact and E-regular spaces generalizing concepts of ENGELKING and MROWKA ([3], [10]). These concepts were fundamental for starting the theory of epireflective subcategories. At the same time the author [11] examined the class of ℝ-connected spaces, i.e. spaces on which each real-valued continuous function is constant. The striking similarities to the class of all connected spaces led to the concept of E-connectedness. In 1967, the author discovered the relations between E-compactness and E-connectedness [12]. In 1970, he used the concept of E-connectedness in order to prove the relationship between separation-axioms and non-connectedness of topological spaces [13]. In 1971, COLLINS [2] used generalized connection properties for the factorization theory of maps and thus generalized well-known results of WHYBURN [19] and MICHAEL [9]. 1972 SALICRUP and VAZQUEZ [15] began their investigations on connection categories. (In 1967, the author [12] introduced the similiar concept of disjoint component class K and developed the theory of locally K-spaces.) In 1974, STRECKER [18] found further factorizations of maps by means of connection properties. In 1975, ARHANGEL'SKII and WIEGANDT [1] characterized topological connectednesses and disconnectedness (in the sense of the author) axiomatically. In 1977, PREUSS [14] introduced relative connectednesses and disconnectednesses in topological categories (in the sense of HERRLICH [7]) and characterized them axiomatically. It turned out that a relative disconnectedness in a topological category $\underline{C}$ is nothing else than an extremal epireflective (i.e. quotient reflective) subcategory of $\underline{C}$. Thus, in topological categories the concept of a relative disconnectedness coincides with the concept of a disconnection subcategory in the sense of SALICRUP and VAZQUEZ [16].

Now, let us show that the categorical generalization of connectedness and disconnectedness leads to factorizations of morphisms in "nice" categories corresponding to the concordant-dissonant factorization of COLLINS [2] and to the submonotone-superlight factorization of STRECKER [18] respectively. Especially, the results are studied in topological categories. By the way, we will obtain characterizations of $\underline{E}$-extendable quotient maps by means of $\underline{E}$-quasicomponents.

§ 1 Preliminaries

1.1 Definitions: Let $\underline{C}$ be a category and let $\underline{P}$ be a subclass of its object class $|\underline{C}|$. Then we define the operators C, D and Q as follows:

$$C\underline{P} = \{X \in |\underline{C}| : \text{every } \underline{C}\text{-morphism } f:X\to P \text{ is constant for each } P \in \underline{P}\}$$

$$D\underline{P} = \{X \in |\underline{C}| : \text{every } \underline{C}\text{-morphism } f:Y\to X \text{ is constant for each } Y \in C\underline{P}\}$$

QP = {X ∈|C|: for any two distinct morphisms $\alpha,\beta:Z\to X$ there exist some P ∈ P and some $f:X\to P$ such that $f\circ\alpha \neq f\circ\beta$}.

The elements of CP are called P-connected.

The elements of DP are called totally P-disconnected.

The elements of QP are called totally P-separated.

From now on, let C denote a complete, cocomplete, wellpowered and cowellpowered category.

If M is a class of C-morphisms which contains all C-isomorphisms and if E is a class of epimorphisms in C which contains all C-isomorphisms then C is called an (E,M)-category provided that E and M are closed under composition and every C-morphism f has a unique (E,M)-factorization (i.e. (1) f = m∘e for some e ∈ E and for some m ∈ M, and (2) f = m∘e = m'∘ e' where e, e' ∈ E and m, m' ∈ M implies that there exists an C-isomorphism j such that the diagram

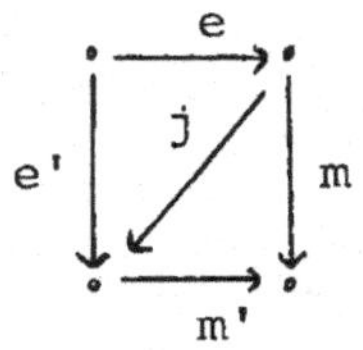

commutes).

For every class E of C-epimorphisms we denote by M (E) the class of all morphisms f in C such that whenever

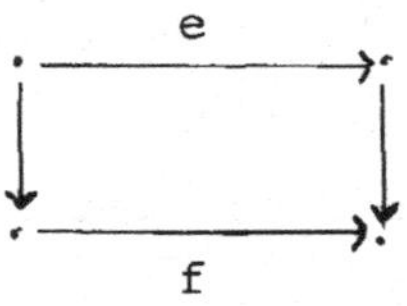

commutes and e ∈ $\underline{E}$ then there exists a morphism l such that

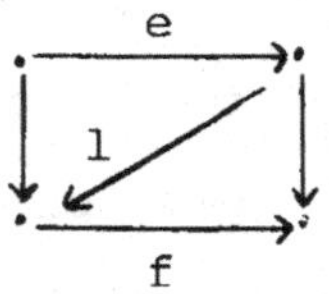

commutes.

Especially, let $\underline{K}$ be a class of objects in $\underline{C}$ and let us consider the class $E(\underline{K})$ of all $\underline{K}$-extendable $\underline{C}$-epimorphisms (i.e. an $\underline{C}$-epimorphism e: $X \to Y$ belongs to $E(\underline{K})$ if and only if for each $K \in \underline{K}$ and each $f: X \to K$ there exists a $\bar{f}: Y \to K$ with $f = \bar{f} \circ e$). Then $M(E(\underline{K}))$ is the class of all <u>$\underline{K}$-perfect morphisms</u> (cf. [5]).

<u>1.2. Theorem:</u> (HERRLICH [6]): For any class $\underline{E}$ of $\underline{C}$-epimorphisms the following conditions are equivalent:

(1) $\underline{C}$ is an $(\underline{E}, M(\underline{E}))$-category.

(2) (a) $\underline{E}$ is closed under composition and contains all $\underline{C}$-isomorphisms

(b) $\underline{E}$ is closed under pushouts in $\underline{C}$, i.e. if

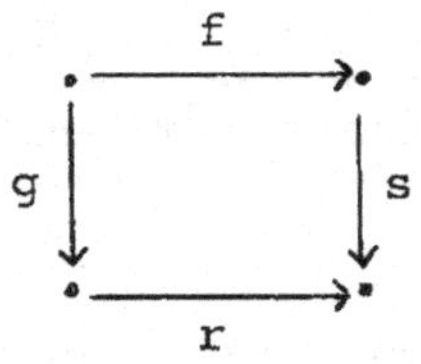

is a pushout square in C and f ∈ E then r ∈ E.

(c) E is closed under cointersections (=multiple pushouts).

1.3. Remarks: ① Especiallly, the results of 1.2. may be applied to topological categories [7], because every topological category is complete, cocomplete, wellpowered and cowellpowered.

② Let P ⊂ |C| where C is a topological category, and (X,ζ) ∈ |C|. A ⊂ X is called P-connected with respect to X, if for each P ∈ P and for each C-morphism f: X → P, f|A is constant. (X,ζ) may be decomposed into maximal subsets M such that M is P-connected with respect to X, the so-called P-quasicomponents (cf.[14]). Further, (X,ζ) ∈ QP if and only if the P-quasicomponents of (X,ζ) consist at most of a singleton. Correspondingly, (X,ζ) may be decomposed into maximal P-connected subsets, the so-called P-components of X (a subset A of X is called P-connected provided that (A,ζ_A) is P-connected, where ζ_A denotes the initial structure on A with respect to the inclusion map i: A → X). Obviously, (X,ζ) ∈ DP if and only if the P-components of X consist at most of a singleton.

§ 2 The (P-extendable extremal epi, relative P-light)-factorization and the (DP-extendable extremal epi, P-superlight)-factorization

2.1. Definition: Let $\underline{P} \subset |\underline{C}|$, and let us denote by $E^*(\underline{P})$ the class of all $\underline{P}$-extendable extremal epimorphisms in $\underline{C}$. The elements of $M(E^*(\underline{P}))$ are called relative $\underline{P}$-light morphisms, and the elements of $M(E^*(D\underline{P}))$ are called $\underline{P}$-superlight morphisms.

2.2. Remarks: ① Every $\underline{P}$-perfect $\underline{C}$-morphism is a relative $\underline{P}$-light morphism.

② $M(E^*(\underline{P}))$ and $M(E^*(D\underline{P}))$ contain all monomorphisms and are closed under composition, pullbacks, multiple pullbacks, products and left-cancellation(cf.[5; Prop.5,p.190])

2.3. Proposition: Let $\underline{P} \subset |\underline{C}|$. Then $E^*(\underline{P})$ fulfills the conditions (2) (a)-(c) in 1.2.

Proof: Since $\underline{C}$ is complete and wellpowered, it is an (extremal epi, mono)-category, and therefore $\underline{C}$ fulfills the (extremal epi, mono)-diagonalization property. Thus, applying the dual of [8;34.2,p.256], the class $\underline{E}$ of all extremal epimorphisms in $\underline{C}$ has the properties (2) (a)-(c). On the other hand, it follows immediately from the definitions that the class of all $\underline{P}$-extendable $\underline{C}$-morphisms has the desired properties too.

2.4. Theorem: $\underline{C}$ is an $(E^*(\underline{P}), M(E^*(\underline{P})))$-category and an $(E^*(D\underline{P}), M(E^*(D\underline{P})))$-category. Thus, each $\underline{C}$-morphism has a unique ($\underline{P}$-extendable extremal epi, relative $\underline{P}$-light)-factorization and a unique ($D\underline{P}$-extendable extremal epi, $\underline{P}$-superlight)-factorization.

Proof: Apply 1.2. and 2.3.

2.5. Theorem: Let $\underline{C}$ be a topological category and let $f:(X,\zeta)\to(Y,\mu)$ be a quotient map*) in $\underline{C}$. Then the following conditions are equivalent for each $\underline{P} \subset |\underline{C}|$:

(1) f is $\underline{P}$-extendable.

(2) f is $Q\underline{P}$-extendable.

(3) $\underline{Q}(f)$ is an isomorphism, provided that $\underline{Q}$ denotes the extremal epireflector $\underline{C} \to \underline{K}$ where $\underline{K}$ is the full and isomorphism-closed subcategory of $\underline{C}$ defined by $|\underline{K}| = Q\underline{P}$.

(4) For each $y \in Y$, $f^{-1}(y)$ is $\underline{P}$-connected with respect to X.

(5) For each subset A of Y, which is $\underline{P}$-connected with respect to Y, $f^{-1}[A]$ is $\underline{P}$-connected with respect to X.

(6) For each $\underline{P}$-quasicomponent K of Y, $f^{-1}[K]$ is a $\underline{P}$-quasicomponent of X.

---

*) that means $f: X \to Y$ is surjective and μ is the final $\underline{C}$-structure with respect to $((X,\zeta),f,Y)$. In topological categories, quotient maps coincide with extremal epimorphisms.

Proof: Let us define a full subcategory $\underline{A}$ of $\underline{C}$ by $|A| = \underline{P}$,
and let us denote by $Q_{\underline{C}}\underline{A}$ the extremal epireflective hull
of $\underline{A}$ in $\underline{C}$. Then the object class $|Q_{\underline{C}}\underline{A}|$ of $Q_{\underline{C}}\underline{A}$ coincides
with $Q\underline{P}$ (cf.[14;3.4.]).

Thus, (1)-(3) are equivalent according to [8;37.9. and 37 F.].

(6) $\Rightarrow$ (4): Since each $y \in Y$ is contained in a
$\underline{P}$-quasicomponent K_y, the $\underline{P}$-quasicomponent $f^{-1}[K_y]$ contains
$f^{-1}(y)$. Thus, $f^{-1}(y)$ is $\underline{P}$-connected with respect to X.

(4) $\Rightarrow$ (1): Let h: $X \to P$ be a $\underline{C}$-morphism with $P \in \underline{P}$.
$\bar{h}$: $Y \to P$ is defined by the property that it makes the diagram

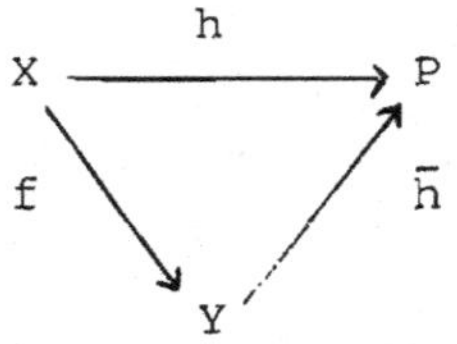

commutative. Thus, $\bar{h}$ is well-defined (let $f(x) = f(x')$; then
x and x' belong to $f^{-1}(f(x))$ which is $\underline{P}$-connected with
respect to X, and consequently, $h|f^{-1}(f(x))$ is constant, i.e.
$h(x) = h(x'))$, and it is a $\underline{C}$-morphism according to the
definition of final structure on Y.

(1) $\Rightarrow$ (5): Let $A \subset Y$ be $\underline{P}$-connected with respect to Y and
let h: $X \to P$ be a $\underline{C}$-morphism with $P \in \underline{P}$. Further, let
$a,b \in f^{-1}[A]$.

There exists a $\underline{C}$-morphism $\bar{h}:Y \to P$ with $\bar{h} \circ f = h$ because
f is $\underline{P}$-extendable. Then $\bar{h}[A]$ is $\underline{P}$-connected with respect to
X (cf.[14;2.1.(3)(b)])and contains $h(a)=\bar{h}(f(a))$ and

$h(b) = \bar{h}\,(f(b)$. This implies $h(a) = h(b)$. Thus, $f^{-1}\,[A]$ is $\underline{P}$-connected with respect to X.

(5) ⇒ (6): Let K be a $\underline{P}$-quasicomponent of Y. According to (5), $f^{-1}\,[K]$ is $\underline{P}$-connected with respect to X, and therefore, $f^{-1}\,[K]$ is contained in a $\underline{P}$-quasicomponent K' of X. Thus, $f\,[K']$ is $\underline{P}$-connected with respect to Y (cf.[14;2.1.(3)(b)]) and contains the $\underline{P}$-quasicomponent $K = f\,[f^{-1}[K]]$ (f is surjective!). Consequently, $K = f\,[K']$. This implies $f^{-1}[K] = f^{-1}[f[K']] \supset K'$, and immediately, $K' = f^{-1}\,[K]$.

<u>2.6. Remarks:</u> ① Let $\underline{P} \subset |\underline{C}|$, and let $\underline{A}$ be a full subcategory of $\underline{C}$ defined by $|\underline{A}| = \underline{P}$. The epireflective hull of $\underline{A}$ in $\underline{C}$ is denoted by $R_{\underline{C}}A$. According to [8;37.9 and 37 F.], (2) (resp. (3)) in 2.5. may be replaced by

(2') f is $|R_{\underline{C}}A|$-extendable (= $R_{\underline{C}}\underline{A}$-extendable).

(resp. (3') $\underline{R}$ (f) is an isomorphism provided that $\underline{R}$ denotes the epireflector $\underline{C} \to R_{\underline{C}}A$).

② Let $\underline{C}$ be the category of topological spaces and continuous maps, and let $\underline{P} \subset |\underline{C}|$ be defined by $|\underline{P}| = \{D_2\}$, where D_2 denotes the two-point discrete space. Then the equivalence of (3) and (4) in the above theorem is nothing else than a theorem of COLLINS on concordant quotient maps (cf.[2; theorem 4,p.588] and [14;3.4.(2)]).

<u>2.7. Theorem:</u> Let $\underline{C}$ be a topological category, and let $\underline{P}$ be a subclass of $|\underline{C}|$. A $\underline{C}$-morphism $f\colon (X,\zeta) \to (Y,\mu)$ is a relative $\underline{P}$-light morphism if and only if for each $y \in Y$ and for each

P-quasicomponent K of X, $f^{-1}(y) \cap K$ consists at most of a singleton.

Proof: 1) " ⇐ ". Let the diagram in C

$$\begin{array}{ccc} A & \xrightarrow{e} & B \\ g\big\downarrow & & \big\downarrow h \\ X & \xrightarrow[f]{} & Y \end{array}$$

be commutative, where e is a P-extendable quotient map. $k: B \to X$ is defined by $k \circ e = g$. Thus, k is well-defined (let $e(a) = e(a') = b$; then $e^{-1}(b)$ is P-connected with respect to A; thus, $g[e^{-1}(b)]$ is P-connected with respect to X, and it is contained in a P-quasicomponent K of X; $K \cap f^{-1}(y)$ with $y = f(g(a)) = h(e(a)) = h(e(a')) = f(g(a'))$ consists at most of a singleton and contains $g(a)$ and $g(a')$; consequently, $g(a) = g(a')$), and according to the definition of final structure on B, it is a C-morphism. Since e is a C-epimorphism, it follows from the commutative diagram above that $fok = h$.

2) " ⇒ ". Let us define an equivalence relation R on X as follows:

$x_1 R x_2 \Leftrightarrow f(x_1) = f(x_2)$ and x_1, x_2 belong to the same P-quasicomponent of X.

Let $\omega: X \to X|R$ be the projection. Now we endow $X|R$ with the final structure with respect to $((X,\zeta),\omega,X|R)$. Hence ω is a P-extendable quotient map. $h: X|R \to Y$, defined by $h \circ \omega = f$, is well-defined, and it is a C-morphism. Thus, the diagram

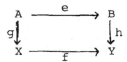

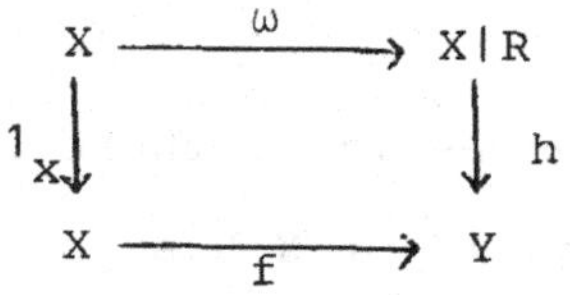

is commutative, and there exists a $\underline{C}$-morphism k such that the diagram

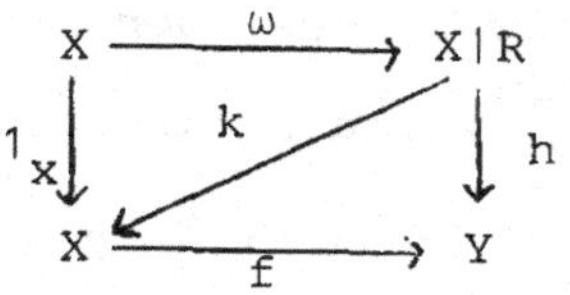

is commutative. Consequently, $k \circ \omega = 1_X$, i.e. ω is an injective map.This implies,that f has the desired property.

<u>2.8. Remarks:</u> ① Let $\underline{C}$ be the category of topological spaces and continuous maps, and let $\underline{P} \subset |\underline{C}|$ be defined by $|\underline{P}| = \{D_2\}$, where D_2 denotes the two-point discrete space. Then the relative $\underline{P}$-light maps coincide with the dissonant maps by COLLINS [2], for in this case the $\underline{P}$-quasicomponents may be identified with the quasicomponents in the sense of HAUSDORFF. Thus, the unique factorization of each continuous map f into a $\underline{P}$-extendable quotient map followed by a relative $\underline{P}$-light continuous map is nothing else than the (concordant quotient, dissonant)-factorization of f by COLLINS.

② Obviously, the factorization of a morphism f in a topological category given in part 2) of the proof of 2.7.is a ($\underline{P}$-extendable quotient,relative $\underline{P}$-light)-factorization

(1. ω is $\underline{P}$-extendable according to 2.5.(4).

2. If h would not be relative $\underline{P}$-light, there would exist a $y \in Y$ and a $\underline{P}$-quasicomponent K of $X|R$ such that $h^{-1}(y) \cap K \supset \{z_1, z_2\}$ with $z_1 \neq z_2$ (cf. 2.7.). Since $f[\omega^{-1}(z_1)] = h[\omega[\omega^{-1}(z_1)]] = h(z_1) = h(z_2) = h[\omega[\omega^{-1}(z_2)]] = f[\omega^{-1}(z2)] = y$, $\omega^{-1}(z_1)$ and $\omega^{-1}(z_2)$ would be contained in $\omega^{-1}[K] \cap f^{-1}(y)$. Since 1. is valid, $\omega^{-1}[K]$ is a $\underline{P}$-quasicomponent of X, and therefore $\omega^{-1}(z_1)$ and $\omega^{-1}(z_2)$ would belong to the same equivalence class which contradicts $z_1 \neq z_2$. According to 2.4. this factorization is unique (up to an isomorphism).

③ Similarly, we obtain from the (D$\underline{P}$-extendable extremal epi, $\underline{P}$-superlight)-factorization of a $\underline{C}$-morphism f the (submonontone quotient, superlight)-factorization of STRECKER provided that $\underline{C}$ is the category top of topological spaces and continuous maps and $\underline{P} = \{D_2\}$, for in this case the $\underline{P}$-components coincide with the components and the following two theorems are valid for each subclass $\underline{P}$ of $|\underline{\text{Top}}|$, especially for $\underline{P} = \{D_2\}$.

Theorem 1: Let $X, Y \in |\text{Top}|$, and let $f: X \to Y$ be a quotient map. Then the following are equivalent:

(1) f is D$\underline{P}$-extendable.

(2) For each $y \in Y$, $f^{-1}(y)$ is contained in a $\underline{P}$-component of X.

(3) For each $\underline{P}$-connected subset A of Y, $f^{-1}(Y)$ is contained in a $\underline{P}$-connected subset of X.

(4) For each $\underline{P}$-component K of Y, $f^{-1}[K]$ is a $\underline{P}$-component of X.

Theorem 2: A continuous map $f:(X,\underline{X}) \to (Y,\underline{Y})$ is a $\underline{P}$-superlight map if and only if for each $y \in Y$ and for each $\underline{P}$-component K of X, $f^{-1}(y) \cap K$ consists at most of a singleton.

Problem: Find nice conditions under which $\underline{P}$-components coincide with $\underline{P}$-quasicomponents in the more general setting of topological categories.

References

[1] ARHANGEL'SKII, A.V. and R. WIEGANDT. Connectednesses and disconnectednesses in topology. General Topology and Appl.5 (1975), 9-33.

[2] COLLINS, P.J. Concordant mappings and the concordant-dissonant factorization of an arbitrary continuous function. Proc.Amer.Math.Soc.27 (1971),387-591.

[3] ENGELKING, R. and S. MROWKA. On E-compact spaces Bull.Acad.Pol.Sci.Math.Astr.Phys.6 (1958),429-436.

[4] HERRLICH, H. E-kompakte Räume.Math.Z.96 (1967),228-255.

[5] ——————. A generalization of perfect maps. General Topology and its Relations to Modern Analysis and Algebra III, Prague 1971,187-191.

[6] ——————. Perfect subcategories and factorizations. Colloquia mathematica societatis Janos Bolyai. 8.Topics in Topology, Keszthely (Hungary).1972,387-403.

[7] ——————. Cartesian closed topological categories. Mathematics Coll.IX,University of Cape Town (1974),1-16.

[8] HERRLICH, H. and G.E. STRECKER. Category Theory. Allyn & Bacon, Boston (1973).

[9] MICHAEL, E. Cuts.Acta Math.111 (1964), 1-36.

[10] MROWKA, S. On universal spaces.Bull.Acad.Pol.Sci.4 (1956),479-481.

[11] PREUSS, G. I-Räume und lokale I-Räume. Diplomarbeit, FU Berlin (1966).

[12] ————. Über den E-Zusammenhang und seine Lokalisation. Diss. FU Berlin (1967).

[13] ————. Trennung und Zusammenhang. Monatsh. Math. 74 (1970), 70-87.

[14] ————. Relative connectednesses and disconnectednesses in topological categories. Quaest. Math. 2 (1977), 297-306.

[15] SALICRUP, G. and R. VAZQUEZ. Categorîas de conexión, An. Inst. Mat. Univ. Nat. Autónoma México 12 (1972), 47-87.

[16] ————————————. Connection and Disconnection. Preprint.

[17] ————————————. Reflectivity and connectivity in topological categories. Preprint.

[18] STRECKER, G.E. Component properties and factorizations. Math. Centre Tracts 52 (1974), 123-140.

[19] WHYBURN, G.T. Analytic Topology. Amer. Math. Soc. Collog. Publ. 28 (1963).

ON PROJECTIVE AND INJECTIVE OBJECTS IN SOME TOPOLOGICAL CATEGORIES.

T.G. Raghavan and I.L. Reilly, University of Auckland, New Zealand.

1. INTRODUCTION. In this paper we characterize the projective and injective objects of certain categories of topological spaces. It is well-known that in the category of Boolean spaces (=zerodimensional compact Hausdorff spaces) the projective objects are exactly the extremally disconnected spaces. Gleason [5] discovered that these spaces are the projective objects in the category of compact Hausdorff spaces. The most general results in this direction were obtained by Banaschewski [1,2]. The principal contributions in this direction of categorical topology were made by Flaschmeyer [4], Gleason [5], Henrikson-Jerison [7], Iliadis [10], Liu [11], Mioduszewski and Rudolf [12,13,14], Ponomarev [15], Purisch [16], Rainwater [17], Strauss [21]. One may refer to the section on projective objects in a survey article by Herrlich [9].

1.1 DEFINITION. Let C be a category and let $\mathcal{P}$ be a class of C-morphisms. An object P of C is called *$\mathcal{P}$-projective* if for each morphism $f:P \to Y$ and for each $\mathcal{P}$-morphism $g:X \to Y$ there exists a morphism $h:P \to X$ such that the following diagram

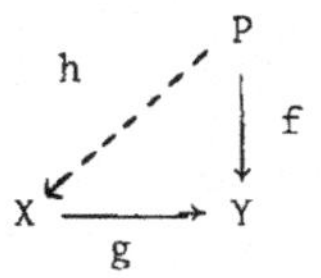

commutes. A morphism f is called *$\mathcal{P}$-essential* if $f \in \mathcal{P}$ and $fg \in \mathcal{P}$ implies $g \in \mathcal{P}$ for each morphism g in C . A morphism $f:P \to X$ is called a *$\mathcal{P}$-projective cover* of X if P is $\mathcal{P}$-projective and f is $\mathcal{P}$-essential.

1.2 DEFINITION. If the $\mathcal{P}$-morphisms in the Definition 1.1 above are epimorphisms i.e. if $\mathcal{P}$ = the class of epimorphisms of C , then P is called *projective*. An object Q of C is called *injective* if and only if it is projective in the dual category of C .

Liu [11] proved that in the category of H-closed spaces and continuous maps the projective objects are the finite topological spaces and the injective objects are singletons. It is also well-known that in the category of compact Hausdorff spaces the injective objects are precisely the retracts of Tychonoff cubes. We modify Liu's technique to obtain characterizations of projective and injective objects in some other categories.

A topological space is called an *HP-space* if and only if it is a Hausdorff space in which every G_δ set is open; i.e., it is both a Hausdorff and P space. A topological space is called *HP-closed* if and only if it is an HP-space and is closed in every embedding into an HP-space. Our attention was drawn to the study of HP-closed spaces when we studied minimal HP-spaces. Indeed every minimal HP-space is HP-closed but not conversely [18]. An HP-space X is HP-closed if and only if every open covering admits a countable subcollection whose union is dense in X [18]. In general, if C is a family of subsets of X, then C is an *almost cover* of X if the collection of closures of members of C covers X. We say C has an *almost subcover* C' of X if $C' \subset C$ and C' is an almost cover of X.

We show that an object in the category of HP-closed spaces and continuous maps is projective if and only if it is a countable topological space, and injective if and only if it is a singleton. We obtain a similar characterization of projective and injective objects in the category of Urysohn P-closed spaces. In the category of Urysohn-closed spaces, the projective objects are finite spaces and the injective objects are singletons.

Let H(1) stand for the property of being Hausdorff and first countable. An H(1)-space is H(1)-closed if and only if it is feebly compact (that is, each countable open filterbase has an adherent point). This result is used in [3] to show that projective objects in the category of H(1)-closed spaces are finite spaces, and injective objects are singletons.

2. PROJECTIVE AND INJECTIVE OBJECTS.

2.1 THEOREM: In the category of HP-closed spaces and continuous maps, X is projective if and only if X is countable.

PROOF: If X is countable, then clearly X is projective.

Conversely if X is projective, we will show that X is discrete, then it will follow X is countable, because X is HP-closed.

Suppose X is not discrete, then there exists a point 'a' in X which is not isolated. Let I be the uncountable set $[0,\Omega)$ and let I have the discrete topology. Let $I^* = [0,\Omega] = I \cup \{\Omega\}$ be its one-point

Lindelof extension, i.e. the neighbourhoods of Ω are sets with countable complements. Let $Y = X - \{a\}$. Let π be an abstract point not in $Y \times I^*$. Define $A = (Y \times I^*) \cup \{\pi\}$. Let $Y \times I^*$ have the product topology. The basic neighbourhoods of π are $(Q \times I) \cup \{\pi\}$ where Q is an open deleted neighbourhood of a, so that $Q \cup \{a\}$ is open in X and $a \notin Q$. We make the following observations. The construction determines a topology T on A which is a P-space.

The space (A,T) is Hausdorff. For points in $Y \times I^*$ can always be separated by disjoint neighbourhoods, because Y and I^* are Hausdorff already. If $(y,\alpha) \in Y \times I^*$ with $\alpha \neq \Omega$ we note that $y \neq a$. Then there are disjoint open sets Q_1 and Q_2 in X such that $y \in Q_1$ and $a \in Q_2$. We note that $U = Q_1 \times \{\alpha\}$ is an open set containing (y,α) and $V = ((Q_2 - \{a\}) \times I) \cup \{\pi\}$ is an open neighbourhood of π and $U \cap V = \phi$. Consider the point $(y,\Omega) \in Y \times I^*$. We note that $U_1 = Q_1 \times (I^* - M)$ and $V_1 = ((Q_2 - \{a\}) \times I) \cup \{\pi\}$ (where Q_1 and Q_2 are described above and M is a countable subset of I) are disjoint open neighbourhoods of (y,Ω) and π in A. Thus A is Hausdorff.

The space A is HP-closed. Let C be an open cover of A. We want to find a countable dense subsystem of C. We may without loss of generality assume C to consist of basis open sets. For $\pi \in A$, we can choose $W_\pi \in C$ such that $\pi \in W_\pi$ and W_π is of form $(Q \times I) \cup \{\pi\}$ where Q is a deleted neighbourhood of a. For each $x \in Y$ consider the set $\{x\} \times I^*$. We choose $W_0(x) \in C$ such that $(x,\Omega) \in W_0(x)$. We can write $W_0(x) = U_0(x) \times [\alpha_x,\Omega]$. For $\beta = 1,2,3,\ldots$ and $\beta < \alpha_x$ there are open neighbourhoods $W_\beta(x)$ of (x,β) where $W_\beta(x) = U_\beta(x) \times \{\beta\}$, and $U_\beta(x)$ is an open neighbourhood of x in Y. Set $U(x) = \{U_\beta(x) \mid 0 \leq \beta < \alpha_x\}$; $U(x)$ is a G_δ-set and hence open. $U(x)$ is open in Y also. Consider the collection $C_1 = \{\{U(x) \mid x \in Y\}, Q \cup \{a\}\}$. C_1 is a subcover of X.

Let the family $\{\{U(x_i) : i=1,2,\ldots, x_i \in Y, Q \cup \{a\}\}$ be a countable subsystem of C_1 which is an almost subcover of X. We define a countable family F as follows: $F = \{(Q \times I) \cup \{\pi\}$, and for $i=1,2,\ldots$ $U(x_i) \times [\alpha_{x_i},\Omega]$, $\{U(x_i) \times \{\beta\} : \beta < \alpha_{x_i}\}\}$. We claim $c\ell(\cup F) = A$. Suppose $z \notin \cup F$ and $z \in A$. Then $z \neq \pi$. So $z = (x,\alpha)$ where $x \notin V = \cup\{U(x_i) : i=1,2,\ldots\}$. Let Q' be an open subset of Y containing x. Then $Q' \times \{\alpha\}$ is an open set containing (x,α), and $Q' \cap (V \cup (Q \cup \{a\})) \neq \phi$.

Two cases arise. Case (i) :- Suppose $Q' \cap U(x_j) \neq \phi$ for some $U(x_j)$ in the collection $\{U(x_i) : i=1,2,3,\ldots\}$. If $\alpha < \alpha_{x_j}$ then $(Q' \times \{\alpha\}) \cap (U(x_j) \times \{\alpha\}) \neq \phi$ where $U(x_j) \times \{\alpha\}$ is a member of F . Therefore $(Q' \times \{\alpha\}) \cap (UF) \neq \phi$. If $\alpha \geqq \alpha_{x_j}$, then $(Q' \times \{\alpha\}) \cap ([U(x_j)] \times [\alpha_{x_j},\Omega]) \neq \phi$. If $\alpha \geqq \alpha_{x_j}$, then $(Q' \times \{\alpha\}) \cap (U(x_j) \times [\alpha_{x_j},\Omega]) \neq \phi$, and therefore $(Q' \times \{\alpha\}) \cap (UF) \neq \phi$.

Case (ii) :- Suppose $Q' \cap (Q \cup \{a\}) \neq \phi$. Then $Q' \cap Q \neq \phi$, so that $(Q' \times \{\alpha\}) \cap [(Q \times I) \cup \{\pi\}] \neq \phi$, and thus $(Q' \times \{\alpha\}) \cap (UF) \neq \phi$.

Thus in each case $(x,\alpha) \in c\ell(UF)$ and $c\ell(UF) = A$, so that A is HP-closed.

We define a new topology T' on $A = (Y \times I^*) \cup \{\pi\}$. Let $Y \times I^*$ have the product topology, and the basic T' neighbourhoods of π be of the form $(Q \times I^*) \cup \{\pi\}$, where Q is an open deleted neighbourhood of a . Now $i : (A,T) \to (A,T')$ is continuous. Hence (A,T') is HP-closed, since (A,T) is. Now define $f : X \to (A,T')$ such that $f(a) = \pi$, and $f(y) = (y,\Omega)$ for each $y \in Y$. Then $f^{-1}[(Q \times I^*) \cup \{\pi\}] = Q \cup \{a\}$ is open. Again $f^{-1}[Q(y) \times (I^* - M)] = Q(y)$, and so is open if $Q(y)$ denotes an open neighbourhood of y . Thus f is continuous.

Since X is projective, there is a continuous function g such that $i \circ g = f$. Clearly $g(a) = \pi$, otherwise $f(a) \neq \pi$. Similarly $g(y) = (y,\Omega)$ for each $y \in Y$. Then $g^{-1}[(Q \times I) \cup \{\pi\}] = \{a\}$, which is not open, and this violates the continuity of g . Thus X is discrete, as desired. □

By a *generalized P-closed space* we mean a P-space (not necessarily Hausdorff) such that every open cover of the space admits a countable almost subcover. If it is Hausdorff, a generalized P-closed space is HP-closed. Using a proof similar to that of the previous theorem, we can prove the following result.

2.2 THEOREM: In the category of generalized P-closed spaces and continuous maps, the projective objects are the countable HP-spaces.

2.3 THEOREM: In the category of HP-closed spaces and continuous maps the injective objects are the singletons.

PROOF: If X is a singleton then X is injective.

To prove the converse, we use an HP-space (X,T) defined as follows.

[In [18, Example 2.8] we show that (X,T) is a minimal HP-space.] Let $A = [2_a,\Omega_a)$, $B_1 = [2_b,\Omega_b)$, $C_1 = [2_c,\Omega_c)$, $B = B_1 \times B_1$, $C = C_1 \times C_1$, and $X = \{0,1\} \cup A \cup B \cup C$. Let x_a, x_b, x_c denote three points having the same ordinal x with $x_a \in A$, $x_b \in B_1$, $x_c \in C_1$, and let P be a countable subset of B_1 and Q a countable subset of C_1. We now define subsets of X as follows: $B(P,x_b) = \{x_b\} \times (B_1 - P)$, $C(Q,x_c) = \{x_c\} \times (C_1 - Q)$, $V(P,Q,x_a) = \{x_a\} \cup B(P,x_b) \cup C(Q,x_c)$, $V(P,0) = \{0\} \cup [(B_1 - P) \times B_1]$, and $V(Q,1) = \{1\} \cup [(C_1 - Q) \times C_1]$. We now define a topology T on X as follows: If $p \in B \cup C$ then $\{p\} \in T$. The T open neighbourhoods of x_a, 0 and 1 are the families $\{V(P,Q,x_a)\}$, $\{V(P,0)\}$, and $\{V(Q,1)\}$ respectively, where P and Q are arbitrary countable subsets of B_1 and C_1 respectively.

Let $Y = A \cup B \cup \{0,1\}$, $X_1 = A \cup B \cup \{0\}$ and $X_2 = \{1\}$. Then X_1 and X_2 are HP-closed, so that $Y = X_1 \cup X_2$ is also.

Let Z be HP-closed and injective, and suppose there are two distinct points a and b in Z. Since Z is Hausdorff, there are disjoint sets U_1 and U_2 in Z with $a \in U_1$ and $b \in U_2$. We define the map $f:Y \to Z$ by $f(X_1) = \{a\}$ and $f(X_2) = \{b\}$. Then f is continuous since X_1 and X_2 are open in Y. Let $i:Y \to X$ be in the inclusion map. Since Z is injective there is a continuous map $g:X \to Z$ such that $g \circ i = f$. Then $g(X_1) = \{a\}$, otherwise $f(X_1) \neq \{a\}$, and $g(X_2) = \{b\}$. Therefore $X_1 = g^{-1}(a) \subset g^{-1}(U_1)$ and $X_2 \subset g^{-1}(U_2)$. Thus $g^{-1}(U_1)$ and $g^{-1}(U_2)$ are disjoint open sets in X containing X_1 and X_2 respectively, which is a contradiction. Hence Z is a singleton. □

Similarly we have

2.4 THEOREM: In the category of generalized P-closed spaces and continuous maps the injective objects are singletons. □

Herrlich [8] has given both filter base and covering characterizations of Urysohn-closed (denoted henceforth U-closed) spaces. We use the following equivalent covering characterization of U-closed spaces given in [6] in the proof of Theorem 2.7 .

2.5 DEFINITION. An open cover C of a subset A of a topological space (X,T) is called a Urysohn open cover if and only if for each $x \in A$ there exists U_1 and $U_2 \in C$ such that $x \in U_1 \subset cl(U_1) \subset U_2$.

2.6 PROPOSITION [6]. A topological space (X,T) is U-closed if and only if every Urysohn open cover C has a finite almost subcover i.e., given a Urysohn cover C of X, there exists $U_i \in C$ $(i=1,2,3,\ldots,n)$ such that $X = \cup \{cl(U_i) \mid i=1,2,3,\ldots,n\}$:

2.7 THEOREM: In the category of U-closed spaces and continuous maps, X is projective if and only if X is finite.

PROOF. It is obvious that if X is finite, then X is projective.

Conversely, if X is projective, we will show that X is discrete so that X is finite, as X is U-closed. If possible, let X have a point 'a' which is not isolated (i.e. $\{a\}$ is not open). Let $Y = X - \{a$ and $A = (Y \times N^*) \cup \{\pi\}$ where π is an abstract point $\notin Y \times N^*$ and $N^* = [1,\omega]$, the one-point compactification of $N = [1,\omega)$. Let a topology τ on A be introduced such that on $Y \times N^*$ the product topology agrees with τ. Also let the basic τ-open neighbourhoods of π be of the form $(U \times N^*) \cup \{\pi\}$ where U is a deleted open neighbourhood of a .

The remainder of the proof consists of showing that A is U-closed, and then obtaining a violation of continuity similar to that in the proof of Theorem 2.1, or in Theorem 5.3 of [11]. □

In the proof of Theorem 5.5 of [11], notice that $Y = \{a_{ij}, c_i, \alpha, \beta \mid i=1,2,3,\ldots ; j=1,2,3,\ldots\}$ is a H-closed Urysohn space so that Y is U-closed. Hence a similar argument with suitable modifications for the Urysohn property gives the following result.

2.8 THEOREM. In the category of U-closed spaces and continuous maps X is injective if and only if X is a singleton.

Suitable modifications of the proofs of Theorems 2.1 and 2.3 yield the following result.

THEOREM: Let C be the category of UP-closed spaces and continuous maps. Then

(i) X is a projective object of C if and only if X is countable.

(ii) X is an injective object of C if and only if X is a singleton.

REFERENCES

[1] Banaschewski, B., Projective covers in categories of topological spaces, Proc. 2nd Prague Symp., 1966, 52-55.

[2] Banaschewski, B., Projective covers in categories of topological spaces and topological algebras, McMaster University, 1968.

[3] Daniel Thanapalan, P.T., and Raghavan, T.G., On H(1)-closed spaces, Bull. Calcutta Math. Soc., (To appear).

[4] Flachsmeyer, J., Topologische Projektivräume, Math. Nachr., 26(1963), 57-66.

[5] Gleason, A.M., Projective topological spaces, Illinois J. Math., 2(1958), 482-489.

[6] Goss, G., and Viglino, G., Some topological properties weaker than compactness, Pacific J. Math., 35(1970), 635-638.

[7] Henriksen, M. and Jerison, M., Minimal projective extensions of compact spaces, Duke Math. J., 32(1965), 291-295.

[8] Herrlich, H., T_ν-Abgeschlossenheit and T_ν-Minimalitat, Math. Z., 88(1965), 675-686.

[9] Herrlich, H., Categorical Topology, General Top. and its Appl., 1(1971), 1-15.

[10] Iliadis, S., Absolutes of Hausdorff spaces, Sov. Math. Dokl., 4(1963), 295-298.

[11] Liu, C.T., Absolutely closed spaces, Trans. Amer. Math. Soc., 130(1968), 86-104.

[12] Mioduszewski, I., and Rudolf, L., A formal connection between projectiveness for compact and not necessarily compact completely regular spaces, Proc. 2nd Prague Symp., 1966.

[13] Mioduszewski, I., and Rudolf, L., On projective spaces and resolutions in categories of completely regular spaces, Colloq. Math., 18(1967), 185-196.

[14] Mioduszewski, I., and Rudolf, L., H-closed spaces and projectiveness, Contr. to Extension theory of topological spaces, Berlin, 1969.

[15] Ponomarev, V., On the absolute of a topological space, Sov. Math. Dokl., 4(1963), 299-301.

[16] Purisch, S., Projectives in the Category of ordered spaces, Studies in Topology, Proc. of Conf. held at Charlotte, North Carolina, (1975), 467-478.

[17] Rainwater, J., A note on Projective resolutions, Proc. Amer. Math. Soc., 10(1959), 734-735.

[18] Raghavan T.G. and Ivan L. Reilly, HP-minimal and HP-closed spaces, (Preprint).

[19] Raghavan T.G. and Ivan L. Reilly, A note on the category of Urysohn-closed spaces, (Preprint).

[20] Stephenson Jr., R.M., Minimal first countable topologies, Trans. Amer. Math. Soc., 138(1969), 115-127.

[21] Strauss, D.P., Extremally Disconnected Spaces, Proc. Amer. Math. Soc., 18(1967), 305-309.

University of Auckland,
NEW ZEALAND.

AN EMBEDDING CHARACTERIZATION OF COMPACT SPACES

S. Salbany, Cape Town

Introduction

Every topological space is initial with respect to the canonical map e into a product of copies of the two point space {0,1} with {0} as the only nontrivial open set. It seems to us remarkable that certain classes of spaces can be neatly described in terms of their *situation* in the canonical product. Almost realcompact and almost compact T_2 spaces where characterized as maximal T_2 subspaces of the closure of e[X] in a canonical product in [1] and [3]. In trying to clarify the role of the maximal T_2 condition, we were led to the notion of a γ^* situated subspace of a canonical product in terms of which we could characterize generalized almost compact spaces (not necessarily T_2). We were also led to consider the dual notion of a γ-situated space which provided the key to the description of compact topological spaces. It is this latter characterization that we now examine.

1. Closed ultrafilters on X and the canonical embedding

For any topological space X , the set of closed ultrafilters wX has a topology with basic closed sets of the form F* , consisting of all closed ultrafilters on X which contain the closed set F . When X is a T_1 space, this is Wallman's topology [6]. As in the Wallman case, $(F_1 \cup F_2)^* = F_1^* \cup F_2^*$, $(F_1 \cap F_2)^* = F_1^* \cap F_2^*$ and wX is a compact T_1 space. When X is an R_0 space, there is an initial map $w: X \to wX$ given by $F \in w(x)$ if $x \in F$. When X is a T_1 space w provides an embedding of X into a dense subspace of the Wallman compactification wX [6].

Consider now a standard space $J \subset \mathbb{R}$, which is either

$I = [0,1]$ or $D = \{0,1\}$. The open sets in $\mathbb{R}$ in the upper topology are of the form $(-\infty, a)$, whereas the lower topology open sets are of the form $(a, +\infty)$. We shall denote the upper topology on $\mathbb{R}$, J by u and the lower topology on $\mathbb{R}$ and J by ℓ . Let C denote the continuous functions $f: (X,T) \to (J,u)$. Let e denote the canonical map from (X,T) to $(J,u)^C$.

We shall first describe wX as a subspace of $(J,u)^C$ and then characterize compact topological spaces in terms of their *situation* in the canonical product.

Definition (1) Let $A \subset X$ and let (X,P,Q) be a bitopological space. Let $\gamma[A]$ consist of all points α such that

(i) α is in the Q-closure of A .

(ii) For every P-neighbourhood V of α , there is a Q-neighbourhood W of α such that $W \cap A \subset V \cap A$.

(2) A set is called γ-closed if $\gamma[A] \subset A$.

Proposition 1 Let $f: (X,T) \to (J,u)$ be a continuous map and F an ultrafilter of closed sets in X , then the filter base $\{f[F] \mid F \in F\}$ converges in the usual topology for J .

Proof Because J is order complete, there are numbers α, β such that $\alpha = \sup\{\inf f[F] \mid F \in F\}$, $\beta = \inf\{\sup f[F] \mid F \in F\}$. Clearly, $\alpha \leq \beta$. We show that $\alpha = \beta$ from which it follows that F converges to $\alpha (= \beta)$ in the usual topology for J . Suppose $\alpha < \beta$. Let δ be a real number between α and β . Put $F_1 = f^{\leftarrow}[cl_u(\delta) \cap J]$. If F_1 is in F , then $f[F_1] \subset cl_u \delta$, so that $\inf f[F_1] \geq \delta$, contradicting our choice of α . If F_1 is not in F , there is F_0 in F which is disjoint from F_1 . Then $f[F_0] \cap cl_u \delta$ is empty so that $\sup f[F_0] \leq \delta$, contradicting our choice of β . Hence $\alpha = \beta$.

To see that $\{f[F]\}$ converges in the usual topology of J to $\gamma = \alpha = \beta$, let $\varepsilon > 0$ be given. By definition of α , there is F_0 in F such that $\sup f[F_0] < \alpha + \varepsilon$.

By definition of β , there is F_1 in $\mathcal{F}$ such that $\inf f[F_1] > \beta - \varepsilon$. Hence $|f[F] - \gamma| < \varepsilon$ for all F in $\mathcal{F}$ such that $F \subset F_0 \cap F_1$, as required.

Proposition 1 is, in a sense, the dual of the "little surprising lemma" (2) of [3] and 6.1.6 of [1], combined.

Notation. We denote the filter $\{B \mid B \supset f[F] , F \in \mathcal{F}\}$, by $f^*[\mathcal{F}]$ and the limit $\gamma = \alpha = \beta$ by $\lim_s f^*[\mathcal{F}]$.

Proposition 2. There is a homeomorphism between wX and $\gamma[e[X]]$, where γ is associated with the bitopological space $(J,u,\ell)^c$.

Proof. Let $\mathcal{F}$ be an ultrafilter of closed sets on X . Let α be the point in the canonical product, given by $\alpha_f = \lim_s f^*[\mathcal{F}]$. We prove that α is in $\gamma[e[X]]$:

(i) Suppose $c < \alpha_f < d$, then there is F in $\mathcal{F}$ such that $f[F] \subset (c,d)$. Hence $F \subset \overleftarrow{e}[\overleftarrow{\pi}_f(c,d)]$, so that α is in the $\Pi u \vee \Pi\ell$-closure of $e[X]$ in J^c .

(ii) Let P be a Πu-neighbourhood of α . It is easy to see that there is a $\{0,1\}$ valued continuous function $g: (X,T) \to (J,u)$ such that $\overleftarrow{\pi}_g[0]$ is a neighbourhood of α and $\overleftarrow{g}[0] \subset \overleftarrow{e}[P]$. Now $\alpha_g = 0$, so that $\overleftarrow{g}[1]$ is not in $\mathcal{F}$. Let F be in $\mathcal{F}$ such that F is disjoint from $\overleftarrow{g}[1]$. There is a $\{0,1\}$ valued continuous function $f: (X,T) \to (J,u)$ such that $F = \overleftarrow{f}[1]$. Hence $g = 0$ on $\overleftarrow{f}[1]$. Moreover, $\alpha_f = \lim_s f^*[\mathcal{F}] = 1$, since $f[\overleftarrow{f}[1]] = \{1\}$. Thus $Q = \overleftarrow{\pi}_f[1]$ is a $\Pi\ell$-neighbourhood of α such that $\overleftarrow{e}[Q] = \overleftarrow{f}[1] \subset \overleftarrow{g}[0] \subset \overleftarrow{e}[P]$, as required.

Denote by σ the map $\mathcal{F} \mapsto \alpha$ discussed above.

We now consider a point α in $\gamma[e[X]]$. Let $\mathcal{F}$ denote the family of closed sets $\overleftarrow{f}[1]$, where $f: (X,T) \to (J,u)$ is continuous and $\{0,1\}$ valued and $\alpha_f = 1$. Because α is in the $\Pi\ell$-closure of $e[X]$, it follows that $\mathcal{F}$ is a filter base of closed sets on X . In fact, $\mathcal{F}$ is a closed ultrafilter on X , for suppose F is closed and $F \notin \mathcal{F}$. Then there is a

$\{0,1\}$-valued continuous $h = (X,T) \to (J,u)$ such that $F = h^{\leftarrow}[1]$ and $\alpha_h = 0$. By condition (ii), there is a $\Pi\ell$-neighbourhood Q of α, such that $Q \cap e[X] \subset \Pi_h^{\leftarrow}[0]$. Because α is in the $\Pi u \vee \Pi\ell$-closure of $e[X]$, there is a $\{0,1\}$- valued function $k: (X,T) \to (J,u)$ such that $k^{\leftarrow}[1] \subset e^{\leftarrow}[Q]$ and $\alpha_k = 1$, hence $k^{\leftarrow}[1] \subset h^{\leftarrow}[0]$. Hence F does not intersect $k^{\leftarrow}[1]$. Moreover $k^{\leftarrow}[1]$ is in $\mathcal{F}$ because $\alpha_k = 1$, so that $\mathcal{F}$ is a closed ultrafilter on X.

Denote by μ the map $\alpha \mapsto \mathcal{F}$ obtained above.

We now show that $\mu\circ\sigma = 1\!\!1_{wX}$ and $\sigma\circ\mu = 1\!\!1_{\gamma[e[X]]}$:

(a) Given $\mathcal{F}$ in wX, let $\alpha = \sigma(\mathcal{F})$. If $\mathcal{G} = \mu(\sigma(\mathcal{F}) \neq \mathcal{F}$, then there is F in $\mathcal{F}$ which is not in $\mathcal{G}$. Let $f:(X,T) \to (J,u)$ be $\{0,1\}$-valued and such that $F = f^{\leftarrow}[1]$. Then $\alpha_f = 1$ by definition of α, so that F is in $\mathcal{G}$ by definition of $\mu(\alpha)$. Thus $\mathcal{G} = \mathcal{F}$, i.e. $\mu\circ\sigma = 1\!\!1_{wX}$.

(b) Given α in $\gamma[e[X]]$, let $\mathcal{F}$ be $\mu(\alpha)$. Suppose $\beta = \sigma(\mathcal{F}) \neq \alpha$. Then there is $f: (X,T) \to (J,u)$ such that $\beta_f \neq \alpha_f$. If $\beta_f < \alpha_f$, let $h: (J,u) \to (J,u)$ be $\{0,1\}$ valued and such that $h = 0$ on β_f, $h = 1$ on α_f. Because α and β are in the $\Pi u \vee \Pi\ell$-closure of $e[X]$ it follows that $\beta_{h\circ f} = 0$ and $\alpha_{h\circ f} = 1$. For simplicity, let $k = h\circ f$. Now $\alpha_k = 1$ implies $k^{\leftarrow}[1] \in \mu(\alpha) = \mathcal{F}$. Also, $\beta_k = 0$ implies $\lim_s k^*[\mathcal{F}] = 0$, which in turn implies $k[F] = \{0\}$ for some F in $\mathcal{F}$, since k is $\{0,1\}$ valued. But then $k^{\leftarrow}[1]$ and F are disjoints sets in $\mathcal{F}$, which is impossible. An analogous reasoning shows that $\alpha_f < \beta_f$ is also not possible. Hence $\alpha = \beta$.

Finally, the continuity of σ and μ :

Let $F = f^{\leftarrow}[1]$ ($f: (X,T) \to (J,u)$ and $\{0,1\}$ valued) be a closed set in X. A basic closed set in wX is F^* which consists of all closed ultrafilters containing F. Now $F^* = \sigma^{\leftarrow}[\pi_f^{\leftarrow}[1]]$, so that σ is continuous. Also $\sigma[F^*] = \pi_f^{\leftarrow}[1] \cap \gamma[e[X]]$, since σ is a bijection. Hence μ is also continuous, as σ is $\mu^{\leftarrow}$.

As a corollary to the preceeding proposition,a T_1 space X is compact if and only if $e[X]$ is γ-closed in the canonical product. More generally, we have the following.

Proposition 3 A topological space is compact if and only if $e[X]$ is γ-closed in the canonical product $(J,u)^c$.

Proof Suppose $e[X]$ is γ-closed. Let $\mathcal{F}$ be a closed ultrafilter on X and $\alpha = \sigma(\mathcal{F})$. Then α is in $\gamma[e[X]]$ so that $\alpha = e(x_0)$ for some x_0 . Then x_0 is in all F in $\mathcal{F}$. For suppose $F = f^{\leftarrow}[1]$, where $f: (X,T) \to (J,u)$ is $\{0,1\}$ valued, is in $\mathcal{F}$. Then $f[F] = \{1\}$, so that $\alpha_f = 1$. Hence $f(x) = 1$, so that $x \in F$.

Conversely, suppose X is compact. Let $\alpha \in \gamma[e[X]]$. Let $\mathcal{F} = \mu(\alpha)$. Because X is compact $\bigcap_{F \in \mathcal{F}} F = F_0$ is non empty (and closed) and F_0 is in $\mathcal{F}$, since $\mathcal{F}$ is a closed ultrafilter. If $x \in F_0$, then $\mathrm{cl}x = F_0$, otherwise it is a proper subset of F_0 , so that it cannot belong to $\mathcal{F}$, consequently $\mathrm{cl}x \cap F = \phi$ for some F in $\mathcal{F}$ and this contradicts the fact that x is in all members of $\mathcal{F}$. Thus, for any x and y in F_0 we have $\mathrm{cl}x = \mathrm{cl}y = F_0$. Because (J,u) is a T_0 space, it follows that $f(x) = f(y)$. Moreover if $z \in \mathrm{cl}x$, then $z \in F_0$, so that $\mathrm{cl}z = \mathrm{cl}x$ and $f(z) = f(x)$ Thus $f(\mathrm{cl}x) = \{f(x)\}$. We show that $\alpha = e(x)$. Now $\alpha_f = \lim_s f^*[\mathcal{F}] = f(x)$. Hence $\alpha = e(x)$, as required.

It is quite remarkable that a characterization of compactness can be achieved through both the canonical products $\Pi(I,u)$ and $\Pi(D,u)$.

A natural question is whether or not one can replace the canonical product by arbitrary products. The analogous question concerning compact T_2 spaces has a well known answer : a T_2 space X is compact if and only if it is homeomorphic to a closed subspace of an arbitrary product of copies of I . Our question has a negative answer as shown by the following example of a non-compact space which is homeomorphic to a γ-closed subspace of a product of copies of (D,u) .

Example Let $\mathbb{N}$ denote the non negative integers $1,2,3,\ldots,n,\ldots$ with the discrete topology. Clearly $\mathbb{N}$ is not compact. Consider the product D^C, where C consists of the continuous functions f_n given by $f_n(m) = 0$ if $n = m$ and $f_n(m) = 1$ if $n \neq m$. The associated product map e is one to one, continuous and separates points and closed sets since $n \notin F$ implies $f_n(n) = 0$ and $f_n(m) = 1$ for all m in F, so that $e(n)$ is not in the closure of $e[F]$ as $\Pi_{f_n}^{\leftarrow}[0]$ is a neighbourhood of $e(n)$ which does not intersect $e[F]$. Thus $\mathbb{N}$ and $e[\mathbb{N}]$ are homeomorphic. It is easy to see that the $\Pi\ell$-closure of $e[\mathbb{N}]$ consists of $e[\mathbb{N}]$ together with $\underline{0}$ (which is the point such that $\underline{0}_f = 0$ for all f in C). But $\underline{0}$ cannot belong to $\gamma[e[\mathbb{N}]]$ as this requires that $e[\mathbb{N}] \subset P$ for all Πu-neighbourhoods P of $\underline{0}$. Hence $e[\mathbb{N}]$ is γ-closed in the product $(D,u)^C$, $\mathbb{N}$ and $e[\mathbb{N}]$ are homeomorphic and $\mathbb{N}$ is not compact T_1.

2. Compact spaces and maximal separation properties

Definition Let (X,T) be a topological space and $A \subsetneqq X$. We shall say that A is weakly separated with respect to $X - A$, or simply, that A is separated in X, if given x in $X - A$, for every a in A, there is a neighbourhood of a which does not contain x.

We say that A is a maximal separated subset of X if A is separated in X and is not properly contained in any separated subset of X.

There is an interesting relationship between γ-closed sets and certain maximal separated subspaces.

Proposition 4 The following are equivalent

(1) $e[X]$ is γ-closed in the canonical product

(2) $e[X]$ is a maximal separated subspace of its $\Pi\ell$-closure in the canonical product

(3) $e[X]$ is a maximal separated subspace of its $\Pi u \vee \Pi\ell$-closure in the canonical product.

Proof (1) ⇒ (2): Suppose $e[X] \cup \{\alpha\}$ is separated for some α in the $\Pi\ell$-closure of $e[X]$ in $(J,u)^C$. The sets $f^{\leftarrow}[\gamma,1]$, where $\gamma < \alpha_f \leq 1$ then form a filter base on X. Let F_0 be a closed ultrafilter which contains F. Let $\beta = \sigma(F)$. Because $e[X]$ is γ-closed, and $\beta \in \gamma[e[X]]$, we have $\beta = e(x_0)$, for some x_0 in X. Because $f^*[F_0]$ converges to β_f and $F \subset F_0$, it follows that $\beta_f \geq \alpha_f$ for all f in C. Hence $f(x_0) \geq \alpha_f$ for all f. This contradicts our assumption that there is a Πu-neighbourhood of x_0 which misses α.

(2) ⇒ (3) is clear.

(3) ⇒ (1) Let α be in $\gamma[e[X]]$. If $\alpha \notin e[X]$, then $\{\alpha\} \cup e[X]$ is a separated subspace of the $\Pi u \vee \Pi\ell$-closure of $e[X]$: Suppose $x \in X$, there is f in C such that $f(x) \neq \alpha_f$. If $f(x) < \alpha_f$, then $\pi_f^{\leftarrow}[0,\alpha_f)$ is a Πu-neighbourhood of $e(x)$ which misses α. If $\alpha_f < f(x)$, then $P = \pi_f^{\leftarrow}[0,f(x))$ is a Πu-neighbourhood of α, so there is a $\Pi\ell$-neighbourhood Q of α such that $Q \cap e[X] \subset P \cap e[X]$. Because α is in the $\Pi u \vee \Pi\ell$-closure of $e[X]$, there is a $\{0,1\}$ valued function $g: (X,T) \to (J,u)$ such that $\pi_g^{\leftarrow}[1]$ is a $\Pi\ell$-neighbourhood of α and $g^{\leftarrow}[1] \subset e^{\leftarrow}[Q]$. Then $g^{\leftarrow}[1] \subset f^{\leftarrow}[0,f(x))$ so that $g^{\leftarrow}[0]$ is a neighbourhood of x and $\pi_g^{\leftarrow}[0]$ is a neighbourhood of $e(x)$ which misses α. This contradicts our assumption (3).

We now have a characterization of compact spaces which is an immediate consequence of the preceeding results.

Proposition 5 X is compact if and only if $e[X]$ is a maximal separated subspace of its lower closure in the canonical product $(J,u)^C$.

For R_0-spaces the above characterization can be improved.

Proposition 6 Suppose X is an R_0 topological space. The following are equivalent

(1) $e[X]$ is a maximal separated subspace of its lower closure in the canonical product

(2) $e[X]$ is a maximal R_0 subspace of its lower closure in the canonical product

(3) $e[X]$ is a maximal T_1 subspace of its lower closure in the canonical product.

<u>Proof</u> It is clear that (1) implies (2) which in turn implies (3). That (3) implies (2) follows from the fact that $e[X]$ is T_0 and that $T_0 + R_0$ is T_1 . It remains to show that (2) implies (1).

We prove, in fact, that (2) implies that $e[X]$ is γ-closed. Suppose $\alpha \in \gamma[e[X]]$; if $\alpha \notin e[X]$, then $\{\alpha\} \cup e[X]$ will be an R_0 subspace of the lower closure of $e[X]$ in the canonical product, contradicting assumption (2). We show that $\{\alpha\} \cup e[X]$ is R_0 . Let P be a Πu-neighbourhood of α , let Q be a $\Pi\ell$-neighbourhood of α such that $Q \cap e[X] \subset P \cap e[X]$. Now the Πu-closure of α is contained in any $\Pi\ell$-neighbourhood of α , so that it is contained in $P \cap (\{\alpha\} \cup e[X])$. Given x in X , there is f in C such that $f(x) \neq \alpha_f$ (assuming $\alpha \notin e[X]$). If $f(x) > \alpha_f$, then the Πu-closure of $e(x)$ cannot contain α , since $\pi_f[\overline{e(x)}] \subset \overline{f(x)} = [f(x),1]$ and $\alpha_f < f(x)$. If $f(x) < \alpha_f$, let γ be in $(f(x),\alpha)$, then $e^{\leftarrow}[\pi_f^{\leftarrow}[0,\gamma)]$ is a neighbourhood of x . Thus $\overline{x} \subset e^{\leftarrow}[\pi_f^{\leftarrow}[0,\gamma)]$. Let c be the characteristic function of $\overline{x}$, then $\pi_c(\alpha) = 0$ otherwise there is x' such that $e(x') \in \pi_f^{\leftarrow}(\gamma,1] \cap \pi_c^{\leftarrow}[1]$, since α is in the $\Pi u \vee \Pi\ell$-closure of $e[X]$. This is clearly impossible. Thus $\pi_c^{\leftarrow}[0]$ is a neighbourhood of α which does not contain $e(x)$, so that the closure of $e(x)$ does not contain α . Together with the above, this shows that $e[X] \cup \{\alpha\}$ is R_0 .

An interesting corollary is

<u>Corollary</u> A $\begin{Bmatrix} R_0 \\ T_1 \end{Bmatrix}$ space X is compact if and only if $e[X]$ is maximal $\begin{Bmatrix} R_0 \\ T_1 \end{Bmatrix}$ in its lower closure in the canonical product.

Using the fact that a compact T_2 space is normal, an elaboration of the proof of the preceeding proposition shows that

Proposition 7 A T_2 space X is compact if and only if $e[X]$ is maximal T_2 in its lower closure in the canonical product.

3. An embedding characterization of almost compact spaces

Definition A topological space is almost compact if every ultrafilter of open sets converges.

These spaces are called generalized absolutely closed in [3].

In [2], almost compact T_2 spaces were characterized as maximal T_2 subspaces of their (upper) closure in the canonical product $(I,u)^C$. An analogous characterization was obtained in [1] by considering the canonical product $(D,u)^C$ rather than $(I,u)^C$.

In this section we outline an embedding characterization which is dual to the one for compact spaces and retrieve a common generalization of the results of [1] and [2]. Details will be presented elsewhere.

Definition (1) Let $A \subset X$ and (X,P,Q) be a bitopological space. Let $\gamma^*[A]$ consist of all points α such that

(i) α is in the P-closure of A.

(ii) For every Q-neighbourhood V of α, there is a P-neighbourhood W of α such that $W \cap A \subset V \cap A$.

(2) A set is γ^*-closed if $\gamma^*(A) \subset A$.

Proposition 8 A topological space X is almost compact if and only if $e[X]$ is γ^*-closed in the canonical product $(J,u)^C$, where γ^* is associated with the bitopological space $(J,u,\ell)^C$.

Proposition 9 A topological space X is almost compact if and only if $e[X]$ is maximal separated in its (upper) closure in the canonical product $(J,u)^C$.

Proposition 10 A T_2 space is almost compact if and only if it is maximal T_2 in its (upper) closure in the canonical product $(J,u)^C$.

Partly promised in [5] , we have

Proposition 11 The Fomin extension of a Hausdorff space is homeomorphic to a maximal T_2 subspace of the closure of e[X] in the canonical product $(J,u)^C$, containing e[X].

References

1. K. Halpin, H-closed spaces and almost realcompact spaces, M.Sc. Thesis, University of Cape Town, 1974.

2. Z. Frolik and Chen-Tung Liu, An embedding characterization of almost realcompact spaces, Proc. Amer. Math. Soc. 32 (1972) 294-298.

3. Chen-Tung Liu, Absolutely closed spaces, Trans. Amer. Math. Soc. 130 (1968).

4. C.T. Liu and G.E. Strecker, Concerning almost realcompactifications, Czechoslovak Math. J. 22 (1972) 181-190.

5. J.R. Porter, Categorical Problems in minimal spaces, Proc. Conf. on Categorical Topology, Mannheim, 1975, Lecture Notes in Mathematics 540, Springer-Verlag 1976.

6. H. Wallman, Lattices and topological spaces, Ann. of Math. (2) 42 (1941) 687-697.

Grants to the Topology Research Group by the University of Cape Town and the South African Council for Scientific and Industrial Research are acknowledged.

Department of Mathematics,
University of Cape Town,
Rondebosch, 7700.
Republic of South Africa.

CONNECTION AND DISCONNECTION

G. Salicrup and R. Vázquez

0. Introduction.

Several authors have recently investigated connection and disconnection in the category of topological spaces ([AW], [H1], [H2], [P1], [P2], [P3], [P4], [SV1], [SV2], [SV3], [SV4]) as well as in the more general setting of topological categories ([SV5], [P5], [P6], [HSV2]).

The aim of this contribution is to develop a theory for connection and disconnection in a class of categories which includes all topological (in the sense of [H3]) and all well powered subcomplete (in the sense of [D]) abelian categories.

Section 1 is devoted to the definition of the auxiliary notions of fibre and cofilament. In section 2 the definition and several characterizations of a connection subcategory of a category $\underline{K}$ are given, thus bringing forth the relations between connection and coreflectivity. Left constant subcategories are important cases of connection subcategories of a given category $\underline{K}$. In $\underline{Top}$ - as well as in well-powered subcomplete abelian categories - left-constant subcategories can be characterized as connection subcategories $\underline{A}$ such that, if $p: X \to Q$ is a quotient whose fibres are the $\underline{A}$-components of X then Q is totally $\underline{A}$-disconnected, i.e. the $\underline{A}$-components of Q are singletons. ([SV1], 1.25) It turns out that in the general case the class of connection subcategories satisfying this condition is smaller than the class of left-constant subcategories. Theorem 2.20 gives a list of characterizations of this class of left-constant subcategories. Disconnection subcategories are defined in section 3 and several results concerning them - parallel to those given for connection subcategories in section 2 - are stated. Finally, torsion theories on a - not necessarily abelian - category $\underline{K}$ are considered, as well as characterizations of pairs of subcategories of $\underline{K}$ which constitute a torsion (hereditary torsion) theory in $\underline{K}$.

1. Preliminaries.

In this paper all subcategories of a given category are supposed to be full and isomorphism-closed. $\underline{K}$ will always denote a category which satisfies the following conditions:

(a) $\underline{K}$ is supplied with a factorization structure (E,M) (in the sense of [HSV1]), where M denotes the class of all $\underline{K}$-monosources, i.e. each $\underline{K}$-source has an (extremal epimorphisms, monosources) - factorization and $\underline{K}$ has the (extremal epimorphisms, monosources) - diagonalization property.

(b) $\underline{K}$ is supplied with a factorization costructure (M',E'), where E' denotes the class of all $\underline{K}$-episinks.

(c) For each $\underline{K}$-object X and each $\underline{K}$-object Y such that Y is not a non-zero initial $\underline{K}$-object, $\underline{K}(X,Y) \neq \emptyset$.

(d) Every $\underline{K}$-terminal object is $\underline{K}$-projective.

T will always denote a $\underline{K}$-terminal object and, for each $\underline{K}$-object X, $t_X: X \to T$ will stand for the only morphism in $\underline{K}$ with domain X and codomain T. The symbol ϕ will be used to designate a non-zero $\underline{K}$-initial object, and the symbol $\emptyset$ will denote, as usual, the empty set.

1.1 <u>Remark</u>. From (a) and (c) it follows that $\underline{K}$ has terminal objects; (c) implies that every constant $\underline{K}$-morphism $f: X \to Y$ with $X \neq \phi$ can be factorized through T.

1.2 <u>Definitions and notation</u>: (1) $\underline{K}_*$ will denote the comma category of T over $\underline{K}$. A $\underline{K}_*$-object $x: T \to X$ will be denoted by (x,X). For each subcategory $\underline{A}$ of $\underline{K}$, $\underline{A}_*$ will denote the subcategory of $\underline{K}_*$ whose objects are those (x,X) with $X \in \underline{A}$. If F is a class of $\underline{K}$-morphisms, F_* will stand for the class of $\underline{K}_*$-morphisms $f: (x,X) \to (y,Y)$ such that $f: X \to Y$ belongs to F.

(2) If $(f_i: X \to Y_i)_I$ is a $\underline{K}$-source and $c: T \to X$ is a $\underline{K}$-morphism, a pair (F,m) is a <u>fibre</u> of $(f_i)_I$ <u>over</u> c iff there exists a pull-back:

$$\begin{array}{ccc} F & \xrightarrow{t_F} & T \\ {\scriptstyle m}\downarrow & & \downarrow{\scriptstyle ec} \\ X & \xrightarrow[e]{} & Z \end{array}$$

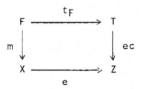

where $(X \xrightarrow{f_i} Y_i) = (X \xrightarrow{e} Z \xrightarrow{m_i} Y_i)$ is the (E,M)-factorization of $(f_i)_I$.

(3) If $(f_i\colon Y_i \to X)_I$ is a $\underline{K}$-sink, a pair (p,Q) is a cofilament of $(f_i)_I$ iff $p\colon X \to Q$ is a $\underline{K}$-morphism which satisfies the following conditions:

(a) $p \circ f_i$ is constant for each $i \in I$.

(b) If $p'\colon X \to Q'$ is a $\underline{K}$-morphism such that $p' \circ f_i$ is constant for each $i \in I$, then there exists a unique $\underline{K}$-morphism $h\colon Q \to Q'$ such that $h \circ p = p'$.

(c) Consider in I the relation $i \sim i'$ iff there exist $c\colon T \to Y_i$ and $c'\colon Y \to Y_{i'}$, such that $f_i \circ c = f_{i'} \circ c'$. Let $\approx$ be the equivalence relation in I generated by $\sim$. If I_j is an equivalence class of I with respect to $\approx$ and $(Y_i \xrightarrow{f_i} X)_{I_j} = (Y_i \xrightarrow{e_i} F \xrightarrow{m} X)_{I_j}$ is the (M',E')-factorization of $(f_i)_{I_j}$, then (F,m) is a fibre of p, or $F=\emptyset$.

(4) $\underline{K}$ is said to have fibres iff for each $\underline{K}$-source $(f_i\colon X \to Y_i)_I$ and each $\underline{K}$-morphism $c\colon T \to X$, the fibre of $(f_i)_I$ over c exists. $\underline{K}$ is said to have cofilaments iff each $\underline{K}$-sink has a cofilament.

1.3 Proposition. The following hold in $\underline{K}$:

(1) If (F,m) is a fibre of a $\underline{K}$-source $(f_i\colon X \to Y_i)_I$ then m is an extremal monomorphism.

(2) If $(f_i\colon X \to Y_i)_I$ is a monosource and (F,m) is a fibre of $(f_i)_I$ then F is a terminal $\underline{K}$-object.

(3) If (F,m) and (F',m') are fibres of a source $(f_i\colon X \to Y_i)_I$ and $g\colon Y \to F$, $g'\colon Y \to F'$ are $\underline{K}$-morphisms such that $m \circ g = m' \circ g'$ and $Y \neq \emptyset$ then there exists an isomorphism $h\colon F' \to F$ such that $m \circ h = m'$.

(4) If $(X \overset{f}{\to} Y) = (X \overset{g}{\to} W \overset{h}{\to} Y)$, (F,m) is a fibre of g over $k\colon T \to X$ and (F',m') is a fibre of f over k then there exists a morphism $s\colon F \to F'$ such that $m's = m$.

(5) If (p,Q) is a cofilament of a $\underline{K}$-sink $(f_i\colon Y_i \to X)_I$ then p is an extremal epimorphism.

(6) If $(f_i\colon (y_i,Y_i) \to (x,X))_I$ is a $\underline{K}_*$-sink then $(f_i)_I$ is a $\underline{K}_*$-episink iff (t_X,T) is a cofilament of $(f_i\colon Y_i \to X)_I$ in $\underline{K}$.

(7) If $(f_i\colon (y_i,Y_i) \to (x,X))_I$ is a $\underline{K}_*$-sink, (p,Q) is a cofilament

of $(f_i : Y_i \to X)_I$ and $(Y_i \xrightarrow{f_i} X)_I = (Y_i \xrightarrow{e_i} Z \xrightarrow{m} X)_{I \neq \emptyset}$ is the (M', E')-factorization of $(f_i)_I$ in $\underline{K}$ then there exists a pull-back and push-out square:

$$\begin{array}{ccc} Z & \xrightarrow{m} & X \\ {\scriptstyle t_Z}\downarrow & & \downarrow{\scriptstyle p} \\ T & \xrightarrow[k]{} & Q \end{array}$$

2. Connection subcategories.

2.1 Definitions. (1) A subcategory $\underline{A}$ of $\underline{K}$ is said to be a connection subcategory iff for each $\underline{K}_*$-episink $(f_i : (a_i, A_i) \to (x, X))_I$ with $A_i \in \underline{A}$ for all i, X belongs to $\underline{A}$.

(2) A subcategory $\underline{A}$ of $\underline{K}$ is said to be weakly coreflective iff $\underline{A}$ contains a $\underline{K}$-object $A \neq \emptyset$ and for each $\underline{K}$-object X there exists a non void $\underline{K}$-sink $(f_i : A_i \to X)_I$ such that the following conditions hold:

(a) For each $i \in I$ A_i belongs to $\underline{A}$.

(b) If $f: A \to X$ is a $\underline{K}$-morphism with $A \in \underline{A}$ and $A \neq \emptyset$ then there exists a unique $i \in I$ such that the $\underline{K}$-morphism f can be factorized through f_i, and this factorization is also unique.

(c) If $X \neq \emptyset$ then $A_i \neq \emptyset$ for each $i \in I$.

2.2 Remark. If $\underline{A}$ is a weakly coreflective subcategory of $\underline{K}$ then every $\underline{K}$-terminal and every $\underline{K}$-initial object belong to $\underline{A}$.

2.3 Proposition. If $\underline{A}_*$ is a monocoreflective subcategory of $\underline{K}_*$, X is a $\underline{K}$-object and $m_i : (a_i, A_i) \to (x_i, X)$ $i = 1,2$ are two $\underline{A}_*$-coreflections such that there exist $\underline{K}$-morphisms $c_i : T \to A_i$ with $m_1 \circ c_1 = m_2 \circ c_2$, then there exists a $\underline{K}$-isomorphism $h: A_1 \to A_2$ such that $m_2 h = m_1$.

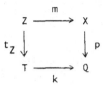

Proof. Let $m: (a, A) \to (m_1 \circ c_1, X)$ be the $\underline{A}_*$-coreflection of $(m_1 c_1, X)$. Since $m_1 : (c_1, A_1) \to (m_1 \circ c_1, X)$ is a $\underline{K}_*$-morphism there exists a $\underline{K}_*$-morphism $h_1 : (c_1, A_1) \to (a, A)$ such that $m \circ h_1 = m_1$, hence $m \circ h_1 \circ a_1 = m_1 \circ a_1 = x_1$ and therefore $m: (h_1 \circ a_1, A) \to (x_1, X)$ is a $\underline{K}_*$-morphism. Let $h_1' : (h_1 \circ a_1, A) \to (a_1, A_1)$ be the $\underline{K}_*$-morphism such that $m_1 \circ h_1' = m$. Hence $m_1 \circ h_1' \circ h_1 = m \circ h_1 = m_1$ in $\underline{K}$ and this implies that $h_1' \circ h_1 = 1_{A_1}$. Therefore h_1' is a $\underline{K}$-monomorphism and a $\underline{K}$-retraction, hence a $\underline{K}$-isomorphism. In the same way one can prove that there exist $\underline{K}$-isomorphisms $h_2: A_2 \to A$ and $h_2': A \to A_2$ such that $m \circ h_2 = m_2$ and $m_2 \circ h_2' = m$. Therefore $h = h_2' \circ h_1$ is a $\underline{K}$-isomorphism such that $m_2 \circ h = m_2 \circ h_2' \circ h_1 = m \circ h_1 = m$.

2.4 Corollary. If $\underline{A}_*$ is a monocoreflective subcategory of $\underline{K}_*$ and $m: (a,A) \to (x,X)$ is an $\underline{A}_*$-coreflection then, for each $\underline{K}$-morphism $c: T \to A$, $m: (c,A) \to (m \circ c, X)$ is an $\underline{A}_*$-coreflection.

2.5 Theorem. If $\underline{A}$ is a subcategory of $\underline{K}$ and $\underline{K}^o$ is the subcategory of non-zero $\underline{K}$-initial objects then the following are equivalent:

(a) $\underline{A}$ is a connection subcategory.

(b) $\underline{A}_*$ is (extremal $\underline{K}_*$-monomorphisms)-coreflective.

(c) $\underline{A}_*$ is coreflective and if $e: X \to Y$ is a $\underline{K}$-epimorphism with $X \in \underline{A}$ then Y belongs to $\underline{A}$.

(d) $\underline{A} \cup \underline{K}^o$ is weakly (extremal $\underline{K}$-monomorphisms)-coreflective.

(e) If $(f_i: A_i \to X)_I$ is a $\underline{K}$-sink such that (t_X, T) is a cofilament of $(f_i)_I$ and A_i belongs to $\underline{A}$ for each $i \in I$, then X belongs to $\underline{A}$.

Proof. (a) $\Leftrightarrow$ (b) Follows from the fact that $\underline{K}_*$ has a factorization costructure $(M_*^!, E_*^!)$ (See [HSV1]).

(b) $\Rightarrow$ (c) Is obvious.

(c) $\Rightarrow$ (b) If $c: (a,A) \to (x,X)$ is an $\underline{A}_*$-coreflection let $((a,A) \overset{c}{\to} (x,X)) = ((a,A) \overset{e}{\to} (z,Z) \overset{m}{\to} (x,X))$ be the $(M_*^!, E_*^!)$-factorization of c. By (c), (z,Z) belongs to $\underline{A}_*$. Hence m is an $\underline{A}_*$-coreflection.

(b) $\Leftrightarrow$ (d) Follows from 2.2 and 2.4.

(b) $\Rightarrow$ (e) Since (t_X, T) is a cofilament of $(f_i)_I$ and $(X, 1_X)$ is a fibre of t_X, it follows from the definition of cofilament that there exists $I_j \subset I$ such that $(f_i: A_i \to X)_{I_j}$ is a $\underline{K}$-episink and such that I_j is a class of equivalence of $\approx$ in I. By 2.4, if $x, x': T \to X$ are $\underline{K}$-morphisms and $m: (a,A) \to (x,X)$, $m': (a',A') \to (x',X)$ are $\underline{A}_*$-coreflections then one can suppose $A = A'$. Hence $(f_i)_{I_j}$ can be factorized through $m: A \to X$. Therefore m is a $\underline{K}$-epimorphism and an extremal mono morphism, which implies that m is an isomorphism. Hence X belongs to $\underline{A}$.

(e) $\Rightarrow$ (a) By 1.3 (6).

2.6 Definitions. (1) It follows from the definition of a connection subcategory of $\underline{K}$ that the intersection of all connection subcategories that contain a given subcategory $\underline{A}$ of $\underline{K}$ is a connection subcategory of $\underline{K}$. Hence the connective hull of $\underline{A}$ is well defined and will be denoted by $L\underline{A}$.

(2) A $\underline{K}$-object X is said to be strongly projective iff for each $\underline{K}$-morphism $f: X \to Y$ and each $\underline{K}$-episink $(f_i: Z_i \to Y)_I$ there exist $i \in I$ and a $\underline{K}$-morphism $g: X \to Z_i$ such that $f_i \circ g = f$.

(3) If $\underline{A}$ is a connection subcategory of $\underline{K}$, X is a $\underline{K}$-object, $x: T \to X$ is a $\underline{K}$-morphism and $m: (a,A) \to (x,X)$ is the $\underline{A}_*$-coreflection of (x,X) then (A,m) is the $\underline{A}$-component of x in X.

2.7 Proposition. If $\underline{A}$ is a subcategory of $\underline{K}$ and X is a $\underline{K}$-object then the following are equivalent:

(a) X belongs to $L\underline{A}$ and $X \neq \emptyset$.

(b) There exists a $\underline{K}$-sink $(f_i: A_i \to X)_I$ such that (t_X,T) is a cofilament of $(f_i)_I$ and A_i belongs to $\underline{A}$ for each $i \in I$.

Proof. (a) $\Rightarrow$ (b) Let $\underline{B}$ be the subcategory of $\underline{K}$ whose objects are those X for which there exists a $\underline{K}$-sink $(f_i: A_i \to X)_I$ with $A_i \in \underline{A}$ for all $i \in I$ and cofilament (t_X,T) and those $X \in \underline{A} \cap \underline{K}^o$. By 2.5, $\underline{A} \subset \underline{B} \subset L\underline{A}$. Suppose that $(g_i: B_i \to X)_I$ is a $\underline{K}$-sink such that (t_X,T) is a cofilament of $(g_i)_I$, $B_i \neq \emptyset$ and $B_i \in \underline{B}$ for each $i \in I$. Then for each $i \in I$ there exists a $\underline{K}$-sink $(f_j: A_j \to B_i)_{J_i}$ such that (t_{B_i},T) is a cofilament of $(f_i)_{J_i}$ and such that $A_j \in \underline{A}$ for each $j \in J_i$. Using the fact that T is strongly projective, one can prove that (t_X,T) is a cofilament of $(g_i \circ f_j: A_j \to X)_{j \in J_i, i \in I}$. Hence X belongs to $\underline{B}$. Therefore $\underline{B}$ is a connection subcategory of $\underline{K}$. This implies that $\underline{B} = L\underline{A}$.

(b) $\Rightarrow$ (a) By 2.5.

2.8 Definitions. (1) If $\underline{B}$ is a subcategory of $\underline{K}$, $I\underline{B}$ will stand for the subcategory of $\underline{K}$ whose objects are those X for which every $\underline{K}$-morphism $f: B \to X$ with $B \in \underline{B}$ is constant. A subcategory $\underline{A}$ of $\underline{K}$ is said to be right-constant iff there exists a subcategory $\underline{B}$ of $\underline{K}$ such that $\underline{A} = I\underline{B}$.

(2) If $\underline{B}$ is a subcategory of $\underline{K}$, $N\underline{B}$ will denote the subcategory of $\underline{K}$ whose objects are those X for which every $\underline{K}$-morphism $f: X \to B$ with $B \in \underline{B}$ is constant. A subcategory $\underline{A}$ of $\underline{K}$ is said to be left-constant iff there exists a subcategory $\underline{B}$ of $\underline{K}$ such that $\underline{A} = N\underline{B}$.

2.9 Remark. If A_1, A_2 are subcategories of $\underline{K}$ such that $\underline{A}_1 \subset A_2$ then $I\underline{A}_2 \subset I\underline{A}_1$, and $N\underline{A}_2 \subset N\underline{A}_1$. For each subcategory $\underline{A}$ of $\underline{K}$, $INI\underline{A} = I\underline{A}$ and $NIN\underline{A} = N\underline{A}$.

2.10 Proposition. If $\underline{A}$ is a left-constant subcategory of $\underline{K}$ then $\underline{A}$ is a connection subcategory of $\underline{K}$.

Proof. Suppose $\underline{A} = N\underline{B}$. If $(f_i: (a_i, A_i) \to (x, X))_I$ is a $\underline{K}_*$-episink with all $A_i \in \underline{A}$ and $g: X \to B$ is a $\underline{K}$-morphism with $B \in \underline{B}$, then $(A_i \xrightarrow{f_i} X \xrightarrow{g} B) = (A_i \xrightarrow{tA_i} T \xrightarrow{k} B)$ for each $i \in I$. But $(f_i: A_i \to X)_I$ is a $\underline{K}$-episink. Therefore g is constant.

2.11 Proposition. If every $\underline{K}$-terminal object is strongly projective and P is a $\underline{K}$-object then the following are equivalent:

(a) For every connection subcategory $\underline{A}$ of $\underline{K}$ different from $\underline{K}$, $\underline{A} \subset NP \subsetneq \underline{K}$.

(b) P satisfies the following conditions:

(i) P is neither terminal nor initial.

(ii) For each object $X \neq \emptyset$ there exists a $\underline{K}$-sink $(f_i: P \to X)_I$ such that (t_X, T) is a cofilament of $(f_i)_I$.

(iii) If $f: X \to P$ is a non-constant $\underline{K}$-morphism then there exists a $\underline{K}$-sink $(f_i: X \to P)_I$ such that (t_P, T) is a cofilament of $(f_i)_I$.

Proof. (a) ⇒ (b) (i) If P were either terminal or initial then NP would be $\underline{K}$. Hence P is neither terminal nor initial.

(ii) By 2.7 all one has to prove is that $LP = \underline{K}$. Suppose that this is not true. By (a) $LP \subset NP$. Hence P belongs to NP. This implies that P is either terminal or initial. Therefore $NP = \underline{K}$.

(iii) If $f: X \to P$ is a non-constant $\underline{K}$-morphism then X does not belong to NP. Hence by (a) and (ii), $LX = \underline{K} = LP$. Hence (iii) follows from 2.7.

(b) ⇒ (a) Suppose $\underline{A}$ is a connection subcategory such that $\underline{A} \not\subset NP$. Therefore there exists a $\underline{K}$-object $X \in \underline{A} \setminus NP$. Hence, by (iii), P belongs to $\underline{A}$ and by (ii) $LP = \underline{K}$. Therefore $\underline{A} = \underline{K}$. Suppose $NP = \underline{K}$. Therefore $P \in NP$. Hence P is either terminal or initial.

2.12 Proposition. If $\underline{A}$ is a subcategory of $\underline{K}$ and X is a $\underline{K}$-object then of the following propositions:

(a) X belongs to $I\underline{A}$.

(b) X belongs to $IL\underline{A}$.

(c) If (A,m) is an $\underline{LA}$-component of X then A is a $\underline{K}$-terminal object.

(c) ⇔ (b) ⇒ (a). If every $\underline{K}$-terminal object is strongly projective then (a) ⇒ (b).

Proof. (a) ⇒ (b) Suppose that $f: W \to X \neq \emptyset$ is a $\underline{K}$-morphism with $W \in \underline{LA}$. By 2.7 there exists a $\underline{K}$-sink $(g_i: A_i \to W)_I$ with all $A_i \in \underline{A}$ and such that (t_W, T) is a cofilament of $(g_i)_I$. Since $f \circ g_i$ has to be constant for each $i \in I$, there exists a $\underline{K}$-morphism $k: T \to X$ such that $k \circ t_X = f$. Therefore f is constant. Hence X belongs to $\underline{ILA}$.

(b) ⇔ (c) and (b) ⇒ (a) is obvious.

2.13 Proposition. If $\underline{A}$ is a subcategory of $\underline{K}$, $\underline{K}$ is (C,D)-factorizable, C is a class of $\underline{K}$-epimorphisms closed under compositions and D is a class of $\underline{K}$-monomorphisms then the following are equivalent:

(a) $\underline{A}$ is left-constant.

(b) $\underline{A} = N\underline{IA}$.

(c) $X \in \underline{A}$ ⇔ For each non-constant C-morphism $e: X \to Y$, $Y \notin \underline{IA}$.

(d) $X \in \underline{A}$ ⇔ For each non-constant C-morphism $e: X \to Y$ there exists a non-constant D-morphism $m: A \to Y$ with $A \in \underline{A}$.

2.14 Proposition. If $\underline{K}$ is (C,D)-factorizable with C and D as in 2.13 and $\underline{A}$ is a subcategory of $\underline{K}$ such that if $e: X \to Y$ is a C-morphism and $X \in \underline{A}$ then $Y \in \underline{A}$, then the following are equivalent:

(a) $X \in N\underline{IA}$.

(b) For each non-constant C-morphism $e: X \to Y$ there exists a non-constant D-morphism $m: A \to Y$ with $A \in \underline{A}$.

2.15 Definitions. Let $\underline{A}$ be a subcategory of $\underline{K}$.

(1) A $\underline{K}$-source $(f_i: X \to Y_i)_I$ is said to be $\underline{A}$-monotone iff for each $\underline{K}$-morphism $c: T \to X$ the fibre (F,m) of $(f_i)_I$ over c exists and F belongs to $\underline{A}$.

(2) A $\underline{K}$-sink $(f_i: Y_i \to X)_I$ is said to be $\underline{A}$-comonotone iff $(f_i)_I$ has a cofilament (p,Q) with $Q \in \underline{A}$.

2.16 Proposition. If $\underline{A}$ is a left-constant subcategory of $\underline{K}$ and $(f_i: A_i \to X)_I$ is an $\underline{A}$-comonotone sink with $A_i \in \underline{A}$ for each $i \in I$ then X belongs to $\underline{A}$.

Proof. By 2.13, $\underline{A} = N\underline{IA}$. If $g: X \to B$ is a $\underline{K}$-morphism with

$B \in I\underline{A}$ then $g \circ f_i$ is constant for each $i \in I$. Hence. if (p,Q) is the cofilament of $(f_i)_I$, there exists a $\underline{K}$-morphism $k: Q \to B$ such that $k \circ p = g$. Since $Q \in \underline{A}$, k is constant. Therefore g is constant. This implies that X belongs to $\underline{A}$.

2.17 Corollary. If $\underline{A}$ is a subcategory of $\underline{K}$ then the following are equivalent:

(a) $\underline{A}$ is left-constant.

(b) If $(f_i: A_i \to X)_I$ is an $NI\underline{A}$-comonotone sink and A_i belongs to $NI\underline{A}$ for each $i \in I$, then X belongs to $\underline{A}$.

2.18 Definitions. (1) A $\underline{K}$-monomorphism $m: X \to Y$ is said to be normal iff there exists a pull-back and push-out square:

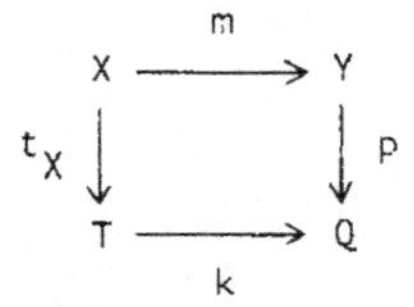

i.e., iff (X,m) is a fibre of p and (p,Q) is a cofilament of m and all the p-fibres (F,m') non equivalent to (X,m) are such that F is terminal.

(2) A $\underline{K}$-epimorphism $e: X \to Q$ is normal iff there exists a $\underline{K}$-sink $(f_i: Y_i \to X)_I$ such that (e,Q) is a cofilament of $(f_i)_I$ and $(f_i)_I$ is the class of all the fibres of e.

Let us consider the following condition that can be satisfied by a weakly coreflective subcategory $\underline{A}$ of $\underline{K}$:

(2.19) If $X \neq \emptyset$ is a $\underline{K}$-object, (p,Q) is a cofilament of the weak $\underline{A}$-coreflection of X and (F,m) is an element of the weak $\underline{A}$-coreflection of Q then there exists a commutative diagram:

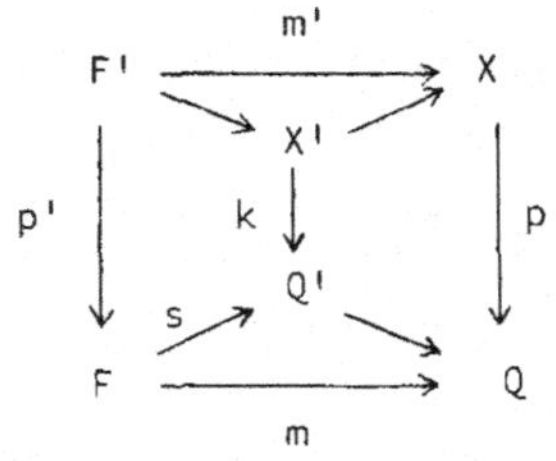

where (F',m') is an element of the weak $\underline{A}$-coreflection of X, s is an $\underline{A}$-comonotone normal monomorphism and k is an $\underline{A}$-monotone normal epimorphism.

2.20 Theorem. If $\underline{K}$ has cofilaments and $\underline{A}$ is a subcategory of $\underline{K}$ and $\underline{K}^{o}$ is the subcategory of non-zero $\underline{K}$-initial objects then the following are equivalent:

(a) $\underline{A}$ is a connection subcategory and for each $\underline{K}$-object X, the sink $(m_j: A_j \to X)_J$ of all $\underline{A}$-components of X is $I\underline{A}$-comonotone.

(b) $\underline{A} \cup \underline{K}^{o}$ is left-constant and satisfies 2.19.

(c) $\underline{A} \cup \underline{K}^{o}$ is weakly coreflective, satisfies 2.19 and if $(f_i: A_i \to X)_I$ is an $\underline{A}$-comonotone sink with $A_i \in \underline{A}$ for each $i \in I$ then X belongs to $\underline{A}$.

(d) $\underline{A}$ is a connection subcategory, satisfies 2.19 and if $e: X \to Y$ is an $\underline{A}$-monotone normal epimorphism with $Y \in \underline{A}$ then X belongs to $\underline{A}$.

Proof. (a) $\Rightarrow$ (b) Clearly $\underline{A} \subset NI\underline{A}$. If $X \in NI\underline{A}$ and $X \neq \emptyset$ let $(m_j: A_j \to X)_J$ be the sink of all $\underline{A}$-components of X. By (a) $(m_j)_J$ has a cofilament (p,Q) with $Q \in I\underline{A}$. Hence p is a constant extremal epimorphism. Therefore, Q is a $\underline{K}$-terminal object and, by the definition of cofilament, and of $\underline{A}$-component, it turns out that each m_j is a fibre of p, hence J is a singleton and m_j is an isomorphism. Thus X belongs to $\underline{A}$. Therefore $\underline{A} \cup \underline{K}^{o} = NI\underline{A}$.

To prove that $\underline{A}$ satisfies 2.19 let $X \neq \emptyset$ be a $\underline{K}$-object, (p,Q) the cofilament of the sink of $\underline{A}$-components of X and (F,m) an $\underline{A}$-component of Q. Since Q belongs to $I\underline{A}$, by 2.12 F is a $\underline{K}$-terminal object. Then, since p is an epimorphism, there exists a morphism $k: F \to X$ such that $p \circ k = m$. Since F belongs to $\underline{A}$, there exist an $\underline{A}$-component (F',m') of X and a $\underline{K}$-morphism $k': F \to F'$ such that $m' \circ k' = k$. Since F' belongs to $\underline{A}$, there exist an $\underline{A}$-component (F'',m'') of Q and a $\underline{K}$-morphism $k'': F' \to F''$ such that $p \circ m' = m'' \circ k''$. Hence $m'' \circ k'' \circ k' = p \circ m' \circ k' = p \circ k = m = m \circ 1_F$. This implies that $(F,m) = (F'',m'')$. Hence in the commutative diagram:

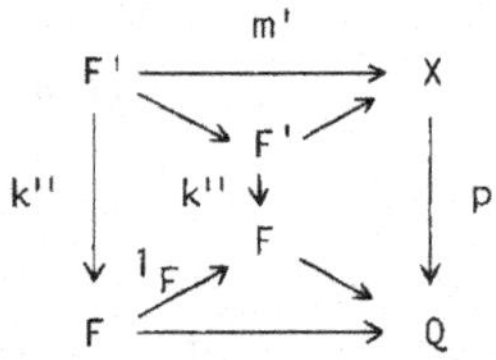

1_F is an A-comonotone normal monomorphism and k'' is an A-monotone normal epimorphism.

(b) ⇒ (c) Follows from 2.5, 2.10 and 2.16.

(c) ⇒ (d) Let $(f_i: A_i \to X)_I$ be a weak A-coreflection and, for each $i \in I$, let $(A_i \xrightarrow{f_i} X) = (A_i \xrightarrow{e} Z \xrightarrow{m} X)$ be the (M', E')-factorization of f_i. If $Z = \emptyset$ then $A_i = \emptyset$ for each $i \in I$ and therefore $Z \in$ A. If $Z \neq \emptyset$, then by 1.3(6) and 2.2 e is an A-comonotone morphism. Therefore, by (c), Z belongs to A. This implies that there exist $i' \in I$ and $g: Z \to A_{i'}$ such that $f_{i'} \circ g = m$. Thus $f_{i'} \circ g \circ e = m \circ e = f_i$. Hence $i = i'$ and $g \circ e = 1_{A_i}$. Therefore e is an isomorphism. Hence f_i is an extremal monomorphism. By 2.5, this implies that A is a connection subcategory. If $e: X \to Y$ is an A-monotone normal epimorphism with $Y \in$ A and $(m_i: A_i \to X)_I$ is the class of e-fibres then $(m_i)_I$ is A-comonotone. Hence X belongs to A.

(d) ⇒ (a) Let $(m_j: F_j \to X)_J$ be the class of A-components of a K-object X. If $X = \emptyset$ then J is empty and $(m_j)_J$ is IA-comonotone. If $X \neq \emptyset$ and (p,Q) is the cofilament of $(m_j)_J$ and (F,m) is an A-component of Q then, since A satisfies 2.19, there exists a commutative diagram:

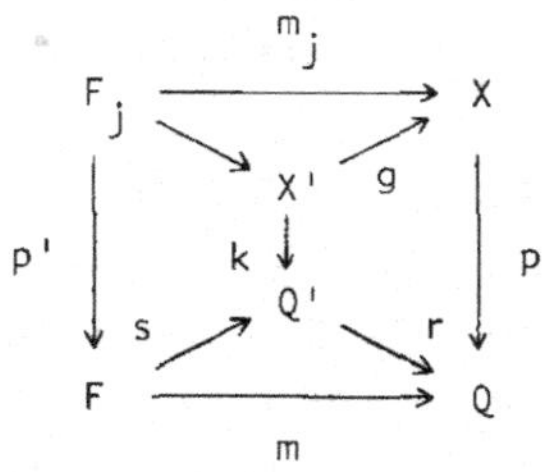

with $j \in J$, s an A-comonotone normal monomorphism and k and A-monotone normal epimorphism. Then, by (d), $Q' \in$ A and $X' \in$ A. Hence there exist an A-component $(F_{j'}, m_{j'})$ of X and a K-morphism $g': X' \to F_{j'}$ such that $m_{j'} \circ g' = g$. Therefore $r \circ k = p \circ g = p \circ m_{j'} \circ g'$ is constant and then, since $(t_{Q'}, T)$ is a cofilament of k, there exists a K-morphism $f: T \to Q$ such that $f \circ t_{Q'} = r$. This implies that r is constant. Therefore m is constant. Thus F is a K-terminal object and, by 2.12, Q belongs to IA.

2.21 Examples. (1) If K is any of the following categories: Top, a topological category in which every quotient is hereditary quotient, a well-powered subcomplete abelian category, then every connection subcategory of $\underline{K}$ satisfies 2.19. Hence, in case $\underline{K}$ is any of the categories mentioned above, condition 2.19 can be omitted in all characterizations of Theorem 2.20. The following are examples of topological categories in which every quotient is hereditary quotient:

(a) The category P-Near of prenearness spaces and nearness preserving maps ([H3]).

(b) The category S-Near of semi-nearness spaces and nearness preserving maps ([H3]), resp. the isomorphic categories of merotopic spaces ([K]) resp. quasi-uniform spaces ([I]).

(c) The category Grill of grill-determined prenearness spaces ([BHR]).

(d) The category Rere ([HSV1]) whose objects are pairs (X,ρ) with X a set and ρ a reflexive relation on X and whose morphisms are relation preserving functions.

(2) If $\underline{K}$ is the topological category whose objects are pairs (X,ξ) with X a set and ξ a subset of PX such that the empty set and X belong to ξ, and whose morphisms $f\colon (X,\xi) \to (Y,\eta)$ are all those functions $f\colon X \to Y$ such that $f^{-1}_{\eta} \subset \xi$, then $\underline{K}$ has a proper class of left-constant subcategories, but only three of them satisfy 2.19, namely: {$\underline{K}$-initial and $\underline{K}$-terminal objects}, {Indiscrete $\underline{K}$-objects} and $\underline{K}$.

3. Disconnection

3.1 Definition. A subcategory $\underline{A}$ of $\underline{K}$ is said to be a disconnection subcategory of $\underline{K}$ iff for each monosource $(f_i\colon X \to A_i)_I$ with $A_i \in \underline{A}$ for each $i \in I$, X belongs to $\underline{A}$.

3.2 Proposition. If $\underline{A}$ is a subcategory of $\underline{K}$ then the following are equivalent:

(a) $\underline{A}$ is a disconnection subcategory.

(b) $\underline{A}$ is (extremal epimorphisms)-reflective.

(c) $\underline{A}$ is reflective and if $m\colon X \to A$ is a monomorphism with $A \in \underline{A}$ then X belongs to $\underline{A}$.

3.3 Notation. If $\underline{A}$ is a subcategory of $\underline{K}$, $G\underline{A}$ will denote the E-reflective hull of $\underline{A}$. It is known ([HSV1]), that $X \in G\underline{A}$ iff there exists a monosource $(f_i\colon X \to A_i)_I$ with all $A_i \in \underline{A}$.

3.4 Proposition. If $\underline{A}$ is a subcategory of $\underline{K}$ and X is a $\underline{K}$-object then the following are equivalent:

(a) X belongs to $N\underline{A}$.

(b) X belongs to $NG\underline{A}$.

(c) If $r: X \to A$ is the $G\underline{A}$-reflection of X then A is a $\underline{K}$-terminal object, or $A = \emptyset$.

3.5 Proposition. If $\underline{A}$ is a right constant subcategory of $\underline{K}$ then $\underline{A}$ is a disconnection subcategory.

3.6 Proposition. If $\underline{K}$ is (C,D)-factorizable, C is a class of $\underline{K}$-epimorphisms, D is a class of $\underline{K}$-monomorphisms closed under compositions and $\underline{A}$ is a subcategory of $\underline{K}$ then the following are equivalent:

(a) $\underline{A}$ is a right-constant subcategory of $\underline{K}$.

(b) $\underline{A} = IN\underline{A}$.

(c) $X \in \underline{A} \Leftrightarrow$ For each non-constant D-morphism $m: X \to Y$, Y does not belong to $N\underline{A}$.

(d) $X \in \underline{A} \Leftrightarrow$ For each non-constant D-morphism $m: X \to Y$ there exists a non-constant C-morphism $e: Y \to A$ such that $A \in \underline{A}$.

3.7 Proposition. If $\underline{A}$ is a right constant subcategory of $\underline{K}$ and $(f_i: X \to A_i)_I$ is an $\underline{A}$-monotone source with $A_i \in \underline{A}$ for each $i \in I$ then X belongs to $\underline{A}$.

3.8 Definition. A class F of $\underline{K}$-morphisms is said to be simple iff every F-morphism with $\underline{K}$-terminal fibres is a monomorphism.

Next, let us consider the following conditions, that can be satisfied by a reflective subcategory $\underline{A}$ of $\underline{K}$.

(3.9) If $r: X \to A$ is an $\underline{A}$-reflection and (F,m) is an r-fibre then there exists a commutative diagram:

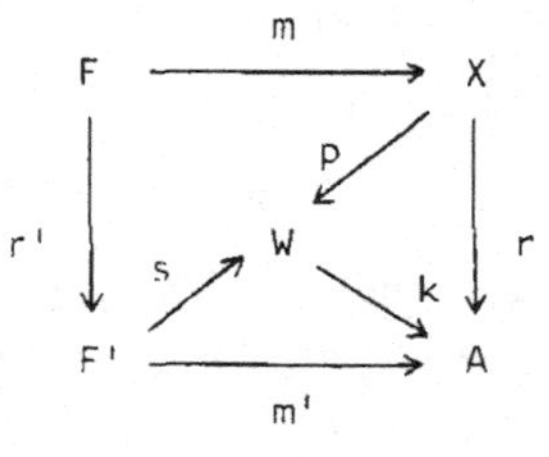

with r' an $\underline{A}$-reflection, s a monomorphism and k an $\underline{A}$-monotone epimorphism.

3.10 Theorem. If $\underline{K}$ has fibres and $\underline{A}$ is a subcategory of $\underline{K}$ then the following are equivalent:

(a) $\underline{A}$ is an ($N\underline{A}$-monotone extremal epimorphisms)-reflective subcategory of $\underline{K}$ and the class of $\underline{A}$-reflections is simple.

(b) $\underline{A}$ is right-constant and satisfies 3.9.

(c) $\underline{A}$ is reflective, satisfies 3.9 and if $(f_i: X \to A_i)_I$ is an $\underline{A}$-monotone source with $A_i \in \underline{A}$ for each $i \in I$, then X belongs to $\underline{A}$.

(d) $\underline{A}$ is reflective, satisfies 3.9 and if $f: X \to Y$ is an $\underline{A}$-monotone morphism with $Y \in \underline{A}$ then X belongs to $\underline{A}$.

(e) $\underline{A}$ is epireflective, satisfies 3.9 and if $e: X \to Y$ is an $\underline{A}$-monotone epimorphism with $Y \in \underline{A}$ then X belongs to $\underline{A}$.

4. Torsion theories

4.1 Definitions. (1) If $\underline{A}$ is a subcategory of $\underline{K}$, $\underline{A}$ is said to be hereditary iff for each extremal monomorphism $m: X \to A$ with $A \in \underline{A}$, X belongs to $\underline{A}$.

(2) A diagram $(Y_i \xrightarrow{m_i} X \xrightarrow{p} Q)_I$ is said to be exact iff $(Y_i, m_i)_I$ is the family of all fibres of p and (p, Q) is a cofilament of $(m_i)_I$.

4.2 Proposition. If $\underline{K}$ has fibres and cofilaments and T and F are two subcategories of $\underline{K}$ such that every non-zero $\underline{K}$-initial object belongs to $T \cap F$ then the following are equivalent:

(a) Every $\underline{K}$-morphism $f: A \to B$ with $A \in T$ and $B \in F$ is constant and for every $\underline{K}$-object $X \neq \emptyset$, there exists an exact diagram $(A_i \xrightarrow{m_i} X \xrightarrow{p} B)_I$ with all $A_i \in T$ and $B \in F$.

(b) F is right-constant, satisfies 3.9, every F-reflection is normal and $T = NF$.

(c) T is left-constant, satisfies 2.19 and $F = IT$.

(d) T is weakly coreflective, F is reflective and for each $\underline{K}$-object $X \neq \emptyset$ the diagram $(A_i \xrightarrow{m_i} X \xrightarrow{p} B)_I$, where $(m_i)_I$ is the weak T-coreflection and p is the F-reflection, is exact.

Proof. (a) $\Rightarrow$ (b). By (a) $T \subset NF$ and $F \subset IT$. If $X \neq \emptyset$ belongs to NF and $(A_i \xrightarrow{m_i} X \xrightarrow{p} B)_I$ is exact with all $A_i \in T$ and $B \in F$ then B belongs to $NF \cap F$. Hence B is a $\underline{K}$-terminal object. Therefore every fibre of p is equivalent to $(X, 1_X)$. This implies that I is a singleton and m_i is an isomorphism. Hence X belongs to T. Thus

$\mathcal{T} = N\mathcal{F}$. Next, let X be a $\underline{K}$-object and let $r: X \to X'$ be the $I\mathcal{T}$-reflection of X. By (a) $r \circ m_i$ is constant for each $i \in I$. Hence there exist $\underline{K}$-morphisms $k: B \to X'$ and $k': X' \to B$ such that $k \circ p = r$ and $k' \circ r = p$. Therefore k is an isomorphism. This implies that $\mathcal{F} = I\mathcal{T}$ and that r is normal. For each $i \in I$, the diagram:

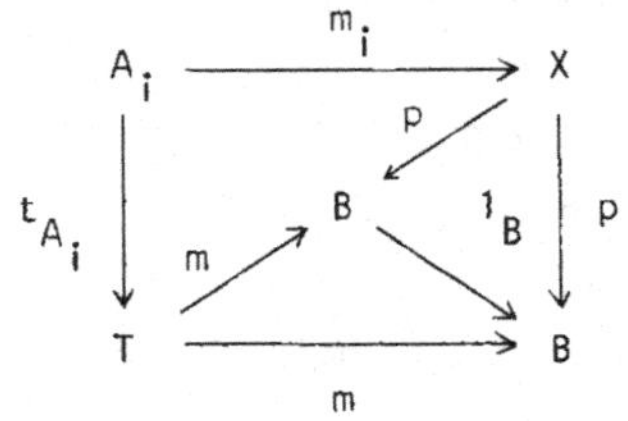

commutes. Hence $\mathcal{T}$ satisfies 3.9.

(b) $\Rightarrow$ (c) By (b), $\mathcal{T}$ is left-constant, $\mathcal{F} = I\mathcal{T}$ and every $\mathcal{F}$-reflection $r: X \to B$ is $\mathcal{T}$-monotone. Let $(m_i: A_i \to X)_I$ be the weak $\mathcal{T}$-coreflection of X and let (p,Q) be a cofilament of $(m_i)_I$. Since $r \circ m_i$ is constant for each $i \in I$, there exists $h: Q \to B$ such that $h \circ p = r$. Since r is a normal epimorphism, (r,B) is a cofilament of the $\underline{K}$-sink $(m'_j: A'_j \to X)_J$ of all r-fibres. Hence, for each $j \in J$ there exist $i \in I$ and $f: A'_j \to A_i$ such that $m_i \circ f = m'_j$. Therefore $p \circ m'_j$ is constant for each $j \in J$. Hence there exists $h': B \to Q$ such that $h' \circ r = p$. This implies that h' is an isomorphism. Thus, by 2.20 $\mathcal{T}$ satisfies 2.19.

(c) $\Rightarrow$ (d) If $X \neq \emptyset$, $(m_i: A_i \rightarrowtail X)_I$ is the weak $\mathcal{T}$-coreflection of X and (p,B) is a cofilament of $(m_i)_I$ then, by (c), B belongs to $\mathcal{F}$ and $(m_i)_I$ is the sink of all the fibres of the $\mathcal{F}$-reflection of X. Therefore p is the $\mathcal{F}$-reflection and the diagram $(A_i \xrightarrow{m_i} X \xrightarrow{p} B)_I$ is exact.

(d) $\Rightarrow$ (a) Let $f: A \to B$ be a $\underline{K}$-morphism with $A \in \mathcal{T}$, and $B \neq \emptyset \in \mathcal{F}$. If $r: A \to A'$ is the $\mathcal{F}$-reflection of A then there exists $h: A' \to B$ such that $h \circ r = f$. By (d) the diagram $A \xrightarrow{1_A} A \xrightarrow{r} A'$ is exact. Hence r is constant. Therefore f is constant.

4.3 <u>Definitions</u>. A <u>torsion theory</u> in $\underline{K}$ is a pair $(\mathcal{T},\mathcal{F})$ of subcategories of $\underline{K}$ such that every non-zero $\underline{K}$-initial object belongs to $\mathcal{T} \cap \mathcal{F}$ and such that $(\mathcal{T},\mathcal{F})$ satisfies any of the equivalent conditions of 4.2. In this case, $\mathcal{T}$ is the <u>torsion</u> subcategory and $\mathcal{F}$ the <u>torsion free</u> subcategory of $(\mathcal{T},\mathcal{F})$. A torsion theory $(\mathcal{T},\mathcal{F})$ is said to be <u>hereditary</u>

iff T is hereditary.

4.4 Proposition. Let $\underline{K}$ be a category that has fibres and cofilaments and such that every pull-back of a $\underline{K}$-epimorphism is a $\underline{K}$-epimorphism. If T is a weakly coreflective subcategory of $\underline{K}$ and F is an epireflective subcategory of $\underline{K}$ then the following are equivalent:

(a) (T,F) is a torsion theory in $\underline{K}$.

(b) If $e: A \to B$ is an F-monotone epimorphism and B belongs to F then A belongs to F, and if $r: X \to X'$ is an F-reflection then the weak T-coreflection $(m_i: A_i \to X)_I$ of X is the sink of all r-fibres.

Proof. (a) $\Rightarrow$ (b) By 4.2.

(b) $\Rightarrow$ (a) If (p,B) is a cofilament of $(m_i)_I$ then there exists $f: B \to X'$ such that $f \circ p = r$. Clearly f is a $\underline{K}$-epimorphism. Let (F,m) be a fibre of f over $c: T \to B$, let $c': T \to X$ be such that $p \circ c' = c$ and let (A_i,m_i) be the fibre of r over c'. Then $(A_i m_i)$ is a fibre of p and in the diagram:

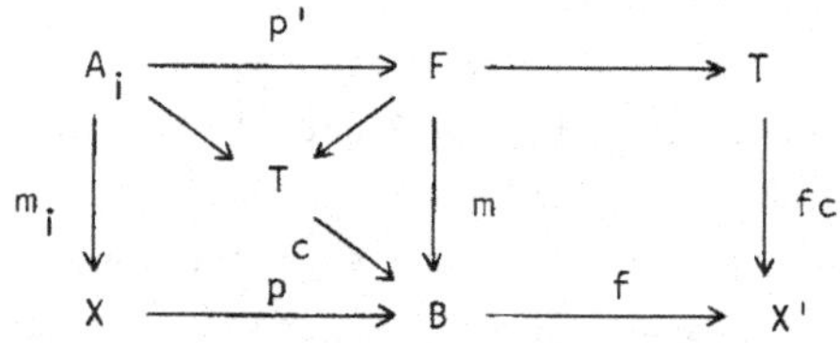

where all squares that contain X are pull-backs, p' is a $\underline{K}$-epimorphism. Since $c \circ t_F \circ p' = c \circ t_{A_i} = p \circ m_i = m \circ p'$, one has $c \circ t_F = m$. Hence F is a $\underline{K}$-terminal object. Therefore f is an F-monotone $\underline{K}$-epimorphism with $X' \in F$. This implies that $B \in F$ and that there exists a $\underline{K}$-morphism $g: X' \to B$ such that $g \circ r = p$. Hence f is an isomorphism. Thus, (T,F) satisfies 4.2(d).

4.5 Lemma. If $\underline{K}$ is like in 4.4 and $\underline{A}$ is a right constant subcategory of $\underline{K}$ then $\underline{A}$ is (normal epimorphisms)-reflective.

Next, consider the following condition that can be satisfied by a reflective subcategory $\underline{A}$ of $\underline{K}$:

(4.6) If $r: X \to A$ is an $\underline{A}$-reflection, (F,m) is an r-fibre and $m_o: Y \to F$ is an extremal monomorphism then there exists a commutative diagram:

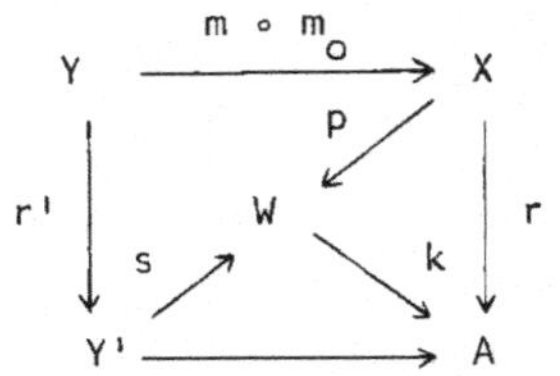

with r' an A-reflection, s a monomorphism and k an A-monotone epimorphism.

4.7 Theorem. Let K be a category that has fibres, cofilaments and such that every pull-back of a K-epimorphism is a K-epimorphism. If A is a subcategory of K then the following are equivalent:

(a) (NA,A) is a hereditary torsion theory in K.

(b) A is right constant and satisfies 4.6.

(c) A is reflective, satisfies 4.6 and if $(f_i : X \to Y_i)_I$ is an A-monotone source with all $Y_i \in$ A then X belongs to A.

(d) A is reflective, satisfies 4.6 and if $f: X \to Y$ is an A-monotone morphism with $Y \in$ A then X belongs to A.

(e) A is epireflective, satisfies 4.6 and if $e: X \to Y$ is an A- monotone epimorphism with $Y \in$ A then X belongs to A.

Proof. (a) ⇒ (b) If $r: X \to A$ is an A-reflection, (F,m) is a fibre of r and $m_o : Y \to F$ is an extremal monomorphism then, by (a) and 4.2, Y belongs to NA. Therefore there exists a commutative diagram:

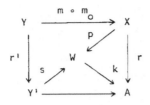

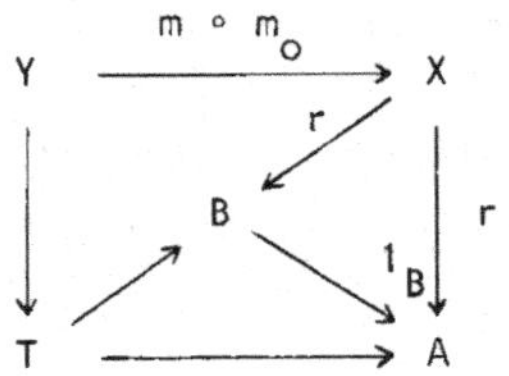

Hence A satisfies 4.6.

Since 4.6 implies 3.9, by 3.10 one has (b) ⇒ (c) ⇒ (d) ⇒ (e).

(e) ⇒ (a) By 3.10 A is right constant. Hence A = INA. By 4.2 and 4.5, NA satisfies condition 2.19. Hence, by 4.2, (NA,A) is a torsion theory. Suppose $m_o : Y \to X$ is an extremal monomorphism with $X \in$ A. Since A satisfies condition 4.6, there exists a commutative diagram:

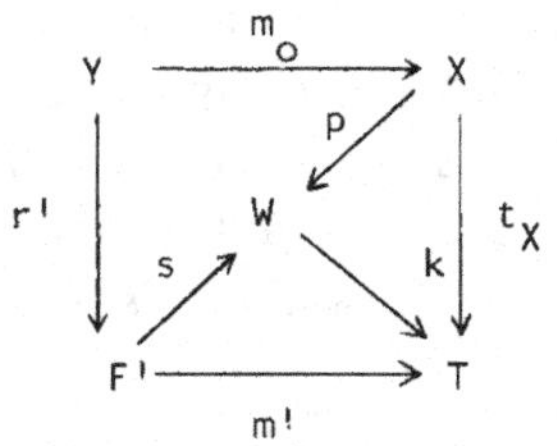

with k an $\underline{A}$-monotone epimorphism, s a monomorphism and r' an $\underline{A}$-reflection. By (e), W belongs to $\underline{A}$. Hence there exists a $\underline{K}$-morphism $k': T \to W$ such that $k' \circ t_X = p$. Therefore $k' \circ m' \circ r' =$ $= k' \circ t_X \circ m_o = p \circ m_o = s \circ r'$. This implies that $k' \circ m' = s$. Hence m' is a monomorphism. Therefore r' is constant. Thus, one has proved that Y belongs to $\underline{NA}$. Hence $\underline{NA}$ is hereditary.

REFERENCES

[AW] A.V. Arhangel'skii and R. Wiegandt, Connectednesses and disconnectednesses in topology, Gen. Top. and its applications. 5 No. 1 (1975), 9-34.

[BHR] H.L. Bentley, H. Herrlich and W.A. Robertson, Convenient categories for topologists, Comment. Math. Univ. Carolinae 17 (1976), 207-227.

[D] S.E. Dickson, A torsion theory for abelian categories, Trans. of the A.M.S. 121 (1966), 223-235.

[H1] H. Herrlich, Topologische Reflexionen und Coreflexionen. Lecture Notes in Math. 78 (1968).

[H2] H. Herrlich, Limit operators and topological coreflections, Trans. of the A.M.S. 146 (1969), 203-210.

[H3] H. Herrlich, Topological structures, Math. Centre Tract 52 (1974), 59-122.

[HS] H. Herrlich and G.E. Strecker, Category Theory, Boston (1973).

[HSV1] H. Herrlich, G. Salicrup and R. Vázquez, Dispersed factorization structures, Preprint.

[HSV2] H. Herrlich, G. Salicrup and R. Vázquez, Light factorization structures, Preprint.

[I] J. R. Isbell, Uniform spaces, Amer. Math. Soc. Math. Surveys 12 (1964).

[K] M. Katetov, On continuity structures and spaces of mappings, Comment. Math. Univ. Carolinae 6 (1965), 257-278.

[P1] G. Preuss, Über den E-Zusammenhang und seine Lokalisation, Thesis, Freie Universität, Berlin (1967).

[P2] G. Preuss, Trennung und Zusammenhang, Monatsh. Math. 74 (1970), 70-87.

[P3] G. Preuss, E-Zusammenhangende Räume, Manuscripta Math. 3 (1970), 331-342.

[P4] G. Preuss, Eine Galois-Korrespondenz in der Topologie, Monatsh. Math. 75 (1971), 447-452.

[P5] G. Preuss, Relative connectednesses and disconnectednesses in topological categories, Quaestiones Math. 2 (1977), 297-306.

[P6] G. Preuss, On factorization of maps in topological categories, Preprint.

[SV1] G. Salicrup and R. Vázquez, Categorías de conexión, An. Inst. Mat. Univ. Nac. Autónoma México 12 (1972), 47-87.

[SV2] G. Salicrup and R. Vázquez, Reflexividad y coconexidad en Top, An. Inst. Mat. Univ. Nac. Autónoma México 14 (1974), 159-230.

[SV3] G. Salicrup and R. Vázquez, Expansiones A-conexas y subespacios A-máximos, An. Inst. Mat. Univ. Nac. Autónoma México 15 No. 1 (1975), 97-111.

[SV4] G. Salicrup and R. Vázquez, Objetos máximos en categorías de conexión en Top, An. Inst. Mat. Univ. Nac. Autónoma México 15 No. 2 (1975), 117-131.

[SV5] G. Salicrup and R. Vázquez, Reflectivity and connectivity in topological categories, Preprint.

[S] B. Stenström, Rings of Quotients, Springer Verlag, Berlin (1975)

Instituto de Matemáticas U. N. A. M.

Ciudad Universitaria, México 20, D.F.

CONNECTIONS BETWEEN CONVERGENCE AND NEARNESS

Friedhelm Schwarz, TU Hannover, Federal Republic of Germany

Abstract

Generalizing usual filter convergence, we introduce stack convergence. This convergence contains nearly each essential concept of generalizing topological and uniform spaces. In particular, the belonging category SCO includes as well the topological spaces as the prenearness spaces. Thus we have solved a problem of Herrlich (1974) and have proved: Convergence contains nearness.

1. Introduction

The category TOP of topological spaces and continuous maps suffers from many deficiencies, e.g. it has no natural function space structures. Hence TOP is inconvenient for many purposes in homotopy theory, topological algebra, functional analysis etc. Consequently, there have been many attempts to replace TOP by supercategories which have the desired properties. In this note, we compare two of the most important concepts of generalizing topological spaces, convergence and nearness.

Herrlich [He 74:1] showed first a connection between convergence and nearness: He proved that the topological nearness spaces (TNEAR) are isomorphic to the R_0-topological spaces (R_0TOP).

Robertson [Ro 75] introduced the categories GRILL and LGRILL (there: R_0Con). These are the nearness categories isomorphic to Katětov's filter merotopic spaces and localized filter merotopic spaces [Ka 65], respectively. Generalizing the topological R_0-condition, Robertson obtained the category R_0LFCO (there: R_0Con) which is a subcategory of Kent's filter convergence spaces [Ke 64] (FCO; there: convergence functions). Robertson succeeded in proving that R_0LFCO and LGRILL are isomorphic, the isomorphism between R_0TOP and TNEAR being a restriction of that between R_0LFCO and LGRILL.

It is now a natural problem if there are isomorphisms between supercategories of R_0LFCO and LGRILL. Indeed, GRILL is isomorphic to the constant filter convergence spaces (ConsFCO; Definition 8). Moreover, generalizing filter convergence, we introduce stack convergence and obtain the stack convergence spaces (SCO; Definition 9). In analogy to the filter situation, the constant stack convergence spaces (ConsSCO; Definition 10) are isomorphic to the category PNEAR introduced in [He 74:2].

Thus we have shown that each nearness category may be regarded as a convergence category. Since only R_0TOP but not TOP is contained in PNEAR (because of the symmetry contained in the concept of nearness), Herrlich asked for a concept including both TOP and PNEAR ([He 74:2] Appendix B). Stack convergence is a simple and natural concept solving this problem.

2. Preliminaries

X and Y are always arbitrary sets. $\mathfrak{P}(X)$ denotes the power set of X, $\mathfrak{P}^2(X) := \mathfrak{P}(\mathfrak{P}(X))$ etc. A function f: $X \longrightarrow Y$ induces a function from $\mathfrak{P}(X)$ to $\mathfrak{P}(Y)$, which we also call f; for $A \subset X$, $\mathfrak{A} \subset \mathfrak{P}(X)$ and $\alpha \subset \mathfrak{P}^2(X)$ we have

$$f(A) = \{ f(a) \mid a \in A \},$$
$$f(\mathfrak{A}) = \{ f(A) \mid A \in \mathfrak{A} \},$$
$$f(\alpha) = \{ f(\mathfrak{A}) \mid \mathfrak{A} \in \alpha \}.$$

$f \restriction A$ denotes the restriction of a function $f: X \longrightarrow Y$ to $A \subset X$.

Now we define some set systems, which are essential for the following, and describe their connections.

Definition 1:

(1) For $\mathfrak{A} \subset \mathfrak{P}(X)$, $\emptyset \neq A \subset X$ and $a \in X$:

$[\mathfrak{A}] := \{ B \mid B \subset X \wedge \exists A(A \in \mathfrak{A} \wedge A \subset B) \}$,

$[A] := [\{A\}]$,

$[a] := [\{a\}]$.

(2) $\mathfrak{S} \subset \mathfrak{P}(X)$ is called stack on X $:\Longleftrightarrow$ $\mathfrak{S} = [\mathfrak{S}]$.

$\mathbb{S}(X)$ denotes the set of all stacks on X.

(3) $\mathfrak{F} \in \mathbb{S}(X)$ is called filter on X $:\Longleftrightarrow$

(F_1) $\mathfrak{F} \neq \emptyset$

(F_2) $A \in \mathfrak{F} \wedge B \in \mathfrak{F} \Rightarrow A \cap B \in \mathfrak{F}$.

$\mathbb{F}(X)$ denotes the set of all filters on X, $\mathbb{U}(X)$ the set of all ultrafilters on X.

(4) $\mathfrak{G} \in \mathbb{S}(X)$ is called grill on X $:\Longleftrightarrow$

(G_1) $\emptyset \notin \mathfrak{G}$

(G_2) $A \cup B \in \mathfrak{G} \Rightarrow A \in \mathfrak{G} \vee B \in \mathfrak{G}$.

$\mathbb{G}(X)$ denotes the set of all grills on X.

For the first time, grills were considered by Choquet [Ch 47], stacks (under the name "Stapel") by Grimeisen [Gr 60]. Choquet showed a duality between filters and grills. From his ideas, the following duality function ch was developed:

$$\mathrm{ch}: \begin{cases} (\mathfrak{P}^2(X),\subset) \longrightarrow (\mathfrak{P}^2(X),\subset) \\ \mathfrak{A} \longrightarrow \mathrm{ch}\,\mathfrak{A} = \{ B \mid B \subset X \wedge X \setminus B \notin \mathfrak{A} \}. \end{cases}$$

ch is an antiisomorphism with $\mathrm{ch} \circ \mathrm{ch} = 1_{\mathfrak{P}^2(X)}$, i.e. in particular a bijection.

Proposition 1:

(1) $\mathrm{ch}\,\mathbb{S}(X) = \mathbb{S}(X)$

(2) $\mathrm{ch}\,\mathbb{F}(X) = \mathbb{G}(X)$

(3) $\mathrm{ch}\,\mathbb{G}(X) = \mathbb{F}(X)$

(4) $\mathrm{ch} \restriction \mathbb{U}(X) = 1_{\mathbb{U}(X)}$.

For more details, we refer to [Gr 60], [Th 73], and [Schw 77].

For $\mathfrak{A} \subset \mathfrak{P}(X)$, $[\mathfrak{A}]$ is the smallest stack which contains $\mathfrak{A}$ and $]\mathfrak{A}[:= \bigcup\{ \mathfrak{T} \mid \mathfrak{T} \in \mathbb{S}(X) \wedge \mathfrak{T} \subset \mathfrak{A} \}$ is the largest stack which is contained in $\mathfrak{A}$. It is possible to describe $]\mathfrak{A}[$ by ch and $[\ ]$: $]\mathfrak{A}[= \mathrm{ch}[\mathrm{ch}\,\mathfrak{A}]$.

A functor and its restrictions are denoted by the same symbol. Subcategories are always assumed to be full and isomorphism-closed.

Definition 2:

For categories $\underline{A},\underline{B}$:

$\underline{A} \cong \underline{B}$ $:\Longleftrightarrow$ $\underline{A}$ and $\underline{B}$ are isomorphic;

$\underline{A}$ bir $\underline{B}$ $:\Longleftrightarrow$ $\underline{A}$ is a bireflective subcategory of $\underline{B}$;

$\underline{A}$ bic $\underline{B}$ $:\Longleftrightarrow$ $\underline{A}$ is a bicoreflective subcategory of $\underline{B}$;

$\underline{A}$ ibir $\underline{B}$ $:\Longleftrightarrow$ $\underline{A}$ is isomorphic to a bireflective subcategory of $\underline{B}$;

$\underline{A}$ ibic $\underline{B}$ $:\Longleftrightarrow$ $\underline{A}$ is isomorphic to a bicoreflective subcategory of $\underline{B}$.

In this paper, we deal only with concrete categories.

Definition 3:

A concrete category is a category $\underline{A}$ with the following properties:

(1) The objects of $\underline{A}$ are structured sets, i.e. pairs (X,φ), where X is a set (the underlying set) and φ a structure on X (the underlying $\underline{A}$-structure). (The class of all $\underline{A}$-structures on X is called the $\underline{A}$-fibre of X and denoted by $\underline{A}(X)$.)

(2) The morphisms $f:(X,\varphi) \longrightarrow (Y,\eta)$ are functions $f: X \longrightarrow Y$ which are, in a certain sense, compatible with the underlying $\underline{A}$-structures.

(3) The composition of morphisms is the usual composition of functions.

By a topological category, we mean a "properly fibred topological category" in the sense of Appendix A to [He 74:2].

Proposition 2:

Let $\underline{A}$ be a topological category. For $\varphi, \eta \in \underline{A}(X)$ define:

$\varphi \leq \eta \;:\Longleftrightarrow\; 1_X : (X,\varphi) \longrightarrow (X,\eta)$ is an $\underline{A}$-morphism.

Then $(\underline{A}(X),\leq)$ is a complete lattice.

3. Convergence

If topology is considered as generalized limit theory, a filter $\mathfrak{F}$ converges to a point a, in symbols: $\mathfrak{F} \longrightarrow a$, iff $\mathfrak{F}$ contains the neighbourhood filter $\mathfrak{U}(a)$ of a. This convergence has three essential properties:

$$(1)\; \mathfrak{F} \longrightarrow a \wedge \mathfrak{G} \supset \mathfrak{F} \;\Longrightarrow\; \mathfrak{G} \longrightarrow a,$$
$$(2)\; [a] \longrightarrow a,$$
$$(3)\; \mathfrak{F} \longrightarrow a \;\Longrightarrow\; \mathfrak{F} \cap [a] \longrightarrow a$$

(where $\mathfrak{F}, \mathfrak{G}$ denote filters). The following categories are induced:

Definition 4:

A function

$$\mathbb{C} : \begin{cases} X \longrightarrow \mathfrak{P}(\mathbb{F}(X)) \\ a \longrightarrow \mathbb{C}(a) \end{cases}$$

which satisfies (1) and (2) (resp. (1),(2), and (3)) is called (localized) filter convergence structure on X. ($\mathbb{C}(a)$ may be regarded as the set of all filters converging to a.)

$(X,\mathbb{C})$ is then called (localized) filter convergence space.

A function $f : (X,\mathbb{C}) \longrightarrow (Y,\mathbb{D})$ between filter convergence spaces is continuous or convergence preserving $:\Longleftrightarrow$

$\mathfrak{F} \in \mathbb{C}(a) \;\Longrightarrow\; [f(\mathfrak{F})] \in \mathbb{D}(f(a))$.

The category of filter convergence spaces and convergence preserving maps is denoted by $\underline{FCO}$, the subcategory of localized filter convergence spaces by $\underline{LFCO}$.

FCO and LFCO are due to Kent (FCO: [Ke 64], there: convergence functions; LFCO: [Ke 67], there: convergence structures).

All categories $\underline{A}$ investigated in this note are topological. Hence, by Proposition 2, $(\underline{A}(X),\leq)$ is always a complete lattice. The partial order of the complete lattice $(FCO(X),\leq)$ may be characterized as follows:

$$\mathbb{C} \leq \mathbb{D} \;\Longleftrightarrow\; \mathbb{C}(a) \subset \mathbb{D}(a).$$

The category TOP of topological spaces is a bireflective subcategory of FCO (and LFCO) if one considers each topology as a filter convergence structure in the following way: $\mathbb{C}(a) = \{\mathfrak{F} \mid \mathfrak{F} \in \mathbb{F}(X) \wedge \mathfrak{F} \supset \mathfrak{U}(a)\}$. In this context, bireflective means that for each $\mathbb{C} \in FCO(X)$, there exists a smallest topology $\mathbb{C}^{TOP}$ which is larger than $\mathbb{C}$, and if $f : (X,\mathbb{C}) \longrightarrow (Y,\mathbb{D})$ is continuous, then so is $f : (X,\mathbb{C}^{TOP}) \longrightarrow (Y,\mathbb{D}^{TOP})$.

FCO and LFCO are cartesian closed; for LFCO, this was shown by Nel [Ne 75]. Until now, nobody has studied the relations between FCO and LFCO.

Theorem 1:

LFCO is as well a bireflective as a bicoreflective subcategory of FCO.

Proof: For $(X,\mathbb{C}) \in FCO$, $1_X : (X,\mathbb{C}) \longrightarrow (X,\mathbb{C}^{LFCO})$ is the bireflection and $1_X : (X,\mathbb{C}_{LFCO}) \longrightarrow (X,\mathbb{C})$ is the bicoreflection, where
$\mathbb{C}^{LFCO}(a) = \{ \mathcal{G} \mid \mathcal{G} \in \mathbb{F}(X) \wedge \exists \mathcal{F} (\mathcal{F} \in \mathbb{C}(a) \wedge \mathcal{F} \cap [a] \subset \mathcal{G}) \}$ and
$\mathbb{C}_{LFCO}(a) = \{ \mathcal{F} \mid \mathcal{F} \in \mathbb{C}(a) \wedge \mathcal{F} \cap [a] \in \mathbb{C}(a) \}$. (For a function $f : X \longrightarrow Y$, $a \in X$ and $\mathcal{F} \in \mathbb{F}(X)$: $[f(\mathcal{F} \cap [a])] = [f(\mathcal{F})] \cap [f(a)]$.)

(At the end of this paper, you can find a diagram containing all considered categories and the isomorphisms, bireflections, and bicoreflections between them.)

4. Nearness

Topology may be described as nearness between points and sets; then proximity is nearness between two sets, contiguity nearness between finite collections of sets, and nearness is nearness between arbitrary collections of sets The last gives rise to the category NEAR of nearness spaces and nearness preserving maps. The following definition includes those nearness categories which are essential for this note.

Definition 5:

(1) $\xi \subset \mathfrak{P}^2(X)$ is called prenearness structure on X $:\Longleftrightarrow$

(N_1) $\mathcal{A} \in \xi \wedge \mathcal{B} \subset \mathcal{A} \Rightarrow \mathcal{B} \in \xi$

$\mathcal{A} \in \xi \Rightarrow [\mathcal{A}] \in \xi$

(N_2) $[a] \in \xi$

(N_3) $\emptyset \neq \xi \neq \mathfrak{P}^2(X)$.

(X,ξ) is then called prenearness space.

A map $f : (X,\xi) \longrightarrow (Y,\eta)$ between prenearness spaces is continuous or nearness preserving $:\Longleftrightarrow f(\xi) \subset \eta$.

The category of prenearness spaces and nearness preserving maps is denoted by PNEAR.

(2) $\xi \in PNEAR(X)$ is called seminearness structure on X $:\Longleftrightarrow$

(N_4) $\mathcal{S}_1 \cap \mathcal{S}_2 \in \xi \Rightarrow \mathcal{S}_1 \in \xi \vee \mathcal{S}_2 \in \xi$ for $\mathcal{S}_1, \mathcal{S}_2 \in \mathbb{S}(X)$

(seminearness space, SNEAR).

(3) $\xi \in$ SNEAR(X) is called nearness structure on X $:\Longleftrightarrow$

(N_5) $cl_\xi \mathfrak{A} = \{ cl_\xi A \mid A \in \mathfrak{A} \} \in \xi \Rightarrow \mathfrak{A} \in \xi$

$(cl_\xi A = \{ a \mid a \in X \wedge \{\{a\},A\} \in \xi \})$

(nearness space, NEAR).

(4) $\xi \in$ NEAR(X) is called topological nearness structure on X $:\Longleftrightarrow$

(N_6) $\mathfrak{A} \in \xi \Rightarrow \bigcap cl_\xi \mathfrak{A} \neq \emptyset$

(topological nearness space, TNEAR).

(5) $\xi \in$ PNEAR(X) is called grill structure on X $:\Longleftrightarrow$

(GRILL) $\mathfrak{A} \in \xi \Rightarrow \exists \mathfrak{G} (\mathfrak{G} \in \xi \cap \mathbb{G}(X) \wedge \mathfrak{A} \subset \mathfrak{G})$

(grill space, GRILL).

(6) $\xi \in$ GRILL(X) is called localized grill structure on X $:\Longleftrightarrow$

(LGRILL) $\mathfrak{G} \in \xi \cap \mathbb{G}(X) \Rightarrow \exists a (a \in X \wedge \mathfrak{G} \cup [a] \in \xi)$

(localized grill space, LGRILL).

The categories PNEAR, SNEAR, NEAR, and TNEAR were introduced and investigated by Herrlich ([He 74:1], [He 74:2] etc.). In particular, he showed the following connections:

SNEAR is a bicoreflective subcategory of PNEAR.

NEAR is bireflective in SNEAR.

TNEAR is bicoreflective in NEAR and isomorphic to the R_0-topological or symmetric topological spaces (R_0TOP) of Shanin [Sh 43] and Davis [Da 61]. (A topological spacc is called symmetric iff it satisfies the following axiom:

(R_0) $a \in cl\{b\} \Rightarrow b \in cl\{a\}$.

R_0TOP is bireflective in TOP.)

The uniform spaces (UNIF), contiguity spaces and proximity spaces are isomorphic to bireflective subcategories of NEAR.

For $\xi \in$ PNEAR(X) and $a,b \in X$, the following condition always holds: $a \in cl_\xi\{b\} \Rightarrow b \in cl_\xi\{a\}$. Hence the nearness concept has a built-in symmetry, which in TOP corresponds to the symmetry axiom (R_0). Therefore not all topological spaces, but only the symmetric ones, may be regarded as prenearness spaces. (On a set with two elements, one has three non-isomorphic topologies, but only two prenearness structures.) This leads to the following problem of Herrlich:

Problem 1: [He 74:2] Appendix B

Is there a concept which includes as well the topological spaces as the prenearness spaces?

The category of merotopic spaces and (merotopically) continuous maps (MER) of Katětov [Ka 65] is isomorphic to SNEAR [He 74:2]. Katětov investigated two subcategories of MER: the filter merotopic spaces (FMER) and the localized filter merotopic spaces (LFMER). By restricting the isomorphism between MER and SNEAR to FMER and LFMER, Robertson [Ro 75] ([BHR 76]) got the nearness categories GRILL and LGRILL (there: R_oCon). He proved:

GRILL is a bicoreflective subcategory of SNEAR.
LGRILL is bicoreflective in GRILL.
TNEAR is bireflective in LGRILL.
Consequently one has: LFMER bic FMER bic MER.

5. Connections between convergence and nearness

In this section, we study the connections between convergence and nearness. Till now we know:

Connection 1: [He 74:1]
TNEAR $\cong$ R_oTOP.

Generalizing the (R_o)-axiom from TOP to LFCO, Robertson [Ro 75] obtained the category R_oLFCO (there: R_oCon) of symmetric localized filter convergence spaces:

Definition 6:
$\mathbb{C} \in$ LFCO(X) is called symmetric $:\Longleftrightarrow$
(R_o) $\mathfrak{F} \cap [a] \in \bigcup\{ \mathbb{C}(b) \mid b \in X \} \implies \mathfrak{F} \in \mathbb{C}(a)$ for $\mathfrak{F} \in \mathbb{F}(X)$.
(Notice: $\bigcup\{ \mathbb{C}(b) \mid b \in X \}$ is the set of all convergent filters.)

Robertson showed that R_oLFCO is bireflective in LFCO. Moreover, we have:

Theorem 2:
R_oLFCO is both a bireflective and bicoreflective subcategory of LFCO and FCO.

Proof: For $(X,\mathbb{C}) \in$ LFCO, $1_X : (X,\mathbb{C}) \longrightarrow (X,\mathbb{C}^{R_o\text{LFCO}})$ is the bireflection and $1_X : (X,\mathbb{C}_{R_o\text{LFCO}}) \longrightarrow (X,\mathbb{C})$ is the bicoreflection, where
$\mathbb{C}^{R_o\text{LFCO}}(a) = \{ \mathfrak{F} \mid \mathfrak{F} \in \bigcup\{\mathbb{C}(b) \mid b \in X\} \wedge \mathfrak{F} \cap [a] \in \bigcup\{\mathbb{C}(b) \mid b \in X \} \}$ and
$\mathbb{C}_{R_o\text{LFCO}}(a) =$
$\mathbb{C}(a) \setminus \{ \mathfrak{G} \mid \mathfrak{G} \in \mathbb{F}(X) \wedge \exists \mathfrak{F} \exists b (\mathfrak{F} \in \mathbb{F}(X) \wedge b \in X \wedge \mathfrak{G} = \mathfrak{F} \cap [b] \wedge \mathfrak{F} \notin \mathbb{C}(b)) \}$.
The application of Theorem 1 completes the proof.

The following generalization of Connection 1 was pointed out by Robertson:

Connection 2: [Ro 75]
LGRILL $\cong$ R_oLFCO.

Connections 1 and 2 give rise to the following problems:

Problem 2:
Are there nearness categories larger than LGRILL which may be regarded as convergence categories?

Problem 3:
Are there convergence categories larger than R_0LFCO which may be considered as nearness categories?

As a first step for the solution of these problems, we introduce the STACK-concept which is equivalent to the nearness concept and turns out to be the natural link between nearness and convergence.

Definition 7:
(1) $\xi \subset \mathfrak{P}^2(X)$ is called stack structure on X $:\Longleftrightarrow$

(S_1) $\mathfrak{A} \in \xi \wedge \mathfrak{A} \subset \mathfrak{B} \Longrightarrow \mathfrak{B} \in \xi$

$\mathfrak{A} \in \xi \Longrightarrow]\mathfrak{A}[\in \xi$

(S_2) $[a] \in \xi$

(S_3) $\emptyset \neq \xi \neq \mathfrak{P}^2(X)$.

(X,ξ) is then called stack space.

A map $f : (X,\xi) \longrightarrow (Y,\eta)$ between stack spaces is continuous $:\Longleftrightarrow$
$[f(\xi)]$ $(= \{[f(\mathfrak{A})] \mid \mathfrak{A} \in \xi\}) \subset \eta$.

The category of stack spaces and continuous maps is denoted by STACK.

(2) $\xi \in$ STACK(X) is called filter structure on X $:\Longleftrightarrow$

(FILTER) $\mathfrak{A} \in \xi \Longrightarrow \exists \mathfrak{F}(\mathfrak{F} \in \xi \cap \mathbb{F}(X) \wedge \mathfrak{F} \subset \mathfrak{A})$

(filter space, FILTER).

By Proposition 1, we obtain:

Proposition 3:
(1) (X,ξ) is a stack space iff $(X, \mathrm{ch}\,\xi)$ is a prenearness space.
(2) (X,ξ) is a filter space iff $(X, \mathrm{ch}\,\xi)$ is a grill space.

Proof: (1) We show that $\xi \subset \mathfrak{P}^2(X)$ satisfies $(S_1),(S_2),(S_3)$ of Definition 7 iff $\mathrm{ch}\,\xi = \{\mathrm{ch}\,\mathfrak{A} \mid \mathfrak{A} \in \xi\}$ fulfills $(N_1),(N_2),(N_3)$ of Definition 5.
$\mathrm{ch}\,\xi$ satisfies (N_1) $\Longrightarrow$ ξ satisfies (S_1):
Let $\mathfrak{A} \in \xi$, $\mathfrak{B} \subset \mathfrak{P}(X)$ with $\mathfrak{A} \subset \mathfrak{B}$. Then we have $\mathrm{ch}\,\mathfrak{A} \in \mathrm{ch}\,\xi$ and $\mathrm{ch}\,\mathfrak{B} \subset \mathrm{ch}\,\mathfrak{A}$. By (N_1), we conclude $\mathrm{ch}\,\mathfrak{B} \in \mathrm{ch}\,\xi$. Hence $\mathfrak{B} = \mathrm{ch}^2\mathfrak{B} \in \mathrm{ch}^2\,\xi = \xi$. Now let us show $]\mathfrak{A}[\in \xi$. By (N_1), we infer from $\mathrm{ch}\,\mathfrak{A} \in \mathrm{ch}\,\xi$: $\mathrm{ch}\,]\mathfrak{A}[= [\mathrm{ch}\,\mathfrak{A}] \in \mathrm{ch}\,\xi$, and thus $]\mathfrak{A}[= \mathrm{ch}^2\,]\mathfrak{A}[\in \mathrm{ch}^2\,\xi = \xi$.
The converse implication is shown analogously.

(S_2) $[a] \in \xi \iff (N_2)$ $[a] \in \mathrm{ch}\,\xi$: Proposition 1 (4)
(S_3) $\emptyset \neq \xi \neq \mathfrak{P}^2(X) \iff (N_3)$ $\emptyset \neq \mathrm{ch}\,\xi \neq \mathfrak{P}^2(X)$: obvious
(2) (1), Proposition 1 (2),(3)

Thus the duality function ch induces the following isomorphisms:

Theorem 3:
(1) STACK $\cong$ PNEAR
(2) FILTER $\cong$ GRILL.

Proof: (1) Proposition 3 (1)
$f : (X,\xi) \longrightarrow (Y,\eta)$ STACK-continuous $\iff$ $f : (X,\mathrm{ch}\,\xi) \longrightarrow (Y,\mathrm{ch}\,\eta)$ PNEAR-continuous: [Schw 77] Satz 5.13 (1),(2)
(2) (1), Proposition 3 (2)

Since GRILL is bicoreflective in PNEAR, FILTER is a bicoreflective subcategory of STACK.

Let us now study the relations between FILTER and FCO. A filter convergence structure $\mathbb{C}$ on X is a function $\mathbb{C} : X \longrightarrow \mathfrak{P}(\mathbb{F}(X))$. $\xi \in$ FILTER(X) is completely determined by $\xi \cap \mathbb{F}(X)$, the set of its filters. $\xi \cap \mathbb{F}(X)$ is an element of $\mathfrak{P}(\mathbb{F}(X))$ and may be regarded as a constant function. This motivates the following definition and Theorem 4 (2).

Definition 8:
$\mathbb{C} \in$ FCO(X) is called constant filter convergence structure on X $:\iff$ $\mathbb{C}$ is a constant function, i.e. each point has the same convergent filters (constant filter convergence space, ConsFCO).

To simplify notation in the following proofs, we define for $\alpha \subset \mathfrak{P}^2(X)$:

$$\uparrow\alpha\uparrow := \{\mathcal{B} \mid \mathcal{B} \subset \mathfrak{P}(X) \wedge \exists \mathfrak{A}(\mathfrak{A} \in \alpha \wedge \mathfrak{A} \subset \mathcal{B})\}.$$

$\uparrow\alpha\uparrow$ is the closure of α with respect to supersystems.

ConsFCO is the natural link between FILTER and FCO:

Theorem 4:
(1) ConsFCO bir FCO
(2) FILTER $\cong$ ConsFCO.

Proof: (1) For $(X,\mathbb{C}) \in$ FCO, $1_X : (X,\mathbb{C}) \longrightarrow (X,\mathbb{C}^{\mathrm{ConsFCO}})$ is the bireflection, where $\mathbb{C}^{\mathrm{ConsFCO}}(a) = \bigcup\{\mathbb{C}(b) \mid b \in X\}$.
(2) For $(X,\xi) \in$ FILTER, define $\mathbb{C}_\xi(a) := \xi \cap \mathbb{F}(X)$. Then $(X,\mathbb{C}_\xi) \in$ ConsFCO, and $\uparrow\bigcup\{\mathbb{C}_\xi(b) \mid b \in X\}\uparrow = \xi$. The coordination $(X,\xi) \longrightarrow (X,\mathbb{C}_\xi)$ induces an isomorphism from FILTER to ConsFCO.

By Theorem 3 (2), we conclude:

Connection 3:
GRILL $\cong$ ConsFCO .

Since LGRILL is bicoreflective in GRILL, we have by Connection 2:
R_oLFCO ibic ConsFCO.

If one considers the essential set systems in FCO and in STACK (or PNEAR), it turns out that a filter convergence structure is described by its convergent filters, while a stack structure (or a prenearness structure) is completely determined by the set of its stacks. From this, we get the natural idea: Replacing in the convergence concept filters by stacks, filter convergence turns to stack convergence.

Definition 9:
A function

$$\mathbb{K} : \begin{cases} X \longrightarrow \mathfrak{P}(\mathbb{S}(X) \setminus \{\emptyset\}) \\ a \longrightarrow \mathbb{K}(a) \end{cases}$$

is called stack convergence structure on X $:\Longleftrightarrow$
(1) $\mathfrak{T} \in \mathbb{K}(a) \wedge \mathfrak{R} \supset \mathfrak{T} \Rightarrow \mathfrak{R} \in \mathbb{K}(a)$ (where $\mathfrak{R}, \mathfrak{T}$ denote stacks)
(2) $[a] \in \mathbb{K}(a)$.
$(X,\mathbb{K})$ is then called stack convergence space.
A map $f : (X,\mathbb{K}) \longrightarrow (Y,\mathbb{L})$ between stack convergence spaces is continuous or convergence preserving $:\Longleftrightarrow \mathfrak{T} \in \mathbb{K}(a) \Rightarrow [f(\mathfrak{T})] \in \mathbb{L}(f(a))$.
The category of stack convergence spaces and convergence preserving maps is denoted by SCO.
(Compare Definition 4.)

The relation between filter convergence and stack convergence is clarified by

Theorem 5:
FCO is isomorphic to a bicoreflective subcategory of SCO.

Proof: For $(X,\mathbb{C}) \in$ FCO, define $\mathbb{K}_{\mathbb{C}}(a) := \uparrow\mathbb{C}(a)\uparrow \cap \mathbb{S}(X)$. Then $(X,\mathbb{K}_{\mathbb{C}}) \in$ SCO . Defining $\mathbb{D}(a) := \mathbb{K}_{\mathbb{C}}(a) \cap \mathbb{F}(X)$, we obtain $\mathbb{D} = \mathbb{C}$. The coordination $(X,\mathbb{C}) \longrightarrow (X,\mathbb{K}_{\mathbb{C}})$ induces an isomorphism from FCO to a subcategory of SCO, denoted by FCO*. For $(X,\mathbb{K}) \in$ SCO , $1_X : (X,\mathbb{K}_{FCO^*}) \longrightarrow (X,\mathbb{K})$ is the bicoreflection with respect to FCO*, where $\mathbb{K}_{FCO^*}(a) = \uparrow\mathbb{K}(a) \cap \mathbb{F}(X)\uparrow \cap \mathbb{S}(X)$.
Thus we have shown: FCO $\cong$ FCO* bic SCO.

Now, in analogy to Definition 8, we define the category ConsSCO of constant stack convergence spaces:

Definition 10:

$\mathbb{K} \in SCO(X)$ is called constant stack convergence structure on X :$\Longleftrightarrow$
$\mathbb{K}$ is a constant function.

SCO and ConsSCO are topological, but not cartesian closed. The categorical relations between ConsSCO and the categories introduced till now may be described as follows:

Theorem 6:

(1) ConsSCO bir SCO
(2) ConsFCO ibic ConsSCO
(3) STACK $\cong$ ConsSCO .

Proof: (1) In analogy to Theorem 4 (1).
(2) In the proof of Theorem 5, replace FCO by ConsFCO, SCO by ConsSCO and FCO* by ConsFCO*.
(3) In analogy to Theorem 4 (2) .

Now Theorem 3 (1) implies:

Connection 4:

PNEAR $\cong$ ConsSCO .

The connections 3 and 4 show very clearly, that the convergence concept is more comprehensive than the nearness concept.

The following three corollaries solve Problem 1,2,3, respectively.

Corollary 1:

SCO contains both TOP and PNEAR.

Corollary 2:

Each nearness category may be regarded as a convergence category.

Corollary 3:

Each subcategory of ConsSCO may be considered as a nearness category.

Thus we have shown: Convergence contains nearness.

In SCO, the usual categorical constructions, such as products, subobjects, coproducts, and quotients, are very elegantly to describe. By using the bireflectors, bicoreflectors, and isomorphisms between the considered categories, one has these constructions for each category of this note.

The following diagram may clarify the relationships among the categories introduced.

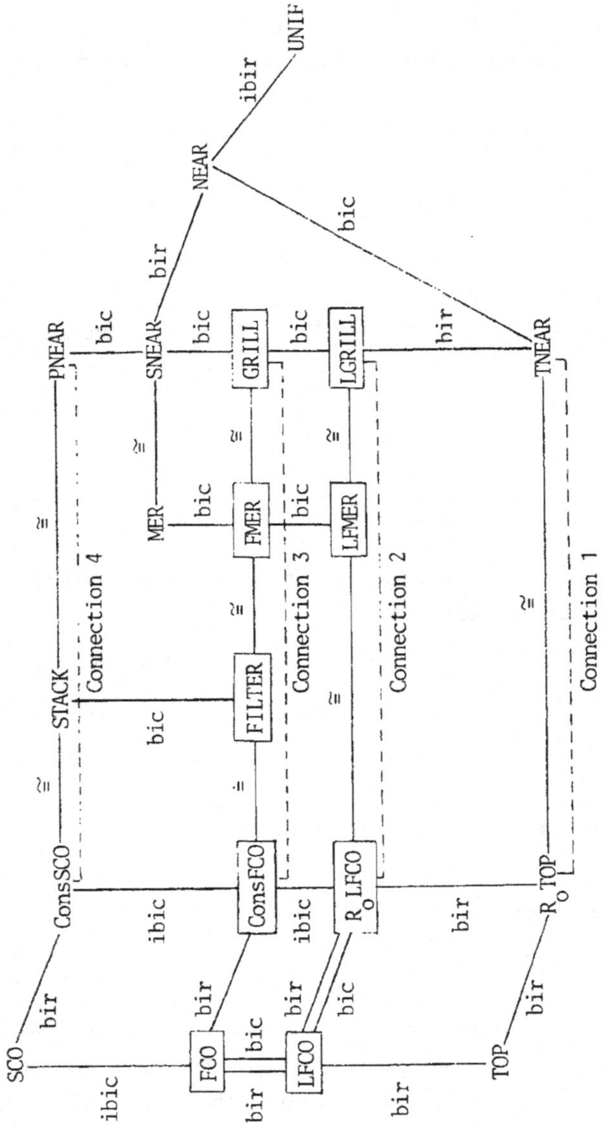

≅, bir, bic, ibir, ibic were introduced in Definition 2.

A category is cartesian closed iff its name is in a small box ☐.

References

[BHR 76] Bentley,H.L., Herrlich,H., Robertson,W.A., Convenient categories for topologists, Comment. Math. Univ. Carolinae 17 (1976), 207-227

[Ch 47] Choquet,G., Sur les notions de filtre et de grille, C.R. Acad. Sci. 224 (1947), 171-173

[Da 61] Davis,A.S., Indexed systems of neighborhoods for general topological spaces, Amer. Math. Monthly 68 (1961), 886-893

[Gr 60] Grimeisen,G., Gefilterte Summation von Filtern und iterierte Grenzprozesse I, Math. Ann. 141 (1960), 318-342

[He 74:1] Herrlich,H., A concept of nearness, Gen. Topol. Appl. 5 (1974), 191-212

[He 74:2] Herrlich,H.,Topological structures, Math. Centre Tract 52 (1974), 59-122

[Ka 65] Katětov,M., On continuity structures and spaces of mappings, Comment. Math. Univ. Carolinae 6 (1965), 257-278

[Ke 64] Kent,D.C., Convergence functions and their related topologies, Fund. Math. 54 (1964), 125-133

[Ke 67] Kent,D.C., On convergence groups and convergence uniformities, Fund. Math. 60 (1967), 213-222

[Ne 75] Nel,L.D., Initially structured categories and cartesian closedness, Can. J. Math. 27 II (1975), 1361-1377

[Ro 75] Robertson,W.A., Convergence as a nearness concept, Thesis, Carleton University Ottawa 1975

[Schw 77] Schwarz,F., Nearnesskategorien und Konvergenzkategorien I: Nearnessstrukturen, Preprint 63, Technische Universität Hannover 1977

[Sh 43] Shanin,N.A., On separation in topological spaces, C.R. (Doklady) Acad. Sci. URSS (N.S.) 38 (1943), 110-113

[Th 73] Thron,W.J., Proximity structures and grills, Math. Ann. 206 (1973), 35-62

Friedhelm Schwarz, Institut für Mathematik, Technische Universität, Welfengarten 1, D-3000 Hannover 1, Federal Republic of Germany

FUNCTORS ON CATEGORIES OF ORDERED TOPOLOGICAL SPACES

by

Z.Semadeni (Warsaw) and H.Zidenberg-Spirydonow (Szczecin)

The topic discussed in this paper has originated in the work by L.Nachbin [8] on ordered topological spaces and that of F.F.Bonsall [1] on semi-algebras of continuous functions.

Nachbin's idea was to generalize various topological concepts to the topology-order case in such a way that for the discrete order (i.e., the order defined as $x \leq y$ iff $x = y$) one gets classical definitions. A preordered topological space X (i.e., a set X provided with a transitive and reflexive relation $\leq$ and with a topology) is said to be monotonically separated iff (the graph of) the relation $\leq$ is closed in $X \times X$ or, equivalently, for any x, y in X the condition $x \not\leq y$ implies that there exist: an open decreasing set V and an open increasing set W such that $x \in V$, $y \in W$ and $V \cap W = \emptyset$ (a subset E of X is called increasing iff the conditions $x \in E$, $x' \geq x$ imply $x' \in E$; E is decreasing iff $X \setminus E$ is increasing). If the preorder on X is discrete, then the above condition simply means that X is a Hausdorff space.

A map $f : X \longrightarrow Y$ (X, Y preordered) is said to be increasing [decreasing] iff $x \leq x'$ implies $f(x) \leq f(x')$ [respectively $f(x) \geq f(x')$]. Note that $E \subset X$ is increasing [decreasing] if and only if its characteristic function $\chi_E : X \longrightarrow \mathbf{2} = \{0,1\}$ is increasing [decreasing]. Moreover, $f : X \longrightarrow Y$ is increasing if and only if for any increasing subset H of Y the counterimage $f^{\leftarrow}(H)$ is increasing. The family $\mathbf{2}^{X,\leqslant}$ of all increasing subsets of X is a topology on X; $f : X \longrightarrow Y$ is increasing if and only if it is $(\mathbf{2}^{X,\leqslant}, \mathbf{2}^{Y,\leqslant})$-continuous.

A preordered topological space X is called <u>monotonically normal</u> iff for every pair of disjoint closed sets E,F such that E is decreasing and F is increasing there exist disjoint open sets V,W such that $E \subset V$, $F \subset W$, V is decreasing and W is increasing. Equivalent conditions in the form of a monotone Urysohn lemma, a monotone between-extension theorem etc. can be found in Nachbin [6], [8], Semadeni and Zidenberg [12], Gh.Bucur [3], H.A.Priestley [9]. If X is monotonically separated, then all the sets

(1) $i(x) = \{y \in X : y \geq x\}$ and $d(x) = \{y \in X : y \leq x\}$

$(x \in X)$ are closed; if X is monotonically normal and all these sets are closed, then X is monotonically separated. X is called <u>monotonically compact</u> if it is a monotonically separated ordered space and, as a topological space, is compact (i.e., compact Hausdorff). Each monotonically compact space is monotonically normal (Nachbin [8], p.48). Moreover, each ordered normed vector space X is monotonically normal; indeed, if E,F are disjoint closed subset of X, E is decreasing and F is increasing, then the distance $\rho(x,E)$ is increasing and continuous with respect to x while the distance $\rho(x,F)$ is decreasing and continuous; consequently, the function

$$f(x) = \frac{\rho(x,E)}{\rho(x,E)+\rho(x,F)} = \frac{1}{1+\dfrac{\rho(x,F)}{\rho(x,E)}}$$

is increasing, continuous, vanishes on E, and equals to 1 on F (cf. Semadeni and Zidenberg [12]).

If X is a preordered topological space, $C^{\uparrow}(X,\leq)$ or shortly $C^{\uparrow}(X)$ will denote the space of bounded nonnegative increasing continuous functions on X. If $\alpha : X \longrightarrow Y$ is a continuous increasing map, then

$$C^{\uparrow}(\alpha) : C^{\uparrow}(Y) \longrightarrow C^{\uparrow}(X)$$

will denote the induced map $C^{\uparrow}(\alpha).g = g \circ \alpha$ for g in $C^{\uparrow}(Y)$.

If A is any subset of the space $C(X)$ of <u>bounded</u> continuous real-valued functions on a topological space X, then the <u>preorder</u> $\leq_A$ <u>determined by</u> A is defined as

$$x \leq_A x' \text{ iff } f(x) \leq f(x') \quad \text{for each } f \text{ in } A.$$

If $\leq$ is any preorder on X, then the set $A = C^{\uparrow}(X,\leq)$ determines, in turn, a preorder $\leq_A$ on X; it is obvious that $x \leq x'$ implies $x \leq_A x'$ and that $C^{\uparrow}(X,\leq) = C^{\uparrow}(X,\leq_A)$. Form the above quoted results it follows that the given preorder $\leq$ on a compact space X coincides with $\leq_A$ if and only if $(X,\leq)$ is monotonically separated (Bonsall [1]).

The converse problem is: given a compact space X and a subset A of the cone $C(X,\mathbf{R}_+)$ of nonnegative functions in $C(X)$, find a condition necessary and sufficient in order that $A = C^{\uparrow}(X,\leq_A)$. Of course, each function in A is increasing with respect to the order $\leq_A$, i.e., $A \subset C^{\uparrow}(X,\leq_A)$. Bonsall [1] (cf. also [2]) has shown that $A = C^{\uparrow}(X,\leq_A)$ if and only if A is a closed semi-algebra of type one with unit, i.e.,

(i) A is a norm-closed convex cone,

(ii) $f,g \in A \Longrightarrow fg \in A$,

(iii) $1_X \in A$,

(iv) A is stable under the (increasing) function $u \longmapsto u(1+u)^{-1}$, i.e., $f \in A \Longrightarrow \frac{f}{1+f} \in A$.

By a <u>Bonsall semi-algebra</u> we shall mean any set $A \subset C(X,\mathbf{R}_+)$ satisfying the conditions (i)-(iv) (X may be any topological space). Thus, in case where X is compact, A is a Bonsall semi-algebra if and only if $A = C^{\uparrow}(X,\leq)$ for some preorder $\leq$ on X.

In particular, $C(X,\mathbf{R}_+)$ itself is a Bonsall semi-algebra; is corresponds to the discrete order on X. If X is the empty space $\mathbf{0}$ (which is monotonically compact), then $C^{\uparrow}(\mathbf{0}) = C(\mathbf{0},\mathbf{R}_+) = \{0\}$, where 0 denotes the empty function $\mathbf{0} \longrightarrow \mathbf{R}_+$, which is identically equal to $1_{\mathbf{0}}$, the unit of $C^{\uparrow}(\mathbf{0})$. The closed half-line $\mathbf{R}_+$ will also be regarded as a Bonsall semi-algebra identified with $C(\mathbf{1},\mathbf{R}_+)$.

Bons will denote the category of Bonsall semi-algebras and Bonsall morphisms, i.e., triples (φ, A, B), written as $\varphi : A \longrightarrow B$, where A, B are Bonsall semi-algebras, $A \subset C(X, \mathbf{R}_+)$, $B \subset C(Y, \mathbf{R}_+)$, and the map $\varphi : A \longrightarrow B$ is additive, positive-homogeneous (i.e., $\varphi(cf) = c\varphi(f)$ for c in $\mathbf{R}_+$), increasing, multiplicative and unit-preserving (i.e., $\varphi(1_X) = 1_Y$). The unique map from $C^{\uparrow}(X)$ to $C^{\uparrow}(\mathbf{0})$ is a Bons-morphism; if $X \neq \mathbf{0}$, there is no Bons-morphism $C^{\uparrow}(\mathbf{0}) \longrightarrow C^{\uparrow}(X)$, because the condition $\varphi(0) = 1_X$ cannot be satisfied.

$\underline{\mathrm{Top}}_{p\leq}$ will denote the category of preordered topological spaces and increasing continuous maps; $\underline{\mathrm{Top}}_{\leq}$ is the full subcategory of ordered topological spaces, and $\underline{\mathrm{Comp}}_{\leq}$ is the full subcategory of monotonically compact spaces. It is clear that

(2) $\quad C^{\uparrow} : \underline{\mathrm{Top}}_{p\leq} \longrightarrow \underline{\mathrm{Bons}}$

is a well-defined contravariant functor. We shall now define another contravariant functor

(3) $\quad \mathfrak{X}^{\uparrow} : \underline{\mathrm{Bons}} \longrightarrow \underline{\mathrm{Comp}}_{\leq}$

as follows: If $A \subset C(X, \mathbf{R}_+)$ is a Bonsall semi-algebra, $\mathfrak{X}^{\uparrow}(A)$ is the set of all Bons-morphisms $\varphi : A \longrightarrow \mathbf{R}_+$ provided with the weak topology $\sigma(\mathfrak{X}^{\uparrow}(A), A)$ and the order

(4) $\quad \varphi \leq \psi \quad$ if $\quad \varphi(f) \leq \psi(f) \quad$ for f in A.

The set $A - A = \{f - g \,\vdots\, f, g \in A\}$ is a subalgebra of $C(X)$ containing 1_X. The uniform closure $\check{A}$ of $A - A$ is a Banach algebra isometrically isomorphic to $C(\mathfrak{X}(\check{A}))$, where $\mathfrak{X}(\check{A})$ is the set of all multiplicative unit-preserving linear functionals $A \longrightarrow \mathbf{R}$, compact in the weak topology $\sigma(\mathfrak{X}(\check{A}), \check{A})$. If $\varphi \in \mathfrak{X}^{\uparrow}(A)$, then φ has a unique extension to a bounded linear functional $\check{\varphi}$ on $\check{A}$; it is clear that $\check{\varphi} \in \mathfrak{X}(\check{A})$ and that the map $\varphi \longmapsto \check{\varphi}$ is a homeomorphism; consequently $\mathfrak{X}^{\uparrow}(A)$ is compact. It is easy to verify that $\mathfrak{X}^{\uparrow}(A)$ with the order (4) is monotonically separated and hence an object of $\underline{\text{Comp}}_{\leq}$.

If $\varphi : A \longrightarrow B$ is a $\underline{\text{B}}$ons-morphism, then the map

$$\mathfrak{X}^{\uparrow}(\varphi) : \mathfrak{X}^{\uparrow}(B) \longrightarrow \mathfrak{X}^{\uparrow}(A),$$

defined as $\mathfrak{X}^{\uparrow}(\varphi).\eta = \eta \circ \varphi$ for η in $\mathfrak{X}^{\uparrow}(B)$, is obviously continuous and increasing. This yields the functor (3).

If X is a preordered topological space, then the map

$$(5) \quad \delta^{(X)} : X \longrightarrow \mathfrak{X}^{\uparrow} C^{\uparrow}(X)$$

is defined as $\delta_x^{(X)}(f) = f(x)$ for f in $C^{\uparrow}(X)$, x in X. If A is a Bonsall semi-algebra, then

$$(6) \quad \Delta^{(A)} : A \longrightarrow C^{\uparrow}\mathfrak{X}^{\uparrow}(A)$$

is defined as $\Delta_f^{(A)}(\varphi) = \varphi(f)$ for f in A, φ in $\mathfrak{X}^{\uparrow}(A)$.

We can now formulate a categorical version of the Bonsall duality between monotonically compact spaces and Bonsall semi-algebras.

<u>Theorem 1</u>. The maps (5) and (6) yield natural equivalences

$$\delta : \underline{\text{Comp}}_{\leq} \longrightarrow \underline{\text{B}}\text{ons}$$

and

$$\Delta : \underline{\text{B}}\text{ons} \longrightarrow \underline{\text{Comp}}_{\leq}.$$

The space $\beta^{\uparrow}(X) = \mathfrak{X}^{\uparrow}C^{\uparrow}(X)$ together with the map (5) will be called the <u>monotone compactification</u> of X. It is characterized by the following property: For each preordered topological space X, each monotonically compact space Y and each increasing continuous map $\xi : X \longrightarrow Y$ there is a unique increasing continuous map $\vartheta : \beta^{\uparrow}(X) \longrightarrow X$ such that the diagram

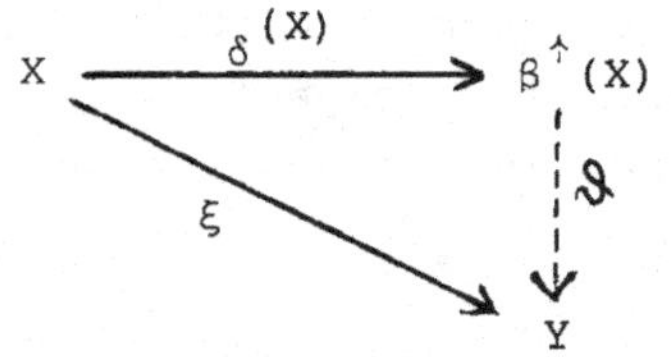

is commutative. The proof of this property is similar to that in the case of the Stone-Čech compactification β (see e.g. [11], §14). If the preorder of X is discrete, then $\beta^{\uparrow}(X)$ may be identified with $\beta(X)$. If it is antidiscrete (i.e., if $x \leq x'$ holds for all x, x' in X) and $X \neq \mathbf{0}$, then $\beta^{\uparrow}(X) \equiv \mathbf{1}$ (here $X \equiv Y$ means that X is homeomorphic to Y; $\mathbf{1}$ is a singleton). If X is a bounded open subset of $\mathbf{R}$ with the usual order and topology and $\mathbf{R} \setminus X$ has no isolated points, then $\beta^{\uparrow}(X)$ is the closure of X in $\mathbf{R}$.

A preordered topological space X is called monotonically pseudo-uniformizable iff the following two conditions are satisfied:

(a) for every x_o in X and every neighborhood V of x_o there exist f, g in $C(X, \mathbf{R}_+)$ such that f is increasing, g is decreasing, $f \leq 1_X$, $g \leq 1_X$, $f(x_o) = 1$, $g(x_o) = 1$, and $\inf(f(x), g(x)) = 0$ for x in $X \setminus V$,

(b) if $x \not\geq y$, then there exists an f in $C^{\uparrow}(X)$ such that $f(x) < f(y)$.

X is called monotonically uniformizable if, additionally,

(c) the preorder of X is an order;

the equivalent conditions are: (c_1) X is a T_o-space, (c_2) X is a $T_{3\frac{1}{2}}$-space, i.e., Hausdorff uniformizable, (c_3) $C^{\uparrow}(X)$ separates the points of X. The importance of these concepts is justified by the following theorem, which is a refinement of results of Nachbin ([7], [8], p.54 and 103):

Each of the following conditions is necessary and sufficient in order that a preordered topological space X be monotonically uniformizable:

(i) the canonical map $x \longmapsto (f(x))_{f \in B}$, where $B = \{h \in C^{\uparrow}(X) : h \leq 1_X\}$, is an order-preserving homeomorphic embedding of X into the product $\mathbf{I}^B$, where $\mathbf{I} = [0,1]$;

(ii) X is a subspace of a monotonically compact space;

(iii) the map (5) is an order-preserving homeomorphism.

There are some sufficient conditions for a space X to be monotonically uniformizable, e.g., it is enough if X is a Hausdorff uniform space and a sub-semilattice with uniformly continuous $(x,y) \longmapsto x \vee y$; a preordered topological Abelian group is monotonically pseudo-uniformizable if (a) for every neighbourhood V of 0 there is a neighbourhood W of 0 such that conditions $x \in W$, $0 \leq y \leq x$ imply $y \in V$; (b) the set $X_+ = \{x \in X : x \geq 0\}$ is closed (Nachbin [8], p. 74 and p.76).

We shall consider the functors shown in the diagram below:

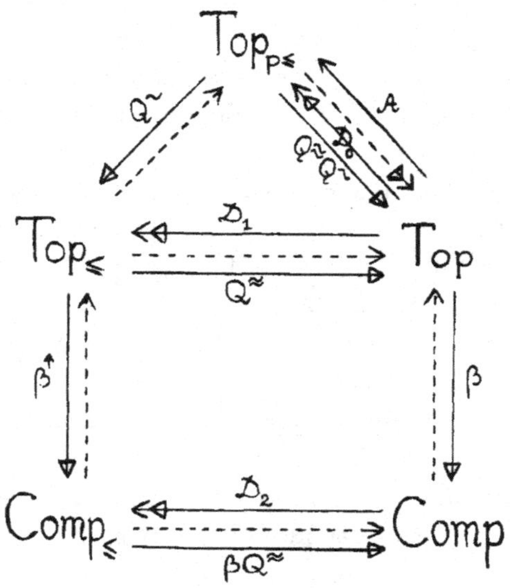

Dotted arrows denote forgetful functors (this concept includes embeddings of subcategories). Tips of the form $\longrightarrow$ denote functors which have left adjoints; tips $\longrightarrow\!\!\triangleright$ denote functors having right adjoints; double tips $\longrightarrow\!\!\triangleright\!\!\triangleright$ denote functors which have both left adjoints and right adjoints.

The functors $\mathcal{D}_0$, $\mathcal{D}_1$, $\mathcal{D}_2$ assign to each topological [resp.

compact] space X the space $\mathcal{D}X$ which is the same topological space with the discrete order. $\mathcal{A}$ assigns to X the same space with the antidiscrete preorder. $Q^{\sim}$ assigns to each preordered topological space X the quotient space $X/\sim$ with respect to the equivalence relation

$$x \sim x' \quad \text{iff} \quad x \leq x' \quad \text{and} \quad x' \leq x.$$

$Q^{\approx}$ assigns to each ordered topological space X the quotient $X/\approx$ with respect to the equivalence relation $\approx$ defined as follows: $x \approx x'$ if there exist $x_1, \ldots, x_{2n}$ such that $x \leq x_1$, $x_1 \geq x_2$, $x_2 \leq x_3, \ldots, x_{2k} \leq x_{2k+1}$, $x_{2k+1} \geq x_{2k+2}, \ldots, x_{2n} \leq x'$. $X/\sim$ and $X/\approx$ are provided with respective quotient topologies. If $\alpha : X \longrightarrow Y$ is a morphism in $\underline{\mathrm{Top}}_{\leq}$, then $Q^{\approx}\alpha$ is the unique continuous map from $X/\approx$ to $Y/\approx$ such that $Q^{\approx}(\alpha)\pi_X = \pi_Y\alpha$, where π_X, π_Y are the respective quotient maps. The remaining functors are familiar: β is the Stone-Čech functor, $\beta Q^{\approx} : \underline{\mathrm{Comp}}_{\leq} \longrightarrow \underline{\mathrm{Comp}}$ is the composition of the embedding functor $\square : \underline{\mathrm{Comp}}_{\leq} \longrightarrow \underline{\mathrm{Top}}_{\leq}$ with $Q^{\approx}$ and β, while $\beta^{\uparrow}$ is the monotone compactification. Note that $\beta^{\uparrow}(Q^{\sim}X) \equiv \beta^{\uparrow}X$ for each object X of $\underline{\mathrm{Top}}_{p\leq}$.

Theorem 2. The functors $\mathcal{D}_0, \mathcal{D}_1, \mathcal{D}_2, Q^{\sim}$, β and $\beta^{\uparrow}$ are left adjoints of the corresponding forgetful functors. The functors $\mathcal{D}_0, \mathcal{D}_1, \mathcal{D}_2$ are also right adjoint of $Q^{\approx}Q^{\sim}$, $Q^{\approx}$ and $\beta Q^{\approx}$, respectively. $\mathcal{A}$ is a right adjoint of the forgetful functor $\square : \underline{\mathrm{Top}}_{p\leq} \longrightarrow \underline{\mathrm{Top}}$.

Proof. For left adjointness of $\mathcal{D}_0, \mathcal{D}_1, \mathcal{D}_2, \beta$ and $Q^{\sim}$ see Herrlich [5], p.15, 57, 80 and 87. The adjointness of $\beta^{\uparrow}$ was formulated above and the adjointness of $\mathcal{A}$ is obvious.

Now, let Y be a $\underline{\mathrm{Top}}$-object and X a $\underline{\mathrm{Top}}_{\leq}$-object. Then each $\underline{\mathrm{Top}}$-morphism $\alpha : Q^{\approx}X \longrightarrow Y$ can be composed with $\pi : X \longrightarrow Q^{\approx}X$ and the continuous map $\tilde{\alpha} = \alpha\pi$ is increasing as a map $X \longrightarrow \mathcal{D}Y$, for if $x_1 \leq x_2$ then $\pi(x_1) = \pi(x_2)$ and $\tilde{\alpha}(x_1) = \tilde{\alpha}(x_2)$. The map $\alpha \longmapsto \alpha\pi$ from the set $\langle Q^{\approx}X, Y\rangle_{\underline{\mathrm{Top}}}$ of $\underline{\mathrm{Top}}$-morphisms $Q^{\approx}X \longrightarrow Y$ into the set $\langle X, \mathcal{D}Y\rangle_{\underline{\mathrm{Top}}_{\leq}}$ of $\underline{\mathrm{Top}}_{\leq}$-morphisms $X \longrightarrow \mathcal{D}Y$ is one-to-one (because π is an epimorphism) and is onto (indeed, if

$\varphi : X \longrightarrow \mathcal{D}Y$ is any increasing continuous map and either $x_1 \leq x_2$ or $x_1 \geq x_2$, then $\varphi(x_1) = \varphi(x_2)$; consequently, $\varphi(x) = \varphi(x')$ whenever $x \approx x'$, and therefore φ can be factored through $Q^{\approx}(X)$. Thus, $Q^{\approx}$ is a left adjoint of $\mathcal{D}_1$ and $Q^{\approx}Q^{\sim}$ is a left adjoint of $\mathcal{D}_o$ (which is the composition of $\mathcal{D}_1$ with the embedding $\square : \underline{\mathrm{Top}}_{\leq} \longrightarrow \underline{\mathrm{Top}}_{p\leq}$). Finally, if Y is compact and X is monotonically compact, we can compose the natural equivalences:

$$\langle X, \mathcal{D}Y\rangle_{\mathrm{Comp}_{\leq}} \cong \langle Q^{\approx}X, Y\rangle_{\mathrm{Top}} \cong \langle \beta Q^{\approx} X, Y\rangle_{\mathrm{Comp}} .$$

We shall now prove that the functors shown in the above diagram have only those adjoints which are listed in Theorem 2. This will follow from Freyd's adjoint functor theorem and the theorem below (see e.g., [11], 12.5.2).

<u>Theorem 3</u>. The categories $\underline{\mathrm{Top}}_{p\leq}$ $\underline{\mathrm{Top}}_{\leq}$, and $\underline{\mathrm{Comp}}_{\leq}$ are complete and cocomplete. The functors $Q^{\sim}$, $Q^{\approx}$, $Q^{\approx}Q^{\sim}$, and $\beta^{\uparrow}$ preserve neither products nor equalizers; $\beta Q^{\approx}$ preserve products but not equalizers. The functor $\mathcal{A}$ preserves coequalizers but not coproducts. The embedding $\square : \underline{\mathrm{Top}}_{\leq} \longrightarrow \underline{\mathrm{Top}}_{p\leq}$ and the forgetful functor $\square : \underline{\mathrm{Top}}_{\leq} \longrightarrow \underline{\mathrm{Top}}$ preserve coproducts but not coequalizers. The analogous functors $\square : \underline{\mathrm{Comp}}_{\leq} \longrightarrow \underline{\mathrm{Top}}_{\leq}$ and $\square : \underline{\mathrm{Comp}}_{\leq} \longrightarrow \underline{\mathrm{Comp}}$ preserve neither coproducts nor coequalizers.

Proof. The constructions of products, coproducts, equalizers and coequalizers in $\underline{\mathrm{Top}}_{p\leq}$, $\underline{\mathrm{Top}}_{\leq}$ and $\underline{\mathrm{Comp}}_{\leq}$ is routine; we shall only note that the empty object is the coproduct of the empty family of objects and this is why we include $\mathbf{0}$ among the objects of $\underline{\mathrm{Comp}}_{\leq}$ and $\{0\}$ among Bonsall semi-algebras.

Let $\mathbf{P}X_t$ be the $\underline{\mathrm{Top}}_{p\leq}$-product of a family $(X_t)_{t\in T}$ of objects and let $\pi_t : X_t \longrightarrow Q^{\sim}(X_t)$ be the quotient maps. By the universal property of $Q^{\sim}$, the morphism $\mathbf{P}\pi_t : \mathbf{P}X_t \longrightarrow \mathbf{P}Q^{\sim}(X_t)$ factors as $\varphi\pi$, where π is the quotient map onto $Q^{\sim}(\mathbf{P}X_t)$. It is clear that $\varphi : Q^{\sim}(\mathbf{P}X_t) \longrightarrow \mathbf{P}Q^{\sim}(X_t)$ is a continuous bijection and both φ and φ^{-1} are increasing. Yet, φ^{-1} need not be continuous. To construct an example note that if $\alpha : Y \longrightarrow Z$ is any quotient map in Top, then defining $y \leq y'$ as equivalent to $\alpha(y) = \alpha(y')$ we infer that $Q^{\sim}Y$ is homeomorphic to Z. Therefore it is enough to take two quotient maps $\pi_i : X_i \longrightarrow Y_i$ in Top $(i = 1,2)$ such that $\pi_1 \times \pi_2 : X_1 \times X_2 \longrightarrow Y_1 \times Y_2$ is not a quotient map (see, e.g., Engelking [4], p.131).

Now, for each $n = 2,3,4,\ldots$ let Z_n be the monotonically compact space $\{1,\ldots,n\}$ with the discrete topology and the order determined by the inequalities $1 < 2$, $2 > 3$, $3 < 4$, etc. Then $\mathsf{P}\, Q^{\approx}(Z_n) \equiv \mathbf{1}$ whereas $Q^{\approx}(\mathsf{P}\, Z_n)$ is infinite. Indeed, let $x = (x_n)$, $y = (y_n)$ belong to $\mathsf{P}\, Z_n$; then $x \approx y$ if and only if the sequence $|x_n - y_n|$ is bounded. Thus, $Q^{\approx}$ does not preserve infinite products (in fact, by modifying the preceding example it is easy to show that $Q^{\approx}$ does not preserve finite products either). However, each set $X/\approx\ = \{y : y \approx x\}$ is dense in $\mathsf{P}\, Z_n$ and therefore $\beta Q^{\approx}(\mathsf{P}\, Z_n)$ is a singleton too. This result shows how to deal with the general case: Let X_t be monotonically compact $(t \in T)$. If $x = (x_t)_{t \in T}$, denote

$$E(x) = \{y \in \mathsf{P}\, X_t : \forall_{t \in T}\ x_t \approxeq y_t\} = \mathsf{P}\, E_t(x_t),$$

where $E_t(x_t) = \{y_t \in X_t : x_t \approxeq y_t\}$ and $x_t \approxeq y_t$ means that $h(x_t/\approx) = h(y_t/\approx)$ for each h in $C(Q^{\approx} X_t)$. Denote $E'_t(x_t)$ $= \{y_t \in X_t : x_t \approx y_t\}$, $E'(x) = \mathsf{P}\, E'_t(x_t)$, $E''(x) = \{y \in \mathsf{P}\, X_t : x \approx y\}$. Let A, A', A'' be the algebra of all functions in $C(\mathsf{P}\, X_t)$ which are constant on each set $E(x)$, $E'(x)$, $E''(x)$, respectively. Obviously $E''(x) \subset E'(x) \subset E(x)$ and hence $A'' \supset A' \supset A$. Moreover, $\mathsf{P}\, \beta Q^{\approx}(X_t)$ may be identified with $\mathfrak{X}(A)$ and $\beta Q^{\approx}(\mathsf{P}\, X_t)$ with $\mathfrak{X}(A'')$. The restriction map $\mathfrak{X}(A'') \longrightarrow \mathfrak{X}(A)$ yields the canonical continuous map ψ from $\beta Q^{\approx}(\mathsf{P}\, X_t)$ onto $\mathsf{P}\, \beta Q^{\approx}(X_t)$. We claim that each set $E''(x)$ is dense in $E'(x)$. Indeed, let $y \in E'(x)$. For each finite subset $F = \{t_1,\ldots,t_k\}$ of T consider $z^i = (z^i_t)$, $1 \leq i \leq k$, where $z^i_{t_j} = y_{t_j}$ for $j = 1,\ldots,i$ and $z^i_t = x_t$ for t in $T \setminus \{t_1,\ldots,t_i\}$. Thus, $x \approx z^1 \approx \ldots \approx z^k$, $z^k \in E''(x)$ and $z^k_t = y_t$ for t in F. Consequently, y belongs to the closure of $E''(x)$. Thus $A'' = A'$. Now, let $f \in A'$. Let $x \in \mathsf{P}\, X_t$ and $s \in T$. Let

$y \in E(x)$ and $y_t = x_t$ for $t \neq s$. If $u \in X_s$, denote $c_s(u) = u$, $c_t(u) = x_t$ for $t \neq s$, and $g(u) = f(c_s(u))$. Then $g \in C(X_s)$ and $u \approx u'$ implies $g(u) = g(u')$; hence $g(u) = h(u/\approx)$ for some h in $C(Q^{\approx}X_s)$. Consequently, $f(y) = h(y_s/\approx) = h(x_s/\approx) = f(x)$. We infer that if $z \in E(x)$ and the set $\{t : z_t \neq x_t\}$ is finite, then $f(z) = f(x)$. Thus, f is constant on a dense subset of $E(x)$. Hence $f \in A$. We have shown that $A = A' = A''$. Therefore, ψ is a bijection.

Since β preserves neither products nor equalizers (cf. e.g. [11], 14.3.3), $\beta^{\uparrow}$ does not preserve them either, as $\beta^{\uparrow}(\mathfrak{D} X) \equiv \beta X$.

To show that $Q^{\sim}$ does not preserve equalizers, consider any maps $\alpha_1 \neq \alpha_2$ from $\mathbf{R}$ to an antidiscrete two-point space; similarly, in the case of $Q^{\approx}$, $Q^{\approx}Q^{\sim}$ and $\beta Q^{\approx}$ consider $\alpha_1 \neq \alpha_2$ form $\mathfrak{D}(\mathbf{I})$ to $\mathbf{I}$, where $\mathbf{I} = [0,1]$.

The functor $\square : \underline{\text{Comp}}_{\leq} \longrightarrow \underline{\text{Comp}}$ does not preserve coequalizers. Indeed, let $X = \{0\}$, $Y = \{0,1,2\}$, and let $\alpha_1(0) = 1$, $\alpha_2(0) = 2$. Then a $\underline{\text{Comp}}$-coequalizer of $\alpha_1, \alpha_2 : X \longrightarrow Y$ has two points, whereas a $\underline{\text{Comp}}_{\leq}$-coequalizer of α_1, α_2 is a singleton.

Now, let $(X_t)_{t \in T}$ be a family of monotonically compact spaces. The canonical surjection

$$\tau : \beta(\textstyle\sum X_t) \longrightarrow \beta^{\uparrow}(\textstyle\sum X_t)$$

from the $\underline{\text{Comp}}$-coproduct onto the $\underline{\text{Comp}}_{\leq}$-coproduct is the restriction map from $\mathfrak{X}(\prod^{\infty} C(X_t))$ onto $\mathfrak{X}(\prod^{\infty} \check{A}_t)$, where $\prod^{\infty}$ denotes the ℓ_{∞}-product ([11], 11.3.1 and 14.3.1), $A_t = C^{\uparrow}(X_t)$ and $\check{A}_t$ is the uniform closure of $A_t - A_t$. By the Stone-Weierstrass theorem, each $A_t - A_t$ is dense in $C(X_t)$; yet $\prod^{\infty} \check{A}_t$ need not fill up all of $\prod^{\infty} C(X_t)$. Consider, e.g., $X_n = \mathbf{I}$ and $f_n(x) = \sin 2^n \pi x$ for x in $\mathbf{I}$, $n = 1,2,\ldots$ Then each f_n can be uniformly approximated by a function of bounded variation $g_n - h_n$ with $g_n, h_n \in C^{\uparrow}(\mathbf{I})$, but the sequence $\|g_n\|$ must be unbounded. Indeed, f_n has 2^{n-1} maxima intertwined with 2^{n-1} minima; if $\|f_n - (g_n - h_n)\| < \varepsilon$, then, by Jordan's theorem, $g_n(1) - g_n(0) > 2^{n+1}(1 - \varepsilon)$. Consequently, in this case, τ is not one-to-one.

We omit easy proofs of the remaining statements of the theorem.

In the following table the symbol + [respectively -] means that the functor listed in the row preserves [does not preserve] the construction listed in the column; $\oplus$ means that the positive answer follows from the adjoint functor theorem (cf. also [10], p. 196).

Functor	product	equalizer	coproduct	coequalizer
$\square : \underline{Comp} \longrightarrow \underline{Top}$	$\oplus$	$\oplus$	-	-
$\square : \underline{Comp}_{\leq} \longrightarrow \underline{Top}_{\leq}$	$\oplus$	$\oplus$	-	-
$\square : \underline{Top}_{\leq} \longrightarrow \underline{Top}_{p\leq}$	$\oplus$	$\oplus$	+	-
$\square : \underline{Top}_{p\leq} \longrightarrow \underline{Top}$	$\oplus$	$\oplus$	$\oplus$	$\oplus$
$\square : \underline{Top}_{\leq} \longrightarrow \underline{Top}$	$\oplus$	$\oplus$	+	-
$\square : \underline{Comp}_{\leq} \longrightarrow \underline{Comp}$	$\oplus$	$\oplus$	-	-
$\mathcal{A} : \underline{Top} \longrightarrow \underline{Top}_{p\leq}$	$\oplus$	$\oplus$	-	+
$\mathcal{D}_0 : \underline{Top} \longrightarrow \underline{Top}_{p\leq}$	$\oplus$	$\oplus$	$\oplus$	$\oplus$
$\mathcal{D}_1 : \underline{Top} \longrightarrow \underline{Top}_{\leq}$	$\oplus$	$\oplus$	$\oplus$	$\oplus$
$\mathcal{D}_2 : \underline{Comp} \longrightarrow \underline{Comp}_{\leq}$	$\oplus$	$\oplus$	$\oplus$	$\oplus$
$\beta : \underline{Top} \longrightarrow \underline{Comp}$	-	-	$\oplus$	$\oplus$
$\beta^{\uparrow} : \underline{Top}_{\leq} \longrightarrow \underline{Comp}_{\leq}$	-	-	$\oplus$	$\oplus$
$Q^{\sim} : \underline{Top}_{p\leq} \longrightarrow \underline{Top}_{\leq}$	-	-	$\oplus$	$\oplus$
$Q^{\approx} : \underline{Top}_{\leq} \longrightarrow \underline{Top}$	-	-	$\oplus$	$\oplus$
$Q^{\approx}Q^{\sim} : \underline{Top}_{p\leq} \longrightarrow \underline{Top}$	-	-	$\oplus$	$\oplus$
$\beta Q^{\approx} : \underline{Comp}_{\leq} \longrightarrow \underline{Comp}$	+	-	$\oplus$	$\oplus$

References

[1] F.F.Bonsall, Semi-algebras of continuous functions, Proc. London Math. Soc. 10(1960), 122-140.

[2] F.F.Bonsall, Stability theorems for cones and wedges of continuous functions, J.Functional Analysis 4(1969), 135-145.

[3] Gh.Bucur, Couples normaux de quasi-topologies. Théorèmes de prolongement, Rev.Roumaine Math. Pures Appl. 14(1969), 1395-1422.

[4] R.Engelking, General topology, PWN, Warszawa 1977.

[5] H.Herrlich, Topologische Reflexionen und Coreflexionen, Springer Verlag, Lecture Notes in Math. 78(1968).

[6] L.Nachbin, Sur les espaces topologiques ordonnés, C.R.Acad. Sci. Paris 226(1948), 381-382.

[7] L.Nachbin, Sur les espaces uniformisables ordonnés, ibid. p.547.

[8] L.Nachbin, Topology and order, Van Nostrand, Princeton 1965.

[9] H.A.Priestley, Separation theorems for semicontinuous functions on normally ordered topological spaces, J.London Math. Soc. (2) 3(1971), 371-377.

[10] Z.Semadeni, Categorical approach to extension problems. Contributions to extension theory of topological structures (Proc.Symposium Berlin 1967), VEB Deutscher Verlag der Wissenschaften, Berlin 1969, 193-198.

[11] Z.Semadeni, Banach spaces of continuous functions, vol.1, PWN, Warszawa 1971.

[12] Z.Semadeni and H.Zidenberg, On preordered topological spaces and increasing semicontinuous functions, Prace Mat. (Comm. Math.) 41(1968), 313-316.

Addresses of the authors:

Instytut Matematyczny
Polskiej Akademii Nauk
ul.Śniadeckich 8, skr.poczt.137
00-950 Warszawa, Poland

Zakład Matematyki
Wyższej Szkoły Pedagogicznej
ul.Wielkopolska 15
70-387 Szczecin, Poland

ON THE COPRODUCT OF THE TOPOLOGICAL GROUPS $\mathbb{Q}$ AND $\mathbb{Z}_2$

by

Barbara V. Smith Thomas

Since 1950, when Graev first established the existencc of coproducts of topological groups, it has been impossible to obtain an explicit construction of this coproduct topology. It was shown by Morris, Ordman, and Thompson in 1973 that if G and H are both k_ω-spaces then $G \amalg H$ is homeomorphic to $G \times H \times F_G(G \wedge H)$, where $F_G(G \wedge H)$ denotes the Graev free topological group over $G \wedge H = G \times H/(G \times \{e_H\}) \cup (\{e_G\} \times H)$ [M-O-T]. It was hoped that his homeomorphism would hold in general. Later Fay and Thomas showed that for arbitrary topological groups the topological identification of $G \amalg H$ with $G \times H \times F_G(G \wedge H)$ was equivalent to the continuity of a certain function $\Psi: G \times H \times F_G(G \wedge H) \longrightarrow F_G(G \wedge H)$ [F-T].

In this note we see that even for such nice groups as $G = \mathbb{Q}$ and $H = \mathbb{Z}_2$ this function Ψ fails to be continuous. We use results and notation of [M-O-T], of [F-T], and of a paper of Fay, Ordman, and Thomas on the Graev free topological group over $\mathbb{Q}$ [F-O-T].

We will need the following identifications: In this context $F_G(G \wedge H)$ is algebraically the "reciprocal commutator subgroup" C of $G \amalg H$, which is generated by commutators of the form $ghg^{-1}h^{-1}$. We use $[g,h]$ to denote both such commutators and elements of $G \wedge H$. The function Ψ is simply conjugation of elements of C by two letter words xy. Written as an element of C, $xy[g,h]y^{-1}x^{-1} = \Psi(x,y,[g,h]) = [x,y][xg,y]^{-1}[xg,yh][x,yh]^{-1}$. Now, if $y = e_H$ this becomes $\Psi(x,e_H,[g,h]) = [xg,h][x,h]^{-1}$. For a product $[g_1,h_1][g_2,h_2]^{-1}$ in C we have

$$\Psi(x,e_H,[g_1,h_1][g_2,h_2]^{-1})$$

$$= [xg_1,h_1][x,h_1]^{-1}([xg_2,h_2][x,h_2]^{-1})^{-1}$$

$$= [xg_1,h_1][x,h_1]^{-1}[x,h_2][xg_2,h_2]^{-1}.$$

So if $h_1 = h_2$

$$\Psi(x,e_H,[g_1,h][g_2,h]) = [xg_1,h][xg_2,h]^{-1}. \qquad (*)$$

For the particular case of $G = \mathbb{Q}$ and $H = \mathbb{Z}_2$ we can make a further identification: $\mathbb{Q} \wedge \mathbb{Z}_2$ is homeomorphic with $\mathbb{Q}$. Thus $F_G(\mathbb{Q})$ and $F_G(\mathbb{Q} \wedge \mathbb{Z}_2)$ are isomorphic topological groups via the correspondence $g_1^{\pm 1}g_2^{\pm 1}\cdots g_n^{\pm 1} \sim [g_1,1]^{\pm 1}[g_2,1]^{\pm 1}\cdots[g_n,1]^{\pm 1}$. In this case we can write (*) as

$$\Psi(x,0,g_1g_2^{-1}) = (x+g_1)(x+g_2)^{-1} \qquad (**)$$

with $\Psi: \mathbb{Q} \times \mathbb{Z}_2 \times F_G(\mathbb{Q}) \longrightarrow F_G(\mathbb{Q})$. To show that Ψ is not continuous it now suffices to find a net in $\mathbb{Q}$ which converges to 0 and a net in $F_G(\mathbb{Q})$ which clusters at $e = 0$ whose image under Ψ fails to cluster at $e = 0$. The net in $F_G(\mathbb{Q})$ which clusters at $e = 0$ is the net $(n + \frac{1}{j})\cdot(n)^{-1}$ of Example 3.7 of [F-O-T]; $\mathbb{N} \times \mathbb{N}$ carries the lexicographic order. We now use the following technical lemma, whose proof is postponed to the end of this note.

Lemma: *There exists a net* $\{q_{n,j}\}$, $(n,j) \in \mathbb{N} \times \mathbb{N}$, *which converges to* 0, *and such that* $\{n + q_{n,j}\}$ *and* $\{n + \frac{1}{j} + q_{n,j}\}$ *are disjoint, closed subsets of* $\mathbb{Q}$.

Using (**) we see that

$$\Psi(q_{n,j}, 0, (n+\tfrac{1}{j})(n)^{-1}) = (n+\tfrac{1}{j}+q_{n,j})(n+q_{n,j})^{-1}.$$

Since $\{n + \frac{1}{j} + q_{n,j}\}$ and $\{n + q_{n,j}\}$ are closed and disjoint there is a continuous function $f:\mathbb{Q} \longrightarrow \mathbb{R}$ such that

$$f(\{n + q_{n,j}\} \cup \{0\}) = 0 \quad \text{and} \quad f(\{n + \tfrac{1}{j} + q_{n,j}\}) = 1.$$

This function lifts to a continuous group homomorphism $\hat{f}:F_G(\mathbb{Q}) \longrightarrow \mathbb{R}$ making the following diagram commute.

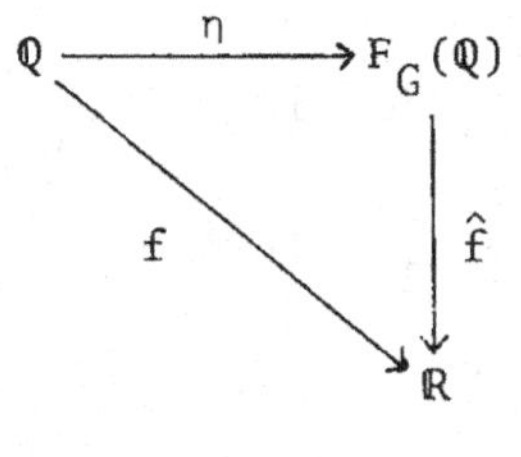

Consider the effect of $\hat{f}$ on $(n + \frac{1}{j} + q_{n,j})\cdot(n + q_{n,j})^{-1}$:

$$\hat{f}((n + \tfrac{1}{j} + q_{n,j})\cdot(n + q_{n,j})^{-1})$$

$$= f(n + \tfrac{1}{j} + q_{n,j}) - f(n + q_{n,j}) = 1 .$$

Thus, while $\hat{f}$ is continuous its composition with Ψ is not. It follows that $\mathbb{Q} \amalg \mathbb{Z}_2$ and $\mathbb{Q} \times \mathbb{Z}_2 \times F_G(\mathbb{Q} \wedge \mathbb{Z}_2)$ do not have the same topology.

Proof of Lemma: The net $\{q_{n,j}\}$ will be chosen inductively so that for each n the sequence $(q_{n,j})$ converges, in $\mathbb{R}$, to $\pi/n+5$. The resulting net on $\mathbb{N} \times \mathbb{N}$ with the lexicographic order will thus converge in $\mathbb{Q}$ to 0. Also, the choice of an irrational limit for each $(q_{n,j})$ will make $\{n + q_{n,j}\}$ and $\{n + \frac{1}{j} + q_{n,j}\}$ closed in $\mathbb{Q}$ so we only need concern ourselves with insuring that they are disjoint.

Choose any $q_{1,1} \in (\pi/6,\pi/5)$; choose $q_{1,2} \in (\pi/6,q_{1,1}) \cap (\pi/6,\pi/6 + 1/2) \setminus \{q_{1,1} - \frac{1}{2}\}$, thus $q_{1,2} < q_{1,1}$ and $q_{1,2} + \frac{1}{2} \neq q_{1,1}$. Choose $q_{1,3} \in (\pi/6,q_{1,2}) \cap (\pi/6,\pi/6 + 1/3) \setminus \bigcup_{j=1}^{2}\{q_{1,j} - \frac{1}{3}\}$; so $q_{1,3} + \frac{1}{3} \notin \{q_{1,1}, q_{1,2}\}$. Inductively choose $q_{1,i+1} \in (\pi/6,q_{1,i}) \cap (\pi/6,\pi/6 + 1/i+1) \setminus \bigcup_{j=1}^{i}\{q_{1,j} - \frac{1}{i+1}\}$

To start the second sequence choose $q_{2,1} \in (\pi/7,\pi/6) \setminus\{q_{1,k} + \frac{1}{k} - 1\}$; this can be done since $q_{1,k} + \frac{1}{k} - 1$ converges to $\pi/6 - 1$. Choose $q_{2,2} \in (\pi/7,q_{2,1}) \cap (\pi/7,\pi/7 + 1/2) \setminus(\{q_{1,k} + \frac{1}{k} - 1\} \cup \{q_{2,1} - \frac{1}{2}\})$; inductively choose $q_{2,i+1} \in (\pi/7,q_{2,i}) \cap (\pi/7,\pi/7 + 1/i+1) \setminus (\{q_{1,k} + \frac{1}{k} - 1\} \cup \bigcup_{j=1}^{i}\{q_{2,j} - \frac{1}{i+1}\})$.

Following this pattern we inductively choose a general

$q_{n+1,1} \in (\pi/n+5, \pi/n+4) \setminus \{q_{n,k} + \frac{1}{k} - 1\}$, and a general

$q_{n+1,i+1} \in (\pi/n+5, q_{n+1,i}) \cap (\pi/n+5, \pi/n+5 + 1/i+1)$

$\setminus (\{q_{n,k} + \frac{1}{k} - 1\} \cup \bigcup_{j=1}^{i} \{q_{n+1,j} - \frac{1}{i+1}\})$.

Now, take any $n + q_{n,j}$ and $m + \frac{1}{k} + q_{m,k}$. If $m > n$ also $m > n + q_{n,j}$; if $m < n - 1$, $m + \frac{1}{k} \leq n - 1$ so $m + \frac{1}{k} + q_{m,k} < n$; so we need only consider the possibilities $m = n$ and $m = n - 1$. If $m = n$ we are comparing $q_{n,k} + \frac{1}{k}$ with $q_{n,j}$; if $k \leq j$ then $q_{n,j} \leq q_{n,k} < q_{n,k} + \frac{1}{k}$, and if $k > j$ then $q_{n,k}$ was chosen so that $q_{n,k} \neq q_{n,j} - \frac{1}{k}$. Finally, suppose that $m = n - 1$; $q_{m,k} < \frac{\pi}{5} < \frac{3}{4}$ so if $k > 4$ $m + \frac{1}{k} + q_{m,k} < m + 1 = n < n + q_{n,j}$. But for $k = 1, 2,$ or 3 $q_{n,j}$ was carefully chosen so that $q_{n,j} \neq q_{m,k} + \frac{1}{k} - 1$.

References

[F-O-T] T.H. Fay, E.T. Ordman, and B.V.S. Thomas, "The free topological group over the rationals", General Topology and its Applications, to appear.

[F-T] T.H. Fay and B.V.S. Thomas, "Remarks on the free product of two Hausdorff groups", to appear.

[M-O-T] S.A. Morris, E.T. Ordman, and H.B. Thompson, "The topology of free products of topological groups", Proc. Second Internat. Conf. Theory of Groups, Canberra (1973) 504-515.

LIFTING SEMIFINAL LIFTINGS

Walter Tholen, Hagen

Abstract.

Let the following commutative diagram of functors be given:

$$(*)\qquad \begin{array}{ccc} \underline{A} & \xrightarrow{\quad U \quad} & \underline{C} \\ {\scriptstyle P}\downarrow & & \downarrow{\scriptstyle T} \\ \underline{X} & \xrightarrow[\quad V \quad]{} & \underline{Z} \end{array}$$

Then a general criterion is proved which allows to show the existence of U-semifinal liftings provided V admits the formation of certain V-semifinal liftings. Several results appear as corollaries, in particular sufficient (and partly necessary) conditions to get the implications

V right adjoint => U right adjoint

V semi-cofibration => U semi-cofibration

V semi-topological => U semi-topological.

Furthermore, a cancellability theorem and the following existence theorem for semi-topological functors can be derived from the general lifting theorem: U is semi-topological iff U is faithful, U preserves all limits and $\underline{A}$ contains certain (large) limits.

1. Semifinal liftings.

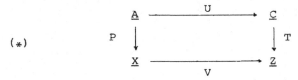

Let $U : \underline{A} \to \underline{C}$ be a functor and let $\gamma : U \circ D \to \Delta C$ be a U-cocone, i.e. a triple (D, γ, C) consisting of a functor $D : \underline{D} \to \underline{A}$ ($\underline{D}$ may be empty or large), an object C of $\underline{C}$ and a natural transformation $\gamma : U \circ D \to \Delta C$ with $\Delta C : \underline{D} \to \underline{C}$ being the constant functor. A U-semifinal lifting of $\gamma : U \circ D \to \Delta C$ is a triple (A, p, α) consisting of an object A of $\underline{A}$, a morphism $p : C \to UA$ of $\underline{C}$ and a natural transformation $\alpha : D \to \Delta A$ with $(\Delta p)\gamma = U \circ \alpha$, such that for any other suitable triple (B,c,β) with $(\Delta c)\gamma = U\circ\beta$ there is a unique $t : A \to B$ with $(Ut)p = c$ and $(\Delta t)\alpha = \beta$.

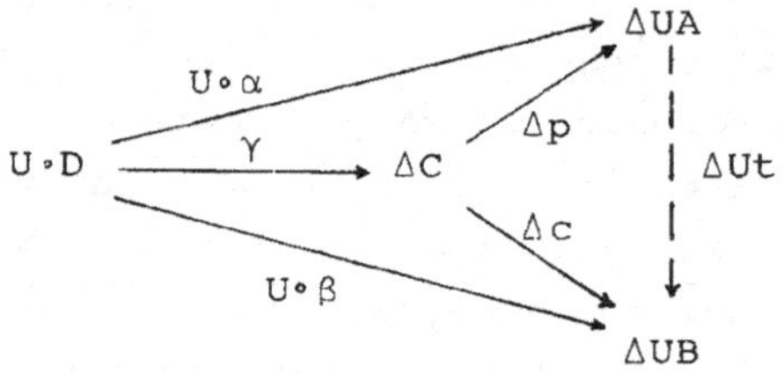

2. Special cases, definitions.

a. In the case of $\underline{D}$ being empty the U-cocone γ is just the object C. Then a U-semifinal lifting consists only of a U-universal map $p : C \to UA$. Therefore, U has a left adjoint functor, iff U admits U-semifinal liftings of all empty U-cocones.

b. Let $\underline{D}$ be the terminal category $\underline{1}$ containing precisely one morphism. Then all natural transformations are morphisms of $\underline{A}$ or $\underline{C}$, and U-semi-final liftings are just U-proclusion pairs in the sense of [Wyler '73]. (Cf. also [Ehresmann '67].) In the following, a functor U admitting the formation of U-proclusion pairs for all U-comorphisms (i.e. U-cocones over $\underline{1}$) is said to be a semi-cofibration.

c. If U admits U-semifinal liftings of all small U-cocones (i.e. U-cocones over small categories $\underline{D}$), then U is a semi-identifying functor in the sense of [Hoffmann '76]. In the following we prefer the notion small semi-topological functor.

d. If U admits U-semifinal liftings of all discrete U-cocones (i.e. U-sinks), U is said to be semi-topological [Trnkova '75, Wischnewsky '76, Hoffmann '77, Tholen '76 and '77]. Then U is, in particular, faithful (for a general proof cf. [Börger-Tholen '78a]) and, therefore, U admits U-semifinal liftings of all non-discrete U-cocones, too.

3. Examples, properties, characterizations.

Top categories [Wyler '71], topological functors (= initially complete functors [Brümmer '71]), (E,M)-topological functors [Herrlich '74a], regular functors [Herrlich '74b], topologically algebraic functors [Hong '74] are semi-topological (but not vice versa : for a concrete counter-example in the case of topologically algebraic functors cf. [Börger-Tholen '78b]). Compositions of semi-topological functors and their restictions to arbitrary full reflective and certain full coreflective subcategories are semi-topological. Every semi-topological functor is the restriction of a topological functor to a full reflective subcategory. A semi-topological functor $U : \underline{A} \to \underline{C}$ lifts the existence of (a fixed not necessarily small type of) limits and colimits from $\underline{C}$ to $\underline{A}$. For $\underline{C}$ being cocomplete and for $\underline{A}$ being cowellpowered U is semi-topological, iff U is faithful and right adjoint and $\underline{A}$ is cocomplete. (All proofs can be found in [Tholen '77].)

Some of these statements are also true for the more general notions of a small semi-topological functor and of a semi-cofibration. It was proved in [Hoffmann '76] that for $\underline{C}$ being cocomplete, U is small semi-topological, iff U is right adjoint and $\underline{A}$ is cocomplete. Therefore, up to completeness and smallness conditions, one has

small semi-topological + faithful = semi-topological.

Any functor $\underline{A} \to \underline{1}$ with $\underline{A}$ being cocomplete but not a preordered class is small semi-topological, but not semi-topological.

Examples of semi-cofibrations which are not (small) semi-topological are given, for instance, by the underlying set functor of a variety together with surjective homomorphisms.

Finally it should be mentioned that to every functor U one can assign a (meta-) functor $\check{U}$ such that

U semi-topological $\Leftrightarrow$ $\check{U}$ semi-cofibration

holds (cf. [Tholen-Wischnewsky '77]).

4. The general lifting problem for semifinal liftings.

Let the commutative square (*) and a U-cocone $\gamma : U \circ D \to \Delta C$ with $D : \underline{D} \to \underline{A}$ be given. In order to construct a U-semifinal lifting of γ we consider the V-cocone $T \circ \gamma : V \circ (P \circ D) \to \Delta TC$ and assume that there exists a V-semifinal lifting (X, q, ξ) of $T \circ \gamma$.

For $\underline{D}$ being small we can form the category $\tilde{\underline{D}}$ the objects of which are triples (B, c, β) as in 1. and the morphisms

$b : (B,c,\beta) \to (B',c',\beta')$ are morphisms $b : B \to B'$ of $\underline{A}$ fulfilling $(Ub)c = c'$ and $(\Delta b)\beta = \beta'$. One gets a projection $\tilde{D} : \underline{\tilde{D}} \to \underline{A}$ which is called the <u>diagram of $\underline{A}$ induced by</u> γ and a (U-)cone $\tilde{\gamma} : \tilde{\Delta}C \to U \circ \tilde{D}$ with $\tilde{\gamma}(B,c,\beta) = c$. Furthermore, there exists a unique morphism $\tilde{\xi}(B,c,\beta) : X \to PB$ rendering commutative the following diagram:

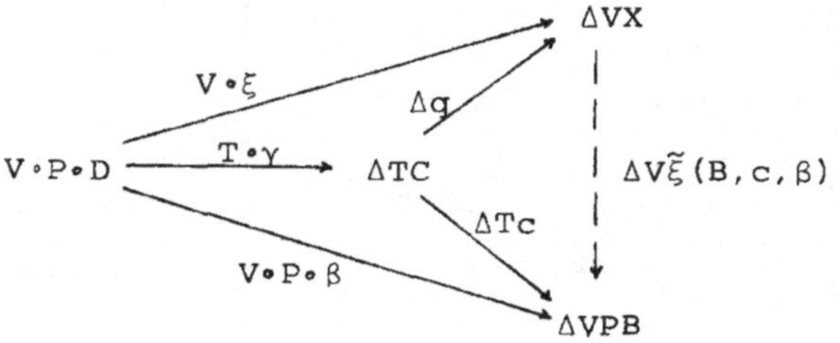

In this way one defines a natural transformation $\tilde{\xi} : \tilde{\Delta}X \to P \circ \tilde{D}$ which is called the P-<u>cone induced by</u> γ. Finally, for each object d of $\underline{D}$ one has a natural transformation $\tilde{\beta}_d : \tilde{\Delta}Dd \to \tilde{D}$ with $\tilde{\beta}_d(B,c,\beta) = \beta d$. Then the equations

$$\tilde{\gamma}(\tilde{\Delta}\gamma d) = U \circ \tilde{\beta}_d, \quad (V \circ \tilde{\xi})(\tilde{\Delta}q) = T \circ \tilde{\gamma}, \quad \tilde{\xi}(\tilde{\Delta}\xi d) = P \circ \tilde{\beta}_d$$

are easily proved.

For $\underline{D}$ not being small one has the same constructions, provided U is faithful. In this case $\underline{\tilde{D}}$ is defined to be the full subcategory of the comma-category $\langle C,U \rangle$ with objects the pairs (B,c), such that there is a $\beta : D \to \Delta B$ with $(\Delta c)\gamma = U \circ \beta$.

Before we can state the main result we have still to introduce the following notions, in which we refer to $\tilde{\xi} : \tilde{\Delta}X \to P \circ \tilde{D}$ as defined above, although an arbitrary $\tilde{\xi}$ can be substituted.

<u>5. Double factorizations of P-cones.</u>

A <u>double factorization</u> of a P-cone $\tilde{\xi} : \tilde{\Delta}X \to P \circ \tilde{D}$ consists of objects Z of $\underline{Z}$, A of $\underline{A}$, morphisms $e : X \to Z$, $s : PA \to Z$ and natural transformations $\tilde{\zeta} : \tilde{\Delta}Z \to P \circ \tilde{D}$, $\tilde{\mu} : \tilde{\Delta}A \to \tilde{D}$ with

$$\tilde{\xi} = \tilde{\zeta}(\tilde{\Delta}e), \quad P \circ \tilde{\mu} = \tilde{\zeta}(\tilde{\Delta}s).$$

The double factorization is called

- <u>rigid</u>, iff for all endomorphisms $z : Z \to Z$ and $a : A \to A$ with $ze = e$, $\tilde{\zeta}(\tilde{\Delta}z) = \tilde{\zeta}$, $zs = s(Pa)$, $\tilde{\mu}(\Delta a) = \tilde{\mu}$ one has $z = Z$ and $a = A$,
- P-<u>semiinitial</u>, iff for any object B of $\underline{A}$, any natural transformation $\tilde{\beta} : \tilde{\Delta}B \to \tilde{D}$ and any morphism $x : PB \to X$

with $\tilde{\xi}(\tilde{\Delta}x) = P\circ\tilde{\beta}$ there is a unique $b : B \to A$ with

$$ex = s(Pb), \quad \tilde{\mu}(\tilde{\Delta}b) = \tilde{\beta}.$$

- a <u>factorization</u>, iff s is an isomorphism.

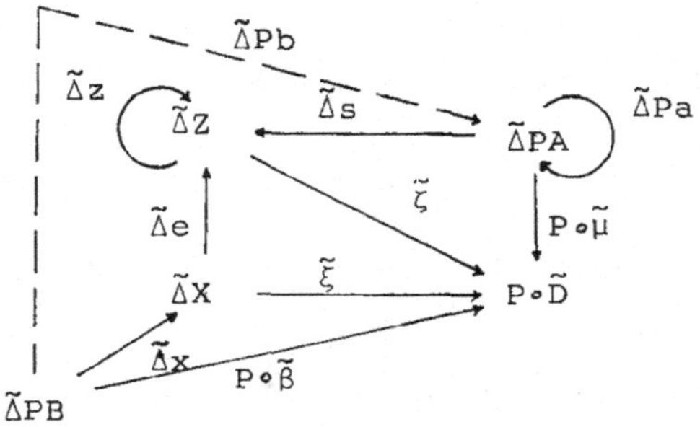

One says that the double factorization is T-<u>semiinitally</u> <u>transformed</u> by the commutative square $(*)$, iff the double factorization

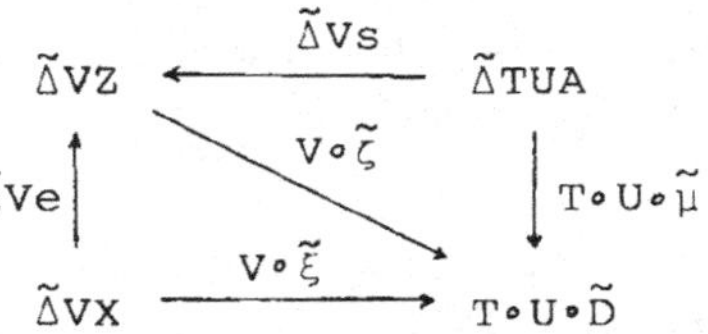

of the T-cone $V\circ\tilde{\xi}$ is T-semiinital.
If P is semi-topological, any P-cone admits a rigid semi-initial factorization. This follows from the duality theorem for semi-topological functors (cf. [Tholen '76 and '77]). But also under the dual assumption, i.e. if P is cosemi-topological any P-cone admits a rigid semiinitial double factorization: Clearly, one chooses e to be the identity.

6. The general lifting theorem for semi-final liftings.

Under the assumptions of 4. the following statements are equivalent:

(i) There exists a U-semifinal lifting of γ.

(ii) The P-cone $\tilde{\xi}$ induced by γ admits a rigid P-semiinital double factorization which is T-semiinitally transformed by $(*)$.

(iii) Same as (ii), but "factorization" instead of "double factorization".

(iv) *The diagram* $\tilde{D}$ *induced by* γ *has a limit which is preserved by* P,U *and* V∘P.

(v) *The diagram* $\tilde{D}$ *induced by* γ *has an absolute limit (i.e. a limit which is preserved by any functor with domain* $\underline{A}$*).*

Proof. (v) => (iv) is trivial.

(iv) => (iii): Let $\tilde{\mu} : \tilde{\Delta}A \to \tilde{D}$ be a limit cone. Then there is a unique morphism $e : X \to PA$ with $\tilde{\xi} = (P\circ\tilde{\mu})(\tilde{\Delta}e)$. This yields the desired factorization as can be easily shown.

(iii) => (ii) is trivial.

(ii) => (i): Let a rigid P-semiinitial double factorization of ξ be given as in 5. Then, because of the P-semiinitiality condition, for each object d of $\underline{D}$ there exists a unique morphism $\alpha d : Dd \to A$ with

$$e(\xi d) = s(P\alpha d), \quad \tilde{\mu}(\tilde{\Delta}\alpha d) = \tilde{\beta}_d.$$

One gets a natural transformation $\alpha : D \to \Delta A$ with

$$(\Delta e)\xi = (\Delta s)(P\circ\alpha), \quad (\Delta\mu(B,c,\beta))\alpha = \beta$$

for all objects (B,c,β) of $\tilde{\underline{D}}$. Since the double factorization is T-semiinitially transformed by (*) one has a unique $p : C \to UA$ with

$$(Ve)q = (Vs)(Tp), \quad (U\circ\tilde{\mu})(\tilde{\Delta}p) = \tilde{\gamma}.$$

Now $(\Delta p)\gamma = U\circ\alpha$ is proved by verifying the equations

$$(Vs)(Tp)(T\gamma d) = (Vs)(TU\alpha d), \quad (U\circ\tilde{\mu})(\tilde{\Delta}p)(\tilde{\Delta}\gamma d) = (U\circ\tilde{\mu})(\tilde{\Delta}U\alpha d)$$

for all d. It remains to show the universal property of U-semifinal liftings. For this let an object (B,c,β) of $\tilde{\underline{D}}$ be given. Then $t := \tilde{\mu}(B,c,\beta)$ fulfills indeed the equations

$$(Ut)p = c, \quad (\Delta t)\alpha = \beta.$$

Vice versa, any such t yields a morphism $t : (A,p,\alpha) \to (B,c,\beta)$ of $\tilde{\underline{D}}$. Therefore, from the naturality of μ we have $t\tilde{\mu}(A,p,\alpha) = \tilde{\mu}(B,c,\beta)$. Hence, in order to prove the uniqueness of t, it suffices to show $\tilde{\mu}(A,p,\alpha) = A$. For this define $z = s\tilde{\xi}(A,p,\alpha)$ and $a = \tilde{\mu}(A,p,\alpha)$. Now, after some calculations, one proves the four equations of the definition of rigidity and, therefore, has the desired result.

(i) => (v): Let (A,p,α) a U-semifinal lifting of γ. For all objects (B,c,β) one has a unique $\tilde{\mu}(B, c,\beta) : A \to B$ with

$(U\tilde{\mu}(B,c,\beta))\,p = c$, $(\Delta\tilde{\mu}(B,c,\beta)\alpha = \beta$. In this way one gets a natural transformation $\tilde{\mu} : \tilde{\Delta}A \to \tilde{D}$ which turns out to be an absolute limit cone.

<u>Remarks.</u> a. One can prove a little more than we have stated in 6: If the P-cone $\tilde{\xi}$ admits a rigid P-semiinitial double factorization which is T-semiinitially transformed by (*), then this double factorization is already an absolute limit factroization, i.e. s is an isomorphism and $\tilde{\mu}$ is an absolute limit cone.
b. $k = s^{-1}e : X \to PA$ is the so called <u>comparison</u> <u>morphism</u> between the V-semifinal lifting (X,q,ξ) of $T\circ\gamma$ and the U-semifinal lifting (A,p,α) of γ, for one has the equations

$$(Vk)q = Tp, \quad (\Delta k)\xi = P\circ\alpha.$$

If k is an isomorphism, for instance, if P is topological, (A,p,α) is said to be a lifting of (X,q,ξ). As a first consequence of the above theorem we state the following corollary:

<u>7. Formal criterion for semi-topologicity.</u>

A functor $U : \underline{A} \to \underline{C}$ *is semi-topological if and only if the following conditions are satisfied:*

a. U *is faithful.*

b. U *preserves all (not necessarily small) limits.*

c. $\underline{A}$ *has limits of diagrams induced by arbitrary* U*-cocones.*

Consider the following diagram:

$$\begin{array}{ccc} \underline{A} & \xrightarrow{U} & \underline{C} \\ \downarrow & & \downarrow \\ \underline{1} & = & \underline{1} \end{array}$$

Then the proof follows immediately from 2.d and 6.(i)<=> (iv). In the same way one also gets a formal criterion for right adjointness (cf.[Mac Lane '71]) and small semi-topologicity.

<u>8. Lifting theorem for right-adjointness, semi-cofibrations and (small) semi-topologicity.</u>

Let P *be semi-topological* <u>*or*</u> *cosemi-topological and let rigid* P*-semiinitial double factorizations be* T*-semiinitially transformed by (*). Then the following implications are valid:*

a. V *right adjoint* => U *right adjoint.*

b. V *semi-cofibration* => U *semi-cofibration.*

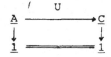

c. *V small semi-topological* => *U small semi-topological.*

d. *V semi-topological, U faithful* => *U semi-topological.*

Again, this is an immediate consequence of 2. and 6. Because of the assumption U faithful cf. the corresponding remark in 4.

In the following we simplify the transformation condition of the above theorem:

Each condition α. *and* β. *implies the validity of the implications a. - d.*

α. *P is an orthogonal* $(\underline{E},\underline{M})$*-functor (cf. [Tholen '76, '77, '78a, '78b]), and U transforms* $\underline{M}$*-cones into T-initial cones.*

β. *P is topologically algebraic (cf. [Hong '74]), and U transforms P-initial cones into T-initial cones.*

The appearing comparison morphisms (cf. 6) belong to $\underline{E}$ (in case α). Vice versa, this condition is necessary for the transformation conditions as stated above. Details can be found in [Tholen'78b and '78c].

9. Cancellability theorem for semi-topological functors.

As a special application of the lifting theorem we finally discuss the cancellability problem for semi-topological functors: Let $U : \underline{A} \to \underline{C}$ and $T : \underline{C} \to \underline{Z}$ be functors such that their composition $T \circ U$ is semi-topological. We try to find conditions which imply the semi-topologicity of U. In general, this implication does not hold: Take $U : \underline{C} \to \{C\}$ to be the full embedding of a single object C without non-identical endomorphisms. Then U is semitopological only if C is terminal in $\underline{C}$, although there is a unique $T : \underline{C} \to \{C\}$ with $T \circ U$ being the identity. We now consider the following diagram:

$$\begin{array}{ccc} \underline{A} & \xrightarrow{\ U\ } & \underline{C} \\ \| & & \big\downarrow T \\ \underline{A} & \xrightarrow[T \circ U]{} & \underline{Z} \end{array}$$

Application of 8 yields:

U is semi-topological if one of the following conditions is satisfied:

α. $\underline{A}$ *is an orthogonal* $(\underline{E},\underline{M})$*-category, U transforms* $\underline{M}$*-cones into T-initial cones and* $T \circ U$ *is semi-topological.*

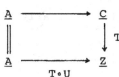

β. $T \circ U$ *is an orthogonal* ($\underline{E}$, $\underline{M}$)*-functor and* U *transforms* $\underline{M}$*-cones into* T*-initial cones.*

γ. $T \circ U$ *is topologically algebraic and* U *transforms* $(T \circ U)$*-initial cones into* T*-initial cones.*

Remark. D. Pumplün pointed out, that under conditions β. and γ. U has already to be topologically algebraic, provided T is faithful

10. Generalizations.

One can try to replace the commutative diagram (*) by a general natural transformation $\psi : T \circ U \to V \circ P$. This was outlined in [Tholen '78c]. Mutatis mutandis one gets the same results as in the special case that ψ is the identity, although the technical details are a little more complicated. If one treats the lifting problem on this general level one necessarily arrives at the notion of a concrete functor (cf. [Tholen '78c]). Although both, semi-topological functors and cosemi-topological functors (cf. the assumptions of the lifting theorem 8.1) are concrete, it is possible to prove the main results for semi-topological functors (duality theorem, existence of a factorization structure, representation as a full restriction of a topological functor) more generally for concrete functors.
Another kind of generalization was outlined in [Tholen-Wischnewsky-Wolff '78] where the lifting problem for adjoint functors is treated in a purely functorial manner which works in any 2-category.

References

Börger-Tholen '78a: Cantors Diagonalprinzip für Kategorien. Math. Z. 160 (1978), 135-138.

Börger-Tholen '78b: Remarks on topologically algebraic functors. Preprint, Hagen 1978 (to appear in Cahiers Topologie Geóm. Différentielle).

Brümmer '71: A categorical study of initiality in uniform topology. Thesis, Cape Town 1971.

Ehresmann '67: Structures quasi-quotient. Math. Ann. 171 (1967), 293-363.

Herrlich '74a: Topological functors. General Topology and Appl. 4 (1974), 125-142.

Herrlich '74b: Regular categories and regular functors. Can. J. Math. 26 (1974), 709-720.

Herrlich-Nakagawa-Strecker-Titcomb '78: Topologically-algebraic and semi-topological functors (are and are not equivalent). Preprint, Manhattan (Kansas) 1978.

Hoffmann '76: Semi-identifying lifts and a generalization of the duality theorem for topological functors. Math. Nachr. '74 (1976), 295-3o7.

Hoffmann '77: Full reflective restrictions of topological functors. Math. Colloq. Univ. Cape Town 11 (1977), 65-88.

Hong '74: Studies on categories of universal topological algebras. Thesis, Hamilton 1974.

Mac Lane '71: Categories for the working mathematician. Springer, Berlin-Heidelberg-New York 1971.

Trnková '75: Automata and categories. Lecture Notes in Computer Science 32 (1975), 132-152.

Tholen '76: M-functors, Mathematik-Arbeitspapiere 7 (Bremen 1976), 178-185.

Tholen '77: Semi-topological functors I. Preprint, Hagen 1977 (to appear in J. Pure Appl. Algebra).

Tholen '78a: On Wyler's taut lift theorem. General Topology and Appl. 8 (1978), 197-206.

Tholen '78b: Zum Satz von Freyd und Kelly. Math. Ann. 232 (1978), 1-14.

Tholen '78c: Konkrete Funktoren. Habilitationsschrift, Hagen 1978.

Tholen-Wischnewsky '77: Semi-topological functors II: External characterizations. Preprint, Hagen 1977 (to appear in J. Pure Appl. Algebra).

Tholen-Wischnewsky-Wolff '78: Semi-topological functors III: Lifting of monads and adjoint functors. Preprint,Hagen 1978.

Wischnewsky '76: A lifting theorem for right adjoints. Preprint, Bremen 1976 (to appear in Cahiers Topologie Geóm. Différentielle).

Wyler '71: On the categories of general topology and topological algebras. Arch. Math. (Basel) 22 (1971), 7-17.

Wyler '73: Quotient maps. General Topology and Appl. 3 (1973), 149-16o.

Added in proof: During this conference Reinhard Börger proved that the cancellability property characterizes semi-topological and topologically algebraic functors completely. For instance, in addition to 9.γ. one has:

P is topologically algebraic, iff for any factorization $P = T \cdot U$ *such that* U *transforms* P*-initial cones into* T*-initial cones* U *has to be right adjoint.*

A similar result holds for semi-topological functors.

Walter Tholen
Fachbereich Mathematik
Fernuniversität
Postfach 940
D-5800 Hagen
West Germany

Normally supercompact spaces and convexity preserving maps.

E. WATTEL Free University Amsterdam.

0 INTRODUCTION

Normally supercompact spaces are spaces with a closed subbase which is both binary and normal. This class of spaces is of importance; for instance :

— The class of compact spaces can be mapped faithfully into it.

— A normally supercompact space has a rich geometrical structure which is mainly reflected by its subbase convexity structure.

A subset of a space with subbase is called closed convex iff it is the intersection of subbase members. This definition fits in with the abstract convexity theory and parallels convexity in linear spaces. The analogue in our theory of a linear function will be a convexity preserving or cp function.

We discuss the category Ω of normally supercompact spaces and cp maps. The objects are characterized as certain subsets of cubes, a CECH type compactification construction is given to obtain normally supercompact spaces from those spaces in which points and subbase members can be separated by cp maps. Finally we obtain absolutes and projective objects in Ω.

1 BASIC NOTIONS.

1.1 A T_1 space X is called normally supercompact iff there exists a closed subbase $\mathcal{S}$ for X which is both normal (i.e. disjoint members of $\mathcal{S}$ can be separated by disjoint complements of members of $\mathcal{S}$) and binary (i.e. if in some subcollection $\mathcal{L}$ of $\mathcal{S}$ every two members meet, then $\bigcap \mathcal{L} \neq \emptyset$.). cf. [12], [8], [5].

1.2 A subset C of a space X with subbase $\mathcal{S}$ is called $\mathcal{S}$-convex iff C is the intersection of some subcollection of $\mathcal{S}$. The collection of all $\mathcal{S}$-convex sets will be denoted by $H(X,\mathcal{S})$ or by $\mathcal{C}$. The collection of $\mathcal{S}$-convex sets seems to be more important than $\mathcal{S}$ itself, since, if X is normally supercompact w.r.t. $\mathcal{S}$ then X is also supercompact w.r.t. $\mathcal{C}$ as a closed subbase.

Proof.

Binarity. If a collection $\mathfrak{L} \subset \mathcal{C}$ is linked then also the collection $\mathfrak{M} = \{ S \mid S \in \mathfrak{S} ; \forall L \in \mathfrak{L} : L \subset S \}$ is linked and $\bigcap \mathfrak{L} = \bigcap \mathfrak{M} \neq \emptyset$ whenever X is normally supercompact w.r.t. $\mathfrak{S}$.

Normality. Suppose that $E \cap F = \emptyset$ for some E and F in $\mathcal{C}$, then $\exists \mathfrak{F}$ and $\exists \mathfrak{E}$ such that $F = \bigcap \mathfrak{F}$ and $E = \bigcap \mathfrak{E}$. Moreover, $\bigcap (\mathfrak{F} \cup \mathfrak{E}) = \emptyset$ and therefore $\mathfrak{F} \cup \mathfrak{E}$ is not linked. We find a F' and an E' such that $F', E' \in \mathfrak{S}$ and $F' \cap E' = \emptyset$. Since E' and F' can be separated, we can find a separation for F and E.

1.3 We define: A mapping is called convexity preserving or shortly cp iff inverse images of $\mathfrak{S}$-convex sets are subbase convex. cf. [7]. We investigate the category of normally supercompact spaces and convexity preserving maps. We denote this category by Ω.

1.4 Let X be a normally supercompact space w.r.t. a subbase $\mathfrak{S}$. The convex hull $I_S(A)$ of a set $A \subset X$ is the intersection of all members of $\mathfrak{S}$ containing A. The convex hull of a pair of points $\{x,y\}$ is called an interval. We also write $I_S(x,y)$ instead of $I_S(\{x,y\})$ cf.[6]. The intervals on three points x,y,z in X intersect in precisely one point which is called the mean of x,y,z.

(i) $\{ m(x,y,z) \} = I_S(x,y) \cap I_S(x,z) \cap I_S(y,z)$. cf.[5].

(ii) The map m is a continuous function from $X * X * X \rightarrow X$.

(iii) Obviously $f(m(x,y,z)) = m(f(x),f(y),f(z))$ for every cp map f between supercompact spaces.

Proof

(i) $\mathfrak{M} = \{ I_S(x,y) , I_S(y,z) , I_S(z,x) \}$ is a linked subcollection of $\mathcal{C}$ and has a non-empty intersection. Suppose $p,q \in \bigcap \mathfrak{M}$. Then p and q can be separated by two sets P and Q in $\mathfrak{S}$ such that $P \cup Q = X$, $p \in P \setminus Q$, $q \in Q \setminus P$. Now either P or Q contains two of the three points x, y and z. Suppose P contains x and y. Then $I_S(x,y) \subset P$ and $q \notin I_S(x,y)$. Contradiction. Therefore $\bigcap \mathfrak{M}$ contains precisely one point called $m(x,y,z)$.

(ii) Choose $S \in \mathfrak{S}$ and x,y,z in X such that $m = m(x,y,z) \notin \mathfrak{S}$. Choose P and Q in $\mathfrak{S}$ such that $S \subset P \setminus Q$ and $m \in Q \setminus P$. Then two of the three points x,y,z are not in P, say x and y. Then $U = (X \setminus P) * (X \setminus P) * X$ is a neighborhood of (x,y,z) in $X * X * X$ with the property that $m[U] \cap S = \emptyset$.

(iii) If $f : (X, \mathfrak{S}) \to (X', \mathfrak{S}')$, then $f^{-1} \mathbf{I}_S(f(x),f(y))$ is convex and hence contained in $\mathbf{I}_S(x,y)$ and now the result is clear from (i).

1.5 A subset M of X is called <u>triple convex</u> iff for every x, y, z in M also $m(x,y,z) \in M$.

1.6 SCHRIJVER [6] showed that a compact space X with closed subbase $\mathfrak{S}$ is supercompact iff there exists a collection of closed intervals $\mathbf{I}(x,y)$ for every pair of points such that

(i) $\forall x,y \in X : x,y \in \mathbf{I}(x,y)$.

(ii) $\forall x,y \in X : \mathbf{I}(x,y) = \mathbf{I}(y,x)$.

(iii) $\forall x,y,u,v \in X : u,v \in \mathbf{I}(x,y) \to \mathbf{I}(u,v) \subset \mathbf{I}(x,y)$.

(iv) $\forall x,y,z \in X : \mathbf{I}(x,y) \cap \mathbf{I}(x,z) \cap \mathbf{I}(y,z) \neq \emptyset$.

(v) $\forall S \in \mathfrak{S} : x,y \in S \to \mathbf{I}(x,y) \subset S$.

A set C is called $\mathbf{I}$-convex iff $\forall x,y \in C$ we have $\mathbf{I}(x,y) \subset C$. It is shown in [7] that the $\mathbf{I}$-convex subsets and the $\mathfrak{S}$-convex subsets coincide for spaces with a binary normal subbase $\mathfrak{S}$.

1.7 The interval structure of a normally supercompact space X can be derived from its mean function because $\mathbf{I}_S(x,y) = \{ z \mid m(x,y,z) = z \}$. A closed subset G of a normally supercompact space X is normally supercompact with respect to the relative subbase iff G is a triple convex subset of X, because it inherits the interval structure. It also follows directly that if $f : X \to Y$ is cp and C is convex in X then $f[C] = D \cap f[X]$ for some convex subset D of Y.

2 CUBES

2.1 Let X be a compact ordered space. Then X is supercompact w.r.t. the subbase $\mathcal{S}$ of all closed halfspaces. The $\mathcal{S}$-convex sets are precisely the closed intervals and for every $x \leq y \leq z$ in X we have that $m(x,y,z) = y$.

2.2 If $\{X_\alpha, \mathcal{S}_\alpha \mid \alpha \in A\}$ is a collection of supercompact spaces then also $\prod_{\alpha\in A} X_\alpha$ is supercompact w.r.t. the subbase

$$\{ \pi_\alpha^{-1} S_\alpha \mid \alpha \in A \ ; \ S_\alpha \in \mathcal{S}_\alpha \}.$$

and the mean of the three points $x = (x_\alpha)$, $y = (y_\alpha)$, $z = (z_\alpha)$ is the point $w = (w_\alpha)$ in which $w_\alpha = m_\alpha(x_\alpha, y_\alpha, z_\alpha)$.

2.3 Let X be a space with closed subbase $\mathcal{S}$. A convexity preserving map from $(X, \mathcal{S})$ onto either the closed unit interval I or onto $\{0,1\} \subset I$ is called an <u>URYSOHN map</u> or shortly <u>ucp map</u>. A space with a binary normal subbase contains sufficiently many ucp maps to separate points and closed sets and to separate disjoint members from $\mathcal{S}$.

This can be shown by a slight modification of the usual proof of URYSOHN's lemma. cf.[8].

2.4 Let X be a TYCHONOV space with normal subbase $\mathcal{S}$. Let $\mathcal{F}$ be the collection of all ucp maps $f : X \to I_f$. (i) Then the evaluation $e : X \to \prod_{f\in\mathcal{F}} I_f$ is a cp embedding and (ii) X is supercompact w.r.t. $\mathcal{S}$ iff $e[X]$ is a closed triple-convex subset of $\prod_{f\in\mathcal{F}} I_f$. cf.[8].

Proof.

(i). The mapping e is cp because for every $f \in \mathcal{F}$ and every subbase member $S_f \subset I_f$, the set $e^{-1}[f^{-1}[S_f]]$ is convex in X and

$$\{ f^{-1}[S_f] \mid f \in \mathcal{F} \ ; \ S_f \in \mathcal{S}_f \}$$

is a subbase for $\prod_{f\in\mathcal{F}} I_f$. The mapping e is one to one because every two points can be separated by some $f \in \mathcal{F}$.

Moreover, let C be convex in X; let $\mathcal{F}'$ be the subcollection of all $f \in \mathcal{F}$ which map C onto O_f. Then $e[C] = \bigcap_{f \in \mathcal{F}'} f^{-1}[O_f]$ and so the image of a convex set in X is convex in e[X].

(ii) If X is normally supercompact w.r.t. $\mathcal{S}$ then we have from 1.4(i) that $m(e(x),e(y),e(z)) = e(m(x,y,z))$ since e is cp. Therefore e[X] has to be triple convex. Since X is compact e[X] is closed in $\prod_{f \in \mathcal{F}} I_f$. Conversely, if e[X] is closed and triple convex then from 1.7 it follows that e[X] is normally supercompact and so is X itself.

2.5 If $\mathcal{S}$ is a normal subbase for a space X and if $\mathcal{F}$ is the collection of all ucp maps from X into I then we can construct a supercompact space X^+ with subbase $\mathcal{S}^+$. Define X^+ to be the smallest closed triple-convex subset of $\prod_{f \in \mathcal{F}} I_f$ which contains e[X]. Then we have the space embedded in a natural way in a normally supercompact space. In [14] we have shown that this space is convexly isomorphic to the superextension of DE GROOT [3], [4], [12] if we use the restriction of the product subbase as $\mathcal{S}^+$. In this construction it is not even necessary to use normal subbases, but subbases which admit sufficiently many ucp maps to separate points and closed sets yield normally supercompact extensions. In such a case the DE GROOT superextension is not always T_2.

3 CATEGORICAL ASPECTS.

3.1 The category Ω of all normally supercompact spaces and convexity preserving maps has

(i) the cp-isomorphic injections as monomorphisms and

(ii) between ordered spaces the epimorphisms are precisely the monotone and onto mappings.

(iii) A cp-map $f:(X,\mathcal{S}) \to (Y,\mathcal{T})$ is an epimorphism iff for each URYSOHN cp-map $g:(Y,\mathcal{T}) \to I$ the composition $g \circ f$ is onto. This has the consequence that between dendrons the epimorphisms are surjections.

Proof.

(i) A mono has to be one to one because the embedding of a singleton is always cp. Now 1.7 implies that each mono is an injection.

(ii) Assume that X is normally supercompact; Y is compact and ordered and $f : X \to Y$ is cp. Suppose that f is not onto. Then there exists a monotone mapping ϑ from Y into [0,1] such that $\vartheta[f[X]] \in \{0,1\}$ and $\exists y \in Y : \vartheta(y) \in (0,1)$. If ψ is any selfmap of [0,1] which is monotone, onto with $\psi(\vartheta(y)) \neq \vartheta(y)$ then $\psi . \vartheta . f = \vartheta . f$ but $\psi . \vartheta \neq \vartheta$ and f would not be epi.

(iii) Cleary for each epi f and every ucp map π the composition $\pi . f$ must be onto.

Suppose conversely that for some mapping f the composition with every ucp map π is onto. Let $g : f[X] \to Z$ be any cp mapping. We now show that if g has an extension h to Y then this extension is unique. Let $\mathcal{F}$ be the collection of all ucp maps from Z to [0,1]. For every $h:Y\to Z$ the mapping $\pi . h$ is a ucp mapping of Y. Let $y \in Y$ be fixed. We show that h(y) does not depend on anything but $h[f[X]] = g[f[X]]$. There exists a point x_π for every $\pi \in \mathcal{F}$ such that $\pi(h(y)) = \pi(h(f(x_\pi))) = \pi(g(f(x_\pi)))$. The space Z is completely determined by its evaluation in the cube $\Pi\{ [0,1]_\pi \mid \pi \in \mathcal{F} \}$ and therefore

$$h(y) = e^{-1}(\Pi \{ \pi . g . f (x_\pi) \mid \pi \in \mathcal{F} \}).$$

which is independent of h. It follows that h is unique.

3.2 Let Ξ be the category of spaces with normal T_1 subbases and cp-maps. Then Ω and the category of TYCHONOV spaces and continuous functions are both full subcategories of Ξ. The subcategory Ω is moreover a bireflective subcategory of Ξ. We obtain the mono-reflection $\lambda : \Xi \to \Omega$ by putting : $\lambda(X)$ is the superextension of 2.5 and for each mapping $f : X \to Y$ we construct a unique mapping $\lambda f : \lambda X \to \lambda Y$ by means of all ucp mappings from Y to [0,1] in the same way as in the proof of 3.1(iii).

This is essentially the same mapping which JENSEN introduced in [4] and which is the unique cp-extension of f although there may be more continuous extensions cf.[7]. In this case λ is the adjoint of the forgetful functor and if $(Y,\mathcal{T})$ is normally supercompact then $\lambda(Y,\mathcal{T}) = (Y,\mathcal{T})$.

3.3 For each extremally disconnected compact space P with collection of closed sets $\mathcal{G}$ the space $\lambda(P,\mathcal{G})$ is a projective object of Ω.

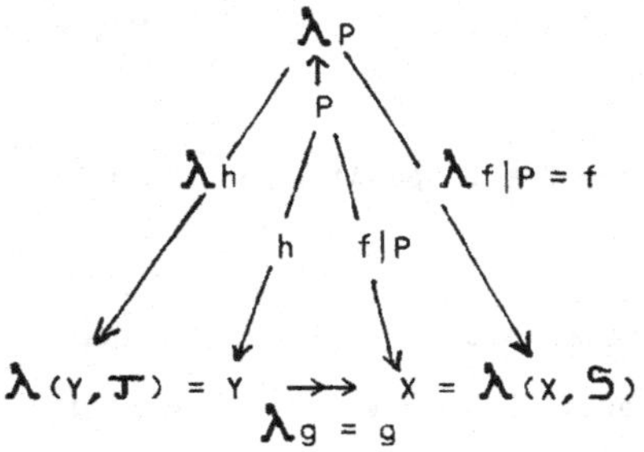

Proof. Let $f: \lambda(P,\mathcal{G}) \to (X,\mathcal{S})$ be any cp mapping and let $g: (Y,\mathcal{T}) \to (X,\mathcal{S})$ be a cp surjection. Assume that $\mathcal{S}$ and $\mathcal{T}$ are closed under intersections.

Since P is projective in Compact HAUSDORFF there exists a mapping $h: P \to Y$ in Compact HAUSDORFF such that $f|P = g \cdot h$. For each of the mappings g, h, $f|P$ there exists a unique cp extension λg, λh, $\lambda f|P$. Since $\lambda(X,\mathcal{S}) = (X,\mathcal{S})$ and $\lambda(Y,\mathcal{T}) = (Y,\mathcal{T})$ we obtain $\lambda g = g$, $\lambda f|P = f$ and finally $g \cdot \lambda h = f$ which shows that λP is a projective object in Ω.

3.4 The category Ω satisfies the requirements of [1] for the existence of projective covers w.r.t. surjective cp-maps. This has the consequence that each object of Ω has a minimal projective cover which could be called the <u>absolute</u> of the object. A mapping $f: X \to Y$ is called <u>triple convex irreducible</u> iff the image of each proper closed triple-convex subset of X is not equal to Y. Absolutes can be recognized by the fact that each triple convex irreducible pre-image is an isomorphic object.

3.5 Let $(X, \mathcal{S})$ be a normally supercompact space. Then X is projective object of Ω w.r.t. all cp-maps iff for each pair of subbase members S_1 and S_2 in $\mathcal{S}$ we have

$$S_1 \cup S_2 = X \iff I_S(X \setminus S_1) \cap I_S(X \setminus S_2) = \emptyset.$$

The requirement stated here is the convex equivalent of extremal disconnectedness.

Proof.

$\Rightarrow$ Suppose that there exist an S_1 and an S_2 is $\mathcal{S}$ such that $S_1 \cup S_2 = X$ and $I_S(X \setminus S_1) \cap I_S(X \setminus S_2) \neq \emptyset$. Then we have for $S_0 = I_S(X \setminus S_2)$ that $S_1 \cup S_0 = X$; $\exists p \in I_S(X \setminus S_1) \cap I_S(X \setminus S_0)$ because $S_0 \subset S_2$ and hence $I_S(X \setminus S_0) \supset I_S(X \setminus S_2)$. Define

$$Y = I_S(X \setminus S_1) * \{1\} \cup I_S(X \setminus S_0) * \{0\} \subset X * \{0,1\}.$$

The projection $f : Y \to X$ is a cp surjection. If X would be projective then there would exist an inverse for f. This is impossible since f is not one to one, since $p*\{0\}$ and $p*\{1\}$ are both mapped onto p.

$\Leftarrow$ Suppose that $S_1 \cup S_2 = X \to I_S(X \setminus S_1) \cap I_S(X \setminus S_2) = \emptyset$ for all $S \in \mathcal{S}$ and suppose that $f : Y \xrightarrow{onto} X$ is tc-irreducible. We assume that Y is embedded in the product of all images under ucp maps. Again we only have to show that f is one to one because it then follows from 1.7 that X is cp-isomorphic with Y. Suppose that f is not one to one and that $f(y) = f(z)$ for two different points y and z. Then there exists a ucp map π from Y to $[0,1]$ with $\pi(y) > \pi(z)$. Let q be a point of $[0,1]$ with $\pi(z) < q < \pi(y)$. Define $Y \setminus \pi^{-1}[\,[0,q)\,] = S_z$; $Y \setminus \pi^{-1}[\,(q,1]\,] = S_y$; $V_z = Y \setminus f^{-1}f[S_z]$ and $V_y = f^{-1}f[S_y]$. $f[S_z] \neq X$ because f is tc-irreducible and S_z is convex and closed. $I_S(V_z) \cup S_z$ is a closed triple-convex subset of Y which is mapped onto X. So this set is equal to Y and $z \in I_S(V_z)$ In the same way $y \in I_S(V_y)$ From 1.7 it follows that $(X \setminus f[S_z]) = f[V_z]$ and therefore we find that $I_S(X \setminus f[S_z]) \cap I_S(X \setminus f[S_y]) \neq \emptyset$ which contradicts our assumption.

4 REFERENCES

[1] **B. BANASCHEWSKI.** Projective covers in categories of topological spaces and topological algebras. Mac Masters University 1968.

[2] **M. BELL.** Not all compact Hausdorff spaces are supercompact. Gen Top & Appl. 8 p 151-155 (1978).

[3] **J. DE GROOT.** Supercompactness and superextensions. Cont. to Ext. Th. of Top. Struct. Symp. Berlin 1967, Deutscher Verlag der Wissenschaften Berlin (1969) p.89-90.

[4] **J. DE GROOT, G. A. JENSEN, A. VERBEEK.** Superextensions. Report ZW 1968-017 Math. Centre Amsterdam.

[5] **J. VAN MILL.** Superextensions and Wallman spaces. Mathematical Centre Tract 85, Mathematical Centre Amsterdam 1977.

[6] **J. VAN MILL & A. SCHRIJVER.** Subbase characterizations of compact topological spaces. To appear in Gen. Top. Appl.

[7] **J. VAN MILL & M. VAN DE VEL.** Convexity preserving maps in subbase convexity theory. Proc. Kon. Ned. Acad. Wet. A-81 (1) p.76-90 (1978).

[8] **J. VAN MILL & E. WATTEL.** An external characterization of spaces which admit binary normal subbases. To appear in the Am. Journ. Math.

[9] **M. STROK & A. SZYMÁNSKI.** Compact metric spaces have binary subbases. Fund. Math. 89 (1975) 81-91.

[10] **M. VAN DE VEL.** Superextensions and Lefschetz fixed point structures. Fund. Math.

[11] **M. VAN DE VEL.** A Hahn-Banach theorem in subbase convexity theory. Rapport 73 Wiskundig Seminarium Free University Amsterdam 1977

[12] **A. VERBEEK.** Superextensions of topological spaces. Mathematical Centre Tract 41 Mathematical Centre Amsterdam 1972.

[13] **E. WATTEL.** Projective objects and k-mappings. 3rd Yugoslav Topological Symposium 1977, Rapport 72 Wiskundig Seminarium Free University Amsterdam.

[14] **E. WATTEL.** Superextensions embedded in cubes. Rapport 75 Wiskundig Seminarium Free University Amsterdam (1978).

STRUCTURE FUNCTORS

- Representation and Existence Theorems -

Manfred B. Wischnewsky

University of Bremen

Introduction

We investigate categories $\underline{A}$ called structured categories resp. functors $S: \underline{A} \to \underline{X}$ called structure functors which represent exactly all full reflective or coreflective restrictions of semitopological functors. The results generalize all fundamental theorems for semitopological functors ([3, 5, 6, 7, 8, 9, 10, 11, 12, 13, 14, 15, 16, 17, 20, 24]). The most wellknown examples for structured categories $\underline{A}$ (over a base category $\underline{X}$) are all full reflective or coreflective subcategories of locally presentable categories ([4]), of algebraic categories (in the sense of Linton) and of topological categories. The main techniques used here are a "Cantor's diagonal lemma for categories" ([2]) and applications of the notion "connectedness with respect to a sequence of functors" ([18]).

§ 0 Notations

Let $S:\underline{A} \longrightarrow \underline{X}$ be a functor. A S-cone is a triple $(X,\psi,D(\underline{A}))$ where X is an $\underline{X}$-object, $D(\underline{A})): \underline{D} \longrightarrow \underline{A}$ is an $\underline{A}$-diagram ($\underline{D}$ may be void or large) and $\psi: \Delta X \longrightarrow SD(\underline{A})$ is a functorial morphism (Δ denotes the "constant" functor into the functor category). We shall abbreviate often $(X,\psi,D(\underline{A}))$ by ψ

Cone(S) denotes the class of all S-cones. If $\underline{D} = \underline{1}$ (i.e. the one point category) then ψ is called a S-morphism denoted by (A,a) where A is an $\underline{A}$-object and $a: X \longrightarrow SA$ is an $\underline{X}$-morphism. The dual notions are S-cocone and S-comorphism. The corresponding classes of S-morphisms, S-cocones and S-comorphisms are denoted by Mor(S), Co-cone(S), Co-Mor(S). Epi(S) denotes the class of all S-epimorphisms $(A,e: X \longrightarrow SA)$ i.e. the class of all S-morphisms (A,e) with the property: for all A-morphisms $p,q : A \rightrightarrows B$ the equation $(Sp)e = (Sq)e$ implies $q = q$.

The dual notion is S-monomorphism. The class of all S-monomorphisms is denoted by Mono(S). Iso(S) denotes the class of all S-isomorphisms i.e. of all objects (A,a) in Mor(S) with a an isomorphism in X. Init(S) denotes the class of all S-initial cones i.e. of all A-cones $\alpha : \Delta A \longrightarrow D(A)$ such that for any A-cone $\beta: \Delta \longrightarrow D(A)$ and any X-morphism $x: SB \longrightarrow SA$ with $(S\beta) = (S\alpha)(\Delta x)$ there exists a unique A-morphisms $a: B \longrightarrow A$ with $\beta = \alpha(\Delta a)$ and $Sa = x$.

§ 1 Structure Functors: Basic Definitions.

Let $S: A \longrightarrow X$ be a functor and let $S = (A \xrightarrow{\tilde{Q}} B \xrightarrow{Q} X)$ be an arbitrary factorization. Let $Id(Q) \subseteq \Phi \subseteq Mor(Q)$ $Id(\tilde{Q}) \subseteq \Gamma \subseteq Co\text{-}Mor(\tilde{Q})$ and $Id(\tilde{Q}) \subseteq ¶ \subseteq Co\text{-}Mor(\tilde{Q})$ be classes of Q-morphisms resp. $\tilde{Q}$-morphisms.[1])

(1.1) DEFINITION ((¶ ,Φ,Γ)-Semifinal Extensions)

Let $SD(A) \xrightarrow{Q\pi} QD(B) \xleftarrow{\phi} D(X) \xrightarrow{\psi} \Delta X$

be a functorial chain with $(D(A), \pi : \tilde{Q}D(A) \longrightarrow D(B))$ being pointwise in ¶ and $(D(B), \phi : D(X) \longrightarrow QD(B))$ being pointwise in Φ.

We call a functorial chain of this type a (¶,Φ)-functorial chain.

---

[1]) Φ, Γ, ¶ fulfill the usual closedness conditions i.e. Φ is closed under composition with B-isomorphisms from the left and Γ and ¶ are closed under composition with A-isomorphisms from the right.

The class of all $(\P,\Phi)$-functorial chains is denoted by $\underline{\mathrm{Chs}}_f(\P,\Phi)$[1]. Let (A,g,B,e,X) be a $(S,\tilde{Q},Q)$-double-morphism with $(A,g: \tilde{Q}A \to B)\varepsilon\Gamma$.

A pair of functorial morphisms

α: $D(\underline{A}) \to \Delta A$ and β: $D(\underline{B}) \to \Delta B$ with $\beta\pi = (\Delta g)(\tilde{Q}\alpha)$ and $(\Delta e)\psi = (Q\beta)\phi$ is called a $(\P,\Phi,\Gamma)$-<u>extension of</u> $(D(\underline{A}),\pi,D(\underline{B}),\phi,\psi)$ <u>by</u> (A,g,B,e,X) (abbreviated by (α,β): $(\pi,\phi,\psi) \to (g,e)$ if there is no confusion).

Let (α,β): $(\pi,\phi,\psi) \to (g,e)$ and (α',β'): $(\pi,\phi,\psi) \to (g',e')$ be $(\P,\Phi,\Gamma)$-extensions of (π,ϕ,ψ).

A morphism from (α,β) to (α',β') is a pair of morphisms a: $A \to A'$ and b: $B \to B'$ such that

$$\alpha' = (\Delta a)\alpha \quad , \quad \beta' = (\Delta b)\beta$$
$$e' = (Qb)e \quad \text{and} \quad bg = g'(\tilde{Q}a).$$

This defines the category of all $(\P,\Phi,\psi)$-extensions over (π,ϕ,ψ).

An initial object (α,β): $(\pi,\phi,\psi) \to (g,e)$ with $(B,e)\varepsilon\Phi$ in the category of all $(\P,\Phi,\Gamma)$-extensions over (π,ϕ,ψ) is called a $(\P,\Phi,\Gamma)$-semifinal extension of (π,ϕ,ψ).

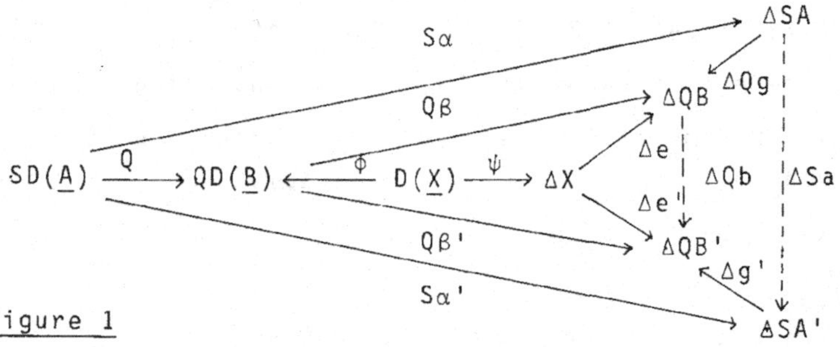

<u>Figure 1</u>

---

[1]) If the domain category $\underline{D} = 1$ then the class of $(\P,\Phi)$-functorial chains is denoted by $\mathrm{Chs}(\P,\Phi)$. In this case we speak about $(\P,\Phi)$ chains.

Let $T(\P,\Phi) \subseteq \underline{Chs}(\P,\Phi)$ be a subclass of $(\P,\Phi)$-chains and denote by $T_f(\P,\Phi)$ the class of all $(\P,\Phi)$-functorial chains being pointwise in $T(\P,\Phi)$.

The functor $S: \underline{A} \to \underline{X}$ is called a ($T(\P,\Phi),\Gamma$)-semifinal-structure functor if every element in $T_f(\P,\Phi)$ has a $(\P,\Phi,\Gamma)$-semifinal extension.

(1.2) DEFINITION. (($\P,\Phi,\Gamma$)-Semiinitial Co-Extensions)

Let $\Phi,\P,\Gamma$, $T(\P,\Phi)$ as above.

Let $(D(\underline{A})$, $\gamma: \tilde{Q}D(\underline{A}) \to D(\underline{B})$, $\rho: \Delta X \to QD(\underline{B}))$ be a double-cone with $(D(\underline{A}),\gamma)$ being pointwise in Γ.

Let $(A,P: \tilde{Q}A \to B,\ f: Y \to QB,\ x: Y \to X)$ be a $T(\P,\Phi)$-chain. A pair of cones $\alpha: \Delta A \to D(\underline{A})$ and $\beta: \Delta B \to D(\underline{B})$ with $\rho(\Delta x) = (Q\beta)(\Delta f)$ and $\beta(\Delta p) = \gamma(\tilde{Q}\alpha)$ is called a ($T(\P,\Phi),\Gamma$)-co-extension of (γ,ρ). A double-morphism $(B,e: X \to QB;\ A,g: \tilde{Q}A \to B)$ with $(B,e)\varepsilon\Phi$ and $(A,g)\varepsilon\Gamma$ together with a pair (α,β) consisting of an $\underline{A}$-cone $\alpha: \Delta A \to D(\underline{A})$ and a $\underline{B}$-cone $\beta: \Delta B \to D(\underline{B})$ with

$$(Q\beta)(\Delta e) = \rho \quad \text{and} \quad \gamma(\tilde{Q}\alpha) = \beta(\Delta g)$$

is called a $(T(\P\ \Phi),\Gamma)$-semiinitial coextension of (γ,ρ) if

SI 1) for all ($T(\P,\Phi),\Gamma$)-coextensions

$(\alpha',\beta'): (A',p',B',f',x) \to (\gamma,\rho)$ of (γ,ρ)

there exists a unique pair (a,b) consisting of an

$\underline{A}$-morphism $a: A' \to A$ and a $\underline{B}$-morphism $b: B' \to B$ such that $\alpha' = \alpha(\Delta a)$, $\beta' = \beta(\Delta b)$,

$ex = (Qb)f'$ and $bp' = g(\tilde{Q}a)$.

SI 2) For any pair of morphisms $s: A \to A$ and $t: B \to B$ the equations $e = (Qt)e$, $\beta = \beta(\Delta t)$, $\alpha = \alpha(\Delta s)$ and $tg = g(\tilde{Q}s)$ imply $s = id(A)$ and $t = id(B)$.

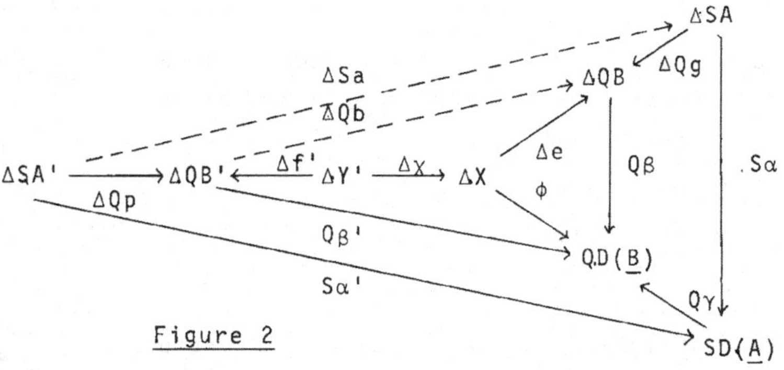

Figure 2

Notation as above.

A functor $S: \underline{A} \to \underline{X}$ is called a $(T(\P,\Phi),\Gamma)$-semiinitial structure functor if every chain (γ,ρ) has a $(T(\P,\Phi),\Gamma)$-semiinitial coextension.

(1.3) Examples

1. Semitopological Functors (Trnkowa [24], Wischnewsky [17], Tholen [13], Hoffmann [8])

are semifinal and semiinitial structure functors. For this special instance we have the following characteristic datas:

$\tilde{Q} = \mathrm{Id}: \underline{A} \to \underline{A}$, $Q = S: \underline{A} \to \underline{X}$, $\underline{\mathrm{Id}}(S) \subseteq \Phi \subseteq \underline{\mathrm{Mor}}(S)$, $\P = \Gamma = \underline{\mathrm{Id}}(\underline{A})$

$T(\P,\Phi)$ can be chosen in many different ways of which we will only note two of them:

(a) $\Phi = \underline{\mathrm{Mor}}(S)$ and $T_1(\P,\Phi) = \underline{\mathrm{Co\text{-}Mor}}(S)$.

These datas give the wellknown semifinal resp. semiinitial characterization of semitopological functors.

(b) $\underline{\mathrm{Id}}(S) \subseteq \Phi \subseteq \underline{\mathrm{Mor}}(S)$ and $T_2(\P,\Phi) \subseteq \underline{\mathrm{Co\text{-}Mor}^2}(S)$ where $(A, SA \xrightarrow{f} Y \xrightarrow{x} X) \varepsilon T_2 \ (\P,\Phi)$ iff $(A,f) \varepsilon \Phi$.

These datas give the locally-orthogonal resp. the left-extension (or Kan-coextension)-characterization of semitopological functors.

The most wellknown examples for semitopological functors are:

[1]) Topological functors.

[2]) Monadic functors provided the domain category has a cone factorization; in particular every monadic functor over Sets is semitopological.

[3]) Full reflective embeddings.

[4]) The composition of semitopological functors.

2. Full Coreflective Restrictions of Semitopological Functors

are semifinal structure functors.

Let $(S: \underline{A} \longrightarrow \underline{X}) = (\underline{A} \xrightarrow{E} \underline{B} \xrightarrow{T} \underline{X})$ be a factorization and assume that T is ψ-locally orthogonal and E is a full coreflective embedding. Take for $\tilde{Q} := E$, $Q := T$, $\Phi = \psi$, $\P = \Gamma$ = the class of all counits of the full coreflective embedding and $T(\P,\Phi) = \underline{Chs}(\P,\Phi)$. Then S is a $(T(\P,\Phi),\Gamma)$-semiinitial and semifinal structure functor.

§ 2 The Duality Theorem for Structure Functors

The duality theorem for structure functors contains as special instances the duality theorems for topological functors (Antoine [1], Roberts [12]), for (E,M)-topological functors, for semitopological functors (Tholen [13]) for locally orthogonal $\underline{Q}$-functors (Tholen [16], Wischnewsky [18], Wolff [20]) and for cosemitopological functors. It is itself a special instance of a much more general duality theorem for structure functor sequences (Wischnewsky [18]).

(2.1) THEOREM (Duality Theorem for Structure Functors - Wischnewsky [18],[19])

Notation as in § 1. Let $S: \underline{A} \to \underline{X}$ be a functor. Then the following assertions are equivalent:

(i) S is a semifinal structure functor (with respect to $(T(\P,\Phi),\Gamma)$).

(ii) S is a semiinitial structure functor (with respect to $(T(\P,\Phi),\Gamma)$).

By specializing the datas $\P, \Phi, \Gamma, T(\P,\Phi)$ we obtain the following corollaries.

(2.2) <u>COROLLARY</u> (<u>Duality for Topological Functors</u>.)
(Antoine [1], Roberts [12])

<u>The following assertions are equivalent for a functor</u> S.

(i) S <u>is topological</u>

(ii) S^{op} <u>is topological</u>

<u>Proof</u>: Take $\tilde{Q}$ = Id, Q = S, ¶= Γ = <u>Id</u>(<u>A</u>), Φ = <u>Id</u>(S) and T(¶,Φ) = <u>Co-Mor</u>(S).

(2.3) <u>COROLLARY</u> (<u>Duality for Semitopological Functors</u> - Tholen [13])

<u>Notation as in</u> [13].

<u>The following assertions are equivalent for a functor</u> S.

(i) S <u>is a semifinal functor</u>.

(ii) S <u>is a semiinitial functor</u>.

<u>Proof:</u> Take for S the characterizing datas in (1.3) 1.a.

(2.4) <u>COROLLARY</u> (<u>Duality for Left-Extension Functors</u> -
Tholen [16], Wischnewsky [18], Wolff [20])

<u>The following assertions are equivalent for a functor</u> S:

(i) S <u>is a Φ-left-extension functor</u>.

(ii) S <u>is a Φ-locally orthogonal functor</u>.

(iii) S <u>is a semitopological functor</u>.

<u>Proof</u>: Take for S the characterizing datas in (1.3) 1.b.

<u>§ 3 The Representation Theorem for Structure Functors.</u>

(3.1) The Canonical Factorization "Induced by an Arbitrary Factorization". ([19])

Let $S = (\underline{A} \xrightarrow{\tilde{Q}} \underline{B} \xrightarrow{Q} \underline{X})$ be a factorization $\underline{Id}(Q) \subseteq \Gamma \subseteq \underline{Co\text{-}Mor}(Q)$ and $\tilde{Q}$ be faithful.

These datas induce in a canonical way a category $\underline{C}$ and severa functors:

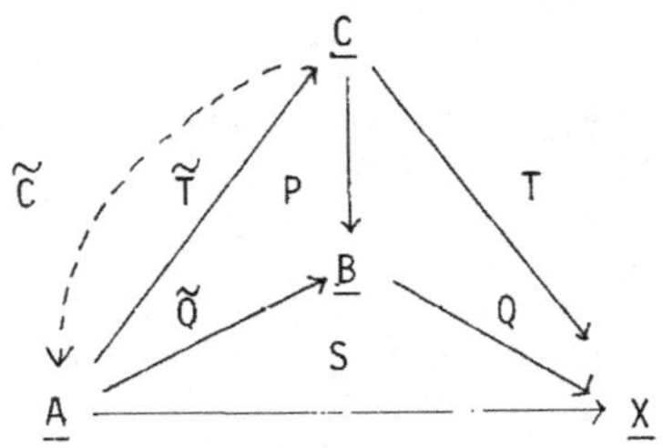

Figure 3

(1) The category $\underline{C}$:

Ob $\underline{C}$ = $\{(A, g: B \leftarrow \tilde{Q}A) \varepsilon \underline{\text{Co-Mor}}(Q);$
$(A,g) \varepsilon \Gamma\}$

Let (A,g) and (A',g') be Γ-objects.

A $\underline{C}$-morphism $(b,a): (A,g) \rightarrow (A',g')$ is a pair consisting of a $\underline{B}$-morphism

$b: B = \text{codom}(g) \longrightarrow \text{codom}(g') = B'$ and a $\underline{A}$-morphism
$a: A \longrightarrow A'$ with $g'(\tilde{Q}a) = bg$.

(2) The functor $\tilde{T}: \underline{A} \rightarrow \underline{C}$ is given by the assignments

$A \longmapsto (A, id(\tilde{Q}A))$

$f \longmapsto (\tilde{Q}f, f)$

$\tilde{T}$ is obviously full and faithfull. ($\tilde{Q}$ is faithful)

(3) The functors $\tilde{C}: \underline{C} \rightarrow \underline{A}$, $P: \underline{C} \rightarrow \underline{B}$ and $T: \underline{C} \rightarrow \underline{X}$ are given by the following assignments

$\tilde{C}: \underline{C} \longrightarrow \underline{A}: (A,g) \longmapsto A$

$P: \underline{C} \longrightarrow \underline{B}: (A,g) \longmapsto \text{codom}(g)$

$T: \underline{C} \longrightarrow \underline{X}: T := QP$

By these definitions we have: $\tilde{Q} = P\tilde{T}$ and $S = T\tilde{T}$

Furthermore we have a functorial morphism
$\gamma: \tilde{Q}\tilde{C} \rightarrow P$ by the assignment $\gamma(A,g) := g: \tilde{Q}\tilde{C}(g) = \tilde{Q}A \rightarrow \text{codom}(g)$
the functorial morphism γ is pointwise in Γ.

(4) $\tilde{T}$: $\underline{A} \longrightarrow \underline{C}$ is coreflective with coreflector $\tilde{C}$.
The counit $\sigma: \tilde{T}\tilde{C} \to Id(\underline{C})$ is given by

$$\sigma(A,g) := (g, id(A)).$$

Given $(\tilde{Q},Q,\Gamma)$. The factorization $S = (\underline{A} \overset{\tilde{T}}{\to} \underline{C} \overset{T}{\to} \underline{X})$ is called the *canonical factorization* induced by $(\tilde{Q},Q,\Gamma)$.

For the rest of this paper we assume that $\Gamma^2 \subseteq T(\P,\Phi)$ where Γ^2 denotes the class of all double-morphisms of type $(A, SA \overset{Qg}{\leftarrow} QB \overset{x}{\to} X)$ where $(A,g) \varepsilon \Gamma$.

(3.2) THEOREM (Representation Theorem for Structure Functors)

Let S: A → X be a functor. Then the following assertions are equivalent:

(i) S *is a Structure Functor*.

(ii) S *is a full coreflective restriction of a semitopological functor*.

Proof: (ii) → (i) 1.3.2

(i) → (ii) Let S be a structure functor with respect to $(\tilde{Q},Q,\P,\Phi,\Gamma,T(\P,\Phi))$. Then S is faithful by [19] Theorem 2.3. Hence $\tilde{Q}$ is faithful. Since $\underline{Id}(\tilde{Q}) \subseteq \Gamma \subseteq \underline{Co\text{-}Mor}(\tilde{Q})$ we obtain a canonical factorization $A \overset{\tilde{T}}{\to} \underline{C} \overset{T}{\to} \underline{X}$ of S where $\tilde{T}$ is a full coreflective embedding. We have to show that T is semitopological.

Let $\psi: \Delta X \to TD(\underline{C})$ be a T-cone. 3.1.3 delivers a functorial morphism $\gamma: \tilde{Q}\tilde{C} \to P$ (notation as in 3.1) being pointwise in Γ . This defines a $(\tilde{Q},Q,S)$-double-cone

$$\Delta X \overset{\psi}{\to} TD(\underline{C}) = QPD(\underline{C}) \xrightarrow{Q\gamma D(\underline{C})} SCD(\underline{C}) \quad \text{with}$$

$\gamma D(\underline{C}): \tilde{Q}\tilde{C}D(\underline{C}) \longrightarrow PD(\underline{C})$ being pointwise in Γ . Hence we obtain a semiinitial $(T(\P,\Phi),\Gamma)$-coextension (α,β):
$(A,g,B,e,X) \longrightarrow (\tilde{C}D(\underline{C}),\ \gamma D(\underline{C}),\ PD(\underline{C}),\psi,X)$.

Then with $C := (A,g) \varepsilon \underline{C}$ $\quad \delta := (\beta,\alpha): \Delta C \to D(\underline{C})$ is a functorial morphism.
Obviously $\psi = (T\delta)(\Delta e)$. Let $(C', x: TC' \to X)$ be a T-comorphism with $C' = (A', g'QA' \to B') \varepsilon \Gamma$ and $\delta': \Delta C' \to D(\underline{C})$ be a functorial morphism with $T\delta' = \psi(\Delta x)$. Hence we have a pair of functorial morphisms $\alpha': \Delta A' \to \tilde{C}D(\underline{C})$ and $\beta': \Delta B' \to PD(\underline{C})$ rendering the

diagram sommutative. Since $SA' \xrightarrow{Qg'} QB' \xrightarrow{x} X \varepsilon T(\P, \Phi)$ by assumption we get unique morphisms $s: A' \to A$ and $A: B' \to B$ and hence a unique $\omega: C' \to C$ such that

$$\delta' = \delta(\Delta\omega) \quad \text{and} \quad ex = T\omega \ .$$

The rest is clear. This completes the proof.

By dualizing and specializing the results in 3.1, 3.2 for semitopological functors S we obtain immediately the following result first proved by W. Tholen and M. Wischnewsky, Oberwolfach 1977 .

(3.3) COROLLARY (Tholen and Wischnewsky - Representation Theorem for Semitopological Functors)

The following assertions are equivalent for a functor S.

(i) S is semitopological.
(ii) S is a full reflective restriction of a topological functor.

Together with (3.2) we obtain.

(3.4) THEOREM (Complete Representation Theorem for Structure Functors)

The following assertions are equivalent for a functor S.

(i) S is a structure functor.

(ii) There exists a factorization of S

$$(S: \underline{A} \to \underline{X}) = (\underline{A} \xrightarrow{C} \underline{B} \xrightarrow{R} \underline{C} \xrightarrow{T} \underline{X})$$

where C ≡ full coreflective embedding
R ≡ full reflective embedding and
T ≡ topological functor.

(3.5) COROLLARY

Let $(S: \underline{A} \to \underline{X}) = (\underline{A} \xrightarrow{\tilde{T}} \underline{B} \xrightarrow{T} \underline{X})$ be a full coreflective restriction of a semitopological functor T where $\tilde{T}$ is the full coreflective embedding (i.e. S is a structure functor).

Let Γ be the class of all counits $g:\tilde{T}A \to B$ of $\tilde{T}$.
If $T(\Gamma) \subseteq Iso(\underline{X})$ we obtain the following assertions:

1. If T is topological then S is topological (Wyler [21]).
2. If T is (E,M)-topological (in the sense of Herrlich [5]) then S is (E,M)-topological.
3. If T is topologically-algebraic in the sense of Y. Hong [1] resp. equivalently a M-functor in the sense of Tholen [13] then S is again a topologically-algebraic functor.
4. If T is semitopological then S is semitopological.

Proof: Let $(S:\underline{A} \to \underline{X}) = (\underline{A} \overset{\tilde{T}}{\to} \underline{B} \overset{T}{\to} \underline{X})$ as above. Then S is a structure functor. Hence S fulfills both of the descriptions in § 1. Since $T(\Gamma) \subseteq Iso(\underline{X})$ everything follows easily from figure 1 resp. figure 2.

(3.6) EXAMPLES:

1. Let Top be the category of all topological spaces (over the category of Sets). Since every full coreflective subcategory of Top (except the trivial one) fulfills the assumption in (3.5) every non-trivial coreflective subcategory of Top is again a topological category (over Sets).

2. The category of compactly-generated spaces is a full coreflective subcategory of the (E,M)-topological category of all Hausdorff spaces fulfilling the assumption in (3.5). Hence it is again an (E,M)-topological category.

Similar results can be obtained for reflective restrictions of coreflective subcategories of topological categories by dualizing the previous results. So for instance we obtain.

(3.7) THEOREM (Complete Representation Theorem for Co-Structure Functors)

The following assertions are equivalent for a functor S.

(i) S is a Co-structure functor. (≡ Co-semifinal resp. Co-semi-initial structure functor).

(ii) There exists a factorization of S

$$(S: \underline{A} \longrightarrow \underline{X}) = (\underline{A} \xrightarrow{R} \underline{B} \xrightarrow{C} \underline{C} \xrightarrow{T} \underline{X})$$

where R ≡ full reflective embedding
C ≡ full coreflective embedding
T ≡ topological functor.

§ 4 Characterization and Existence Theorems for Structure Functors

Let $(S: \underline{A} \to \underline{X}) = (\underline{A} \xrightarrow{\tilde{Q}} \underline{B} \xrightarrow{Q} \underline{X})$ be a factorization of the functor S and $\underline{Id}(Q) \subseteq \Phi \subseteq \underline{Mor}(Q)$ and $\underline{Id}(Q) \subseteq \Gamma \subseteq \underline{Co\text{-}Mor}(Q)$ with the usual closedness conditions.

(4.1) DEFINITION

1) Given $(A,B,SA \xrightarrow{Qg} QB \xleftarrow{f} Y \xrightarrow{x} X)$ with $(B,f)\varepsilon\Gamma$ and $(A,g: \tilde{Q}A \to B)\varepsilon\Gamma$. A semifinal extension of (A,B,g,f,x) is called a (Φ,Γ)-semi-pushout of (A,B,g,f,x).

2) Given a discrete functorial chain

$$(A_i, B_i, SA_i \xrightarrow{Qg_i} QB_i \xleftarrow{e_i} X \xrightarrow{id} X; i\varepsilon I)$$

with $(B_i,e_i)\varepsilon\Phi$ and $(A_i,g_i: \tilde{Q}A_i \to B_i)\varepsilon\Gamma$ for all $i\varepsilon I$.
A semifinal extension of $(A_i,B_i,g_i,e_i,i\varepsilon I)$ is called a multiple (Φ,Γ)-semi-pushout of $(A_i,B_i,g_i,e_i,i\varepsilon I)$.

3) S is called (Φ,Γ)-semi-pushout-complete if every chain (A,B,g,f,x) with $(B,f)\varepsilon\Phi$ and $(A,g)\varepsilon\Gamma$ has a (Φ,Γ)-semi-pushout and if every discrete functorial chain $(A_i,B_i,g_i,e_i,i\varepsilon I)$ with $(B_i,e_i)\varepsilon\Phi$ and $(A_i,g_i)\varepsilon\Gamma$ for all $i\varepsilon I$ has a multiple (Φ,Γ)-semi-pushout.

4) S is called (Φ,Γ)-semi-factorizable if for every functorial chain $(\underline{D}(\underline{A}),D(\underline{B}),\Delta X \longrightarrow QD(\underline{B}) \xleftarrow{Q\gamma} SD(\underline{A}))$ with $(D(\underline{A}),\gamma)$

being pointwise in T there exist at least one chain $(A,B,g,X \xrightarrow{f} QB \xleftarrow{Qg} SA)$ with $(B,f)\varepsilon\Phi$ and $(A,g)\varepsilon\Gamma$ and functorial morphisms $\beta: \Delta B \longrightarrow D(\underline{B})$ and $\alpha: \Delta A \rightarrow D(\underline{A})$ such that $\phi = (Q\beta)\Delta f$ and $\beta\cdot\Delta g = \gamma\cdot(\tilde{Q}\alpha)$.

(4.2) THEOREM (Characterization Theorem for Structure Functors)

Notation as in § 1. Given $S: \underline{A} \rightarrow \underline{X}$ and $(\tilde{Q},Q,\P,\Phi,\Gamma,T(\P,\Phi))$. The following assertions are equivalent:

(i) S is a structure functor (with respect to $(\tilde{Q},Q,\P,\Phi,\Gamma,T(\P,\Phi))$)

(ii) a) S is (Φ,Γ)-semi-factorizable.
b) S is (Φ,Γ)-semi-pushout-complete .

(iii) Let $(S: \underline{A} \rightarrow \underline{X}) = (\underline{A} \xrightarrow{\tilde{T}} \underline{C} \xrightarrow{T} \underline{X})$ be the canonical factorization. Then T is semitopological.

Proof: The equivalence of (i) $\longleftrightarrow$ (iii) follows from (3.1) and (3.2). Hence we have to prove (i) $\longleftrightarrow$(ii). The implication (i) $\rightarrow$ (ii) follows immediately from the semi-final representation of S (Definition (1.1)). (ii) $\rightarrow$ (i): Given a double-cone $(D(\underline{B}),\ \phi: \Delta X \rightarrow QD(B),\ D(\underline{A}),\ \gamma: \tilde{Q}D(\underline{A}) \rightarrow D(\underline{B}))$ with $(D(\underline{A}),\gamma)$ being pointwise in Γ . Since S is (Φ,Γ)-semi-factorizable the class of all chains $(A_i,B_i,g_i;\ X \xrightarrow{f_i} QB_i \xrightarrow{Qg_i} SA_i)$ with $(B_i,f_i)\varepsilon\Phi$ and $(A_i,g_i)\varepsilon\Gamma$ such that there exist functorial morphisms $\beta_i: \Delta B_i \rightarrow D(\underline{B})$ and $\alpha_i: \Delta A_i \rightarrow D(\underline{A})$ with $\gamma = (Q\beta_i)\Delta f_i$ and $\beta_i\Delta g_i = \gamma(\tilde{Q}\alpha_i)$ is nonvoid. The multiple (Φ,Γ)-semi-pushout of the class $(A_i,B_i,g_i,f_i,i\varepsilon I)$ induces a (Φ,Γ)-semi-factorization of the double cone (ϕ,γ). The existence of (Φ,Γ)-semi-pushouts as well as the universal property of the above multiple (Φ,Γ)-semi-pushouts imply that the (Φ,Γ)-semi-factorization constructed above is indeed the semiinitial coextension looked for. This completes the proof.

THEOREM (4.2) has numerous applications by specializing the datas in $(\tilde{Q},Q,\P,\Phi,\Gamma,T(\Phi,\Gamma))$. If $\tilde{Q} = Id_{\underline{A}}$ and $\Gamma = Id(\underline{A})$ we omit Γ in the previous definitions. So for instance we obtain the following corollaries (cp.Tholen I[6]).

(4.3) COROLLARY (Characterization Theorem for Semitopological Functors)

Notation as above. The following assertions are equivalent:

(i) S is semitopological (= Φ locally orthogonal functor)

(ii) a) S is Φ-semi-factorizable
b) S is Φ-semi-pushout-complete ,

(iii) a) S has a left adjoint with a unit being pointwise in Φ .
b) S is Φ-semi-pushout-complete.

(iv) Let $(S: \underline{A} \longrightarrow \underline{X}) = (\underline{A} \xrightarrow{\tilde{T}} \underline{C} \xrightarrow{T} \underline{X})$ be the dual of the canonical construction. Then T is topological.

If one admitts multiple (Φ,Γ)-semi-pushouts for void index-set , S is (Φ,Γ)-semi-factorizable. Hence one can omit in THEOREM (4.2) resp. in COROLLARY (4.3) the conditions (ii) a) resp. (iii) a).

(4.4) COROLLARY (Tholen [16])

The following assertions are equivalent for a category $\underline{A}$.

(i) $\underline{A}$ is a locally orthogonal $(\Phi,\text{Mono-Cone}(\underline{A}))$-category.

(ii) a) $\underline{A}$ is Φ-pushout-complete.
b) $\underline{A}$ has coequalizers
c) Φ is closed under composition with external epimorphisms from the left.

If Φ is closed under composition then $\underline{A}$ is an orthogonal $(\Phi,\text{Mono-Cone}(\underline{A}))$-category.

References

[1] Antoine, P.: Etude élémentaire des catégories d'ensembles structurées. Bull. Soc. Math. Belg. 18 (1966).

[2] Börger, R. and Tholen, W.: Cantors Diagonalprinzip für Kategorien. Preprint (Fernuniversität Hagen 1977).

[3] Börger, R. and Tholen, W.: Is any semitopological functor topologically algebraic. Preprint (Fernuniversität Hagen, 1977).

[4] Gabriel, P. and Ulmer, F.: Lokal präsentierbare Kategorien. Lecture Notes in Math. 221 (Springer,Berlin-Heidelberg-New York 1971).

[5] Herrlich, H.: Topological functors. Gen. Top. Appl. 4 (1974), 125-142.

[6] Herrlich,H., Nakagawa, R., Strecker, E., Titcomb,T.: Topologically-algebraic and semi-topologically functors, Preprint (Bremen, Ibaraki, Manhattan 1977).

[7] Herrlich, H. and Strecker,G.: Semi-universal Maps and Universal Initial Completions. Preprint, (Universität Bremen, Kansas State University, 1977).

[8] Hoffmann, R.-E.: Semi-identifying lifts and a gneralization of the duality theorem for topological functors. Math. Nachr. 74 (1976), 295-307.

[9] Hoffmann, R.-E.: Topological functors admitting generalized Cauchy-completions, Lecture Notes in Math. 540 (Springer, Berlin-Heidelberg-New York 1976), 286-344.

[10] Hong, S.S.: Categories in which every mono-source is initial. Kyungpook Math. J. 15, (1975), 133-139.

[11] Hong, Y.H.: Studies on categories of universal topological algebras. Thesis (MacMaster University, Hamilton 1974).

[12] Roberts, J.E.: A characterization of topological functors. J. Algebra 8 (1968), 131-193.

[13] Tholen, W.: Semi-topological functors I. Preprint (Fernuniversität Hagen 1977). To appear in J. Pure Appl. Algebra.

[14] Tholen, W. and Wischnewsky, M.B.: Semitopological functors II. Preprint (Bremen, Hagen 1977), to appear in J. Pure Appl. Algebra.

[15] Tholen, W., Wischnewsky, M.B., Wolff, H.: Semitopological functors III: Lifting of adjoints. Preprint (Fernuniversität Hagen, Universität Bremen, University of Toledo (1978)

[16] Tholen, W.: Konkrete Funktoren. Habilitationsschrift (Fernuniversität Hagen 1977).

[17] Wischnewsky, M.B.: A lifting theorem for right adjoints. Cahiers de Topo. et Geo.Diff. (1978), Vol. XIX-2, 155-168.

[18] Wischnewsky, M.B.: A Generalized Duality Theorem for Structure Functors. To appear in Cahiers de Topo. et Geo. Diff.

[19] Wischnewsky, M.B.: Topologically algebraic structure functors ≡ full reflective or coreflective subcategories of semi-topological functors. To appear in Cahiers de Topo. et Geo. Diff.

[20] Wolff, H.: External Characterization of Semitopological Functors. Handwritten manuscript (University of Toledo 1977).

[21] Wyler, O.: Top categories and categorical topology. Gei Top. Apll. 1 (1971), 17-28.

[22] Adamēk, J., Herrlich, H., Strecker, G.F.: Least and largest initial completions, Preprint (1978).

[23] Börger, R.: Universal Topological Completions of Semi-topological over Ens need not exist. Preprint (1978).

[24] Trnkowa, V.: Automata and categories. LN Comp. Science 32 (Springer 1975) pp. 132-152.

Bremen, August 1978

M. B. Wischnewsky
FB-Mathematik
Universität Bremen
Kufsteiner Straße
D-2800 Bremen 33
W.-Germany

FUNCTION SPACES IN TOPOLOGICAL CATEGORIES

Oswald Wyler

We do not attempt a comprehensive survey; this paper concentrates on some recent developments in which the author was involved. We shall discuss fine proper function spaces, closed reflective subcategories of cartesian closed topological categories, continuous uniform convergence for uniform convergence spaces, and spaces of closed sets of locally compact spaces.

The title is somewhat inaccurate: we work with initially structured categories as defined in [22], not with the topological categories of [14] and [23]. The list of references is far from comprehensive, but it should be adequate.

1. Fine proper function spaces

In this paper, $\mathcal{C}$ will always denote a concrete category, with a faithful underlying set functor $U : \mathcal{C} \dashrightarrow \mathrm{ENS}$ which transports structures, i.e. if A is an object of $\mathcal{C}$ and $f : UA \dashrightarrow X$ a bijection, then there is exactly one isomorphism $f : A \longrightarrow B$ in $\mathcal{C}$ which lifts f, i.e. such that f is the underlying mapping, and $UB = X$. We use here the common abus de langage which does not distinguish notationally between a map and the underlying mapping.

We say that $\mathcal{C}$ is initially structured if $\mathcal{C}$ and U satisfy:

IS 1. $\mathcal{C}$ has an initial structure for every injective U-source $(f_i : X \dashrightarrow UA_i)$, i.e. the collectively injective f_i always lift to maps $f_i : A \longrightarrow A_i$ in $\mathcal{C}$ such that a mapping $g : UB \dashrightarrow UA$ always lifts to a map $g : B \longrightarrow A$ of $\mathcal{C}$ if all $f_i g$ are maps $f_i g : B \dashrightarrow A_i$ in $\mathcal{C}$.

IS 2. If P is an object of $\mathcal{C}$ with UP empty or a singleton, and B any object of $\mathcal{C}$, then every mapping $f : UP \dashrightarrow UB$ lifts to a map $f : P \dashrightarrow B$ in $\mathcal{C}$.

It follows from IS 1 that U is a topological functor [13], with the nice properties obtained in [13] and [22]. In particular, $\mathcal{C}$ has a one-point terminal object, and an empty initial object. By IS 2, every one-point object of $\mathcal{C}$ is terminal, every empty object of $\mathcal{C}$ initial. These may be the only objects of $\mathcal{C}$. Except for this trivial case, U has a left adjoint which preserves underlying sets, and defines <u>discrete structures</u> for $\mathcal{C}$.

The assumption, not made in [22], that U transports structures is convenient, but not essential. We do not need or assume smallness of U-fibres. 1.4 and 1.5 in [22] may be false if $\mathcal{C}$ has non-initial empty objects.

In an initially structured category $\mathcal{C}$, proper and admissible function spaces $[A,B]$, with underlying set $U[A,B] = \mathcal{C}(A,B)$, can be defined in the usual way. $[A,B]$ is <u>admissible</u> if <u>evaluation</u> $ev : [A,B] \times A \to B$, given at the set level by $ev(f,x) = f(x)$, for $f : A \to B$ and $x \in UA$, is a map of $\mathcal{C}$, and $[A,B]$ is <u>proper</u> if $\hat{u} : C \to [A,B]$, given at the set level by $\hat{u}(z)(x) = u(z,x)$, is a map of $\mathcal{C}$ for every map $u : C \times A \to B$ of $\mathcal{C}$.

Proper and admissible function spaces have the properties which are familiar from topology. In particular, there is always a proper function space, which we denote by $[A,B]_p$, with the initial structure for the mappings $e_x : \mathcal{C}(A,B) \to UA$ given by $e_x(f) = f(x)$, for $x \in UA$ and $f : A \to B$ in $\mathcal{C}$. There is also a <u>fine proper function space</u> which we denote by $[A,B]_f$, with the initial structure for all mappings $id : \mathcal{C}(A,B) \to U[A,B]$ for proper spaces $[A,B]$. On the other hand, admissible function spaces $[A,B]$ need not always exist; this is the case e.g. for uniform spaces. If admissible function spaces exist, there need be no coarsest admissible space $[A,B]$, as the example of TOP shows. Even if there is a coarsest admissible function space $[A,B]$, it may be strictly finer than the fine proper function space $[A,B]_f$.

Fine proper function spaces have not been studied much, but they do have good functorial properties. We state their basic property without proof.

THEOREM 1.1. <u>Fine proper function spaces</u> $[A,B]_f$ <u>in an initially structured category</u> $\mathcal{C}$ <u>define a lifted hom functor</u> $[-,-]_f : \mathcal{C}^{op} \times \mathcal{C} \to \mathcal{C}$, <u>with</u> $U[-,-]_f = \mathcal{C}(-,-)$. <u>The mappings</u> $u \mapsto \hat{u} : \mathcal{C}(C \times A, B) \to \mathcal{C}(C, [A,B]_f)$ <u>lift to maps</u> $[C \times A, B]_f \to [C, [A,B]_f]_f$, <u>natural in</u> C, A, B. <u>These maps are isomorphisms for objects</u> A, B <u>of</u> $\mathcal{C}$ <u>with</u> $[A,B]_f$ <u>admissible</u>.

For topological Hausdorff spaces A and B, the spaces $[A,B]_f$ have the compact-open topology, and the maps $[C \times A, B]_f \to [C, [A,B]]_f$ are embeddings, by a result of J.R. Jackson [18]. These maps are injective for any initially structured category; we do not know whether they always are embeddings.

All spaces $[A,X]_f$ are admissible for an object A if and only if the functor $- \times A$ on $\mathcal{C}$ has a right adjoint $[A,-]_f$. Such objects are of special interest, but we cannot discuss them here.

2. Closed reflective subcategories

We recall that a category $\mathcal{C}$ is called $(\mathcal{E},\mathcal{M})$-factored, for classes $\mathcal{E}$ and $\mathcal{M}$ of morphisms of $\mathcal{C}$, both closed under compositions with isomorphisms, if the following conditions are met.

(1) Every morphism f of $\mathcal{C}$ factors $f = m\,e$ with $e \in \mathcal{E}$ and $m \in \mathcal{M}$.

(2) If $m\,u = e\,v$ in $\mathcal{C}$, with $e \in \mathcal{E}$ and $m \in \mathcal{M}$, then $u = t\,e$ and $v = m\,t$ in $\mathcal{C}$ for a unique morphism t of $\mathcal{C}$.

In this situation, a source $(g_i : A \longrightarrow A_i)$ of $\mathcal{C}$ is called an $\mathcal{M}$-source if for morphisms u of $\mathcal{C}$ and e in $\mathcal{E}$, and factorizations $g_i\,u = v_i\,e$, there is always a unique t in $\mathcal{C}$ such that $t\,e = u$, and $g_i\,t = v_i$ for all i. We say that $\mathcal{C}$ has <u>$(\mathcal{E},\mathcal{M}$-source$)$-factorization</u> if every source (f_i) in $\mathcal{C}$ factors $f_i = g_i\,e$, with e in $\mathcal{E}$ and (g_i) an $\mathcal{M}$-source.

If $\mathcal{C}$ is initially structured, then $\mathcal{C}$ has $(\mathcal{E},\mathcal{M}$-source$)$-factorization for $\mathcal{E}$ all surjective maps of $\mathcal{C}$ and $\mathcal{M}$-sources all initial monosources, and also for $\mathcal{E}$ all quotient maps of $\mathcal{C}$ and $\mathcal{M}$-sources all monosources. If $\mathcal{C}$ is topological, admitting all initial sources, then $\mathcal{C}$ also factors for $\mathcal{E}$ all bijective maps of $\mathcal{C}$, and $\mathcal{M}$-sources all initial sources.

If $\mathcal{C}$ is $(\mathcal{E},\mathcal{M}$-source$)$-factored, with $\mathcal{E}$ consisting of epimorphisms of $\mathcal{C}$, then a full isomorphism-closed subcategory $\mathcal{B}$ of $\mathcal{C}$ is $\mathcal{E}$-reflective, i.e. reflective with all reflections in $\mathcal{E}$, if and only if $\mathcal{B}$ is closed under $\mathcal{M}$-source formation, i.e. if $(g_i : A \to B_i)$ is an $\mathcal{M}$-source in $\mathcal{C}$ with all B_i in $\mathcal{B}$, then A always is in $\mathcal{B}$.

These and other properties of factorizations are well known.

From now on, $\mathcal{C}$ will be cartesian closed, with "function space objects" $[X,Y]$, and $\mathcal{B}$ will be a full and isomorphism-closed reflective subcategory of $\mathcal{C}$. We note first a specialization of B. Day's reflection theorem [7].

THEOREM 2.1. <u>For $\mathcal{C}$ and $\mathcal{B}$ as above, the following two properties of $\mathcal{B}$ are logically equivalent.</u>

(i) $[X,B]$ <u>is in $\mathcal{B}$ for all objects</u> X <u>of</u> $\mathcal{C}$ <u>and</u> B <u>of</u> $\mathcal{B}$.

(ii) <u>The reflector</u> $\mathcal{C} \longrightarrow \mathcal{B}$ <u>preserves finite products</u>.

We say that $\mathcal{B}$ is a <u>closed reflective subcategory</u> of $\mathcal{C}$ if these properties are satisfied; they imply that $\mathcal{B}$ is cartesian closed.

We note two conditions which imply that $\mathcal{B}$ is closed reflective in $\mathcal{C}$.

(1) $\mathcal{B}$ is dense in $\mathcal{C}$, reflective in $\mathcal{C}$, and cartesian closed [7].

(2) $\mathcal{T}$ is initially structured, and $\mathcal{B}$ quotient-reflective in $\mathcal{C}$ [22].

If $\mathcal{C}$ has $(\mathcal{E}, \mathcal{M}$-source)-factorization for a class $\mathcal{E}$ of epimorphisms of $\mathcal{C}$, then the intersection of closed $\mathcal{E}$-reflective subcategories of $\mathcal{T}$ is a closed $\mathcal{E}$-reflective subcategory. Thus we can form the closed $\mathcal{E}$-reflective subcategory of $\mathcal{T}$ <u>generated</u> by a class $\mathcal{L}$ of objects of $\mathcal{T}$, the smallest closed $\mathcal{E}$-reflective subcategory $\mathcal{B}$ of $\mathcal{T}$ such that $\mathcal{L}$ consists of objects of $\mathcal{B}$.

These categories have been studied for $\mathcal{C}$ the category of limit spaces, and for $\mathcal{L} = \{L\}$ a singleton. If $L = \mathbb{R}$, the real numbers, and $\mathcal{M}$-sources are initial monosources, we obtain the c-embedded spaces of Binz [1]. For initial sources, we obtain the c-spaces of Bourdaud [3]. Gazik, Kent and Richardson [12] introduced u-embedded spaces, for $L = \mathbb{R}$ in the category of uniform convergence spaces, and $\mathcal{M}$-sources all initial monosources. The construction of Binz and Bourdaud was generalized by R.S. Lee and the author; see [20] for details. We generalize it further by using arbitrary $\mathcal{L}$.

For a fixed object L of the cartesian closed category $\mathcal{L}$, the contravariant functor $[-,L]$ on $\mathcal{L}$ is self-adjoint on the right. We denote by $a_X : X \longrightarrow [[X,L],L]$ the unit-counit of this self-adjunction, at an object X. We have $a_X(x)(f) = f(x)$ at the set level, for x in X and $f : X \longrightarrow L$ in $[X,L]$, if $\mathcal{C}$ is concrete with constant maps.

THEOREM 2.2. <u>Let $\mathcal{C}$ be cartesian closed, with $(\mathcal{E}, \mathcal{M}$-source)-factorization, where $\mathcal{E}$ consists of epimorphisms of $\mathcal{C}$ and is closed under finite products, with $e \times id_X$ always in $\mathcal{E}$ for $e \in \mathcal{E}$ and an object X of $\mathcal{C}$. If (L_i) is a family of objects of $\mathcal{C}$, then the closed $\mathcal{E}$-reflective subcategory of $\mathcal{C}$ generated by the objects L_i has as objects all objects X of $\mathcal{C}$ for which the source of units $a_X : X \dashrightarrow [[X,L_i],L_i]$ is an $\mathcal{M}$-source.</u>

The proof is given in [20] for a single object L, generalizing Bourdaud's argument in [3]. Generalization to a family of objects L_i is easy. If X is an object of $\mathcal{C}$, then the reflection of X for the closed $\mathcal{E}$-reflective subcategory of $\mathcal{C}$ generated by the objects L_i of $\mathcal{C}$ is the $\mathcal{E}$-factor of the source of units $a_X : X \longrightarrow [[X,L_i],L_i]$ in $\mathcal{C}$.

We note that the conditions for $\mathcal{E}$ in Theorem 2.2 are close to being best possible. They are satisfied for the examples given above if $\mathcal{C}$ is initially structured and cartesian closed.

3. Uniform continuous convergence

Uniform continuous convergence, for spaces of uniformly continuous functions, was first used by Gazik, Kent and Richardson [12]. R.S. Lee [19], [20] showed that uniform convergence spaces form a cartesian closed category, and she obtained other interesting results which I shall review here.

For present purposes, the author's modification [33] of the definition by Cook and Fischer [6] has to be used. Thus a <u>uniform convergence space</u> $(X, \mathcal{J})$ is a set X with a set $\mathcal{J}$ of filters on $X \times X$ such that:

(i) If $x \in X$, then the point filter on (x,x) is in $\mathcal{J}$.

(ii) If $\Phi \in \mathcal{J}$ and Ψ' is finer than Φ, then $\Psi' \in \mathcal{J}$.

(iii) If Φ and Ψ are in $\mathcal{J}$, then $\Phi \cup \Psi$ is in $\mathcal{J}$.

(iv) If Φ and Ψ' are in $\mathcal{J}$, then $\Psi' \circ \Phi$ is in $\mathcal{J}$.

(v) If Φ is in $\mathcal{J}$, then Φ^{-1} is in $\mathcal{J}$.

Here, $\Phi \cup \Psi$ is generated by the sets $A \cup B$ with $A \in \Phi$ and $B \in \Psi$, and the other filter operations are obtained from subset operations in the same way. A <u>uniformly continuous</u> map $f : (X, \mathcal{J}) -> (Y, \mathcal{K})$ is a mapping $f : X -> Y$ such that $(f \times f)(\Phi)$ is in $\mathcal{K}$ for all Φ in $\mathcal{J}$.

EXAMPLES. (1) If Φ is the filter of entourages for a uniform structure of X, then the filters on $X \times X$ finer than Φ form a uniform convergence structure of X. This embeds uniform spaces into uniform convergence spaces.

(2) If X is a locally compact Hausdorff space, then the neighborhood filters of diagonals $\Delta_K = \{(x,x) \mid x \in K\}$ in $X \times X$, for $K \subset X$ compact, together with all filters finer than one of these filters, define a uniform convergence structure of X.

If F is the set of uniformly continuous functions $f : (X, \mathcal{J}) -> (Y, \mathcal{K})$, for uniform convergence spaces $(X, \mathcal{J})$ and $(Y, \mathcal{K})$, then the <u>uniform continuous convergence structure</u> on F consists of all filters $\mathcal{U}$ on $F \times F$ such that $\mathcal{U}(\Phi)$ is in $\mathcal{K}$ for all Φ in $\mathcal{J}$. Here $\mathcal{U}(\Phi)$ is generated by the sets $U(A)$ with $U \in \mathcal{U}$ and $A \in \Phi$, with $U(A)$ consisting of all $(f(x), g(y))$ for $(f,g) \in U$ and $(x,y) \in A$.

THEOREM 3.1 [19]. <u>With uniform continuous convergence for function spaces, uniform convergence spaces form a cartesian closed topological category.</u>

A uniform convergence structure $\mathcal{J}$ on a set X induces a convergence structure on X, with a filter $\mathcal{F}$ on X converging to $x \in X$ whenever the filter $\mathcal{F} \times \dot{x}$, for the point filter $\dot{x}$ on x , is in $\mathcal{J}$.

THEOREM 3.2 [20]. The convergence structure induced by uniform continuous convergence on a function space $[X,Y]$ of uniformly continuous functions is finer than continuous convergence, and coarser than uniform convergence. It is uniform convergence if the principal filter on the diagonal Δ_X is in the structure of X , and uniform convergence on compact sets if X is a locally compact Hausdorff space, with the associated uniform convergence structure.

A uniform convergence space $(X,\mathcal{J})$ is called precompact if every ultrafilter $\mathcal{F}$ on X is a Cauchy filter, i.e. $\mathcal{F} \times \mathcal{F} \in \mathcal{J}$. Cartesian closed categories of precompact uniform spaces were obtained by G. Tashjian [26]; the connection between her results and those of R.S. Lee has not been established.

A pseudouniform space is a uniform convergence space $(X,\mathcal{J})$ such that a filter Φ on $X \times X$ is in $\mathcal{J}$ as soon as every ultrafilter finer than Φ is in $\mathcal{J}$. Pseudouniform spaces form a bireflective subcategory of uniform convergence spaces, and the induced convergence structure of a pseudouniformity is a pseudotopology as defined by G. Choquet.

THEOREM 3.3 [20]. A function space $[X,Y]$ with uniform continuous convergence is a pseudouniform space if Y is a pseudouniform space. Thus pseudouniform spaces form a cartesian closed topological category.

We say that a uniform convergence space X is u-embedded if the map a_X : $X \dashrightarrow [[X,L],L]$ is an embedding, for $L = \mathbb{R}$, the real numbers, or $L = I$, the unit interval, both with the usual uniformity. We say that X is a u-space if a_X is coarse, i.e. an initial source. It turns out that both choices of L define the same spaces [20]. By Theorem 2.2 above, u-spaces and u-embedded spaces define initially structured cartesian closed categories.

THEOREM 3.4 [20]. (a) Every uniform space is a u-space; every separated uniform space is u-embedded. (b) Every u-space is a pseudouniform space. (c) A precompact u-space is a uniform space.

We note that there are precompact pseudouniform spaces which are not uniform spaces, and thus not u-spaces.

4. Compact spaces of closed sets

Closed sets in a topological space X correspond to continuous functions $X \longrightarrow Z$, where Z is the Sierpiński space with two points and three open sets. Thus spaces of closed sets can be considered as function spaces.

We denote by $C(X)$ the space of closed sets of a topological space X , provided with the Scott topology [24] first considered in [9]. In this topology, $G \subset C(X)$ is open whenever (a) G is decreasing, i.e. if $A \in G$ and $B \subset A$, with $B \in C(X)$, then $B \in G$, and (b) if an intersection $\bigcap A_\lambda$ of closed sets is in G , then some finite intersection of sets A_λ is already in G .

We recall that X is called core-compact [15] or quasi-locally compact [30] if for every neighborhood U of a point $x \in X$, there is a neighborhood $V \subset U$ of x such that every ultrafilter $\mathcal{F}$ on X with $V \in \mathcal{F}$ converges to a point of U (and possibly to points not in U). Locally compact spaces, in the proper sense of this term, are core-compact. Using continuous lattice techniques, Hoffman and Lawson have shown in [15] that every sober core-compact space is locally compact, and in [16] that not every core-compact space is locally compact.

Theorem 4.1 [9], [15]. *The following properties of a topological space* X *are logically equivalent.*

(i) X *is core-compact*.

(ii) *The set* $\{(A,x) \in C(X) \times X \mid x \in A\}$ *is closed in* $C(X) \times X$.

(iii) *The functor* $- \times X$ *on* TOP *has a right adjoint* $[X,-]$, *given by proper and admissible function spaces* $[X,Y]$.

(iv) *The complete sup-semilattice* $C(X)$ *is a continuous lattice*.

Many authors prefer open sets in the context of this theorem; closed sets came historically first and seem topologically more natural.

In [27] and [28], L. Vietoris introduced and studied compact spaces of closed sets for compact Hausdorff spaces. We extend his results to locally compact spaces, without requiring T 2. For a locally compact space X , we denote by $V(X)$ the space of all closed sets of X , including $\emptyset$, with a topology which we call the Vietoris topology. This is the coarsest topology for which every segment $\downarrow A = \{B \in V(X) \mid B \subset A\}$ is closed for A closed in X , and every set $K^\perp = \{A \in V(X) \mid A \cap K = \emptyset\}$ is open for K compact in X . We note that the Scott-open sets in $C(X)$ are the decreasing open sets in $V(X)$.

One sees easily that $V(X)$ is a Hausdorff space if X is locally compact. If $SC(X)$ is the Stone space of $C(X)$, the set of all filters of closed sets of X provided with the Stone topology, then it turns out that adherences of filters define a continuous map $adh_X : SC(X) \longrightarrow V(X)$. This map is surjective and $SC(X)$ compact; we conclude that $V(X)$ is compact.

The following theorem extends results of Vietoris [28].

THEOREM 4.2. *If* X *is locally compact and* F *closed in* $V(X)$, *then* $\bigcup F$ *is closed in* X. *The resulting set union map* $V(V(X)) \longrightarrow V(X)$ *is continuous.*

We say that a subset S of a topological space X is *saturated* if S is the intersection of all neighborhoods of S, and we say that a continuous map $f : X \longrightarrow Y$ is *decent* [15] if $f^{-1}(K)$ is compact in X for every saturated compact subset K of Y. For a decent map $f : X \longrightarrow Y$ of locally compact spaces, we denote by $V(f) : V(X) \longrightarrow V(Y)$ the map which assigns to $A \in V(X)$ the closure of the image $f(A)$ in Y.

Every map $f : X \longrightarrow Y$ of compact Hausdorff spaces is decent, and $V(f)$ is the direct image map, restricted to closed sets.

THEOREM 4.3. V *is a functor, from locally compact spaces and decent maps to compact Hausdorff spaces. In the same context, union maps* $V(V(X)) \longrightarrow V(X)$ *and adherence maps* $adh_X : SC(X) \longrightarrow V(X)$ *define natural transformations.*

If X is a compact Hausdorff space, then singletons define a continuous map $s_X : x \longmapsto \{x\} : X \longrightarrow V(X)$, and we have the following result.

THEOREM 4.4 [37]. *The functor* V, *restricted to compact Hausdorff spaces, singleton maps* s_X, *and set union maps* $V(V(X)) \longrightarrow V(X)$, *define a monad on the category of compact Hausdorff spaces. The category of algebras for this* Vietoris monad *is isomorphic to the category of continuous lattices.*

We denote by $A(X)$, for locally compact sober X, the subspace of $V(X)$ consisting of $\emptyset$ and all irreducible closed sets of X. This is the Alexandroff compactification of X if X is locally compact Hausdorff, but not an extension of X if X is not Hausdorff.

THEOREM 4.5 [15]. *For a locally compact sober space* X, *the space* $A(X)$ *is closed in* $V(X)$, *and hence compact, if and only if the intersection of two saturated compact subsets of* X *is always compact in* X.

References

1. Binz, Ernst, Continuous Convergence on C(X). Lecture Notes in Math. 469, 1975.

2. Binz, Ernst, und H.H. Keller, Funktionenräume in der Kategorie der Limesräume. Ann. Acad. Sci. Fenn. Ser. I.A. Mathematica, no. 383, 1966.

3. Bourdaud, Gérard, Some cartesian closed categories of convergence spaces. Categorical Topology -- Mannheim 1975. Lecture Notes in Math. 540 (1976), 93 - 108.

4. Brown, Ronald, Function spaces and product topologies. Quart. J. Math. (Oxford), (2) 15 (1964), 238 - 250.

5. Cook, C.H., and H.R. Fischer, On equicontinuity and continuous convergence. Math. Annalen 159 (1965), 94 - 104.

6. Cook, C.H., and H.R. Fischer, Uniform convergence structures. Math. Annalen 173 (1967), 290 - 306.

7. Day, Brian, A reflection theorem for closed categories. J. Pure Appl. Algebra 2 (1972), 1 - 11.

8. Day, Brian, Limit spaces and closed span categories. Category Seminar, Sydney 1972/73. Lecture Notes in Math. 420 (1974), 65 - 74.

9. Day, B.J., and G.M. Kelly, On topological quotient maps. Proc. Camb. Phil. Soc. 67 (1970), 553 - 558.

10. Edgar, G.A., A cartesian closed category for topology. Gen. Topology Appl. 6 (1976), 65 - 72.

11. Frölicher, Alfred. Kompakt erzeugte Räume und Limesräume. Math. Zeitschr. 129 (1972), 57 - 63.

12. Gazik, R.J., D.C. Kent, and G.D. Richardson, Regular completions of uniform convergence spaces. Bull. Austral. Math. Soc. 11 (1974), 413 - 424.

13. Herrlich, Horst, Topological functors. Gen. Topology Appl. 4 (1974), 125 - 142.

14. Herrlich, Horst, Cartesian closed topological categories. Math. Coll. Univ. Cape Town 9 (1974), 1 - 16.

15. Hofmann, Karl H., and Jimmie D. Lawson, The spectral theory of distributive continuous lattices. Preprint, 1977.

16. Hofmann, Karl H., and Jimmie D. Lawson, Complements to "The spectral theory of distributive continous lattices". SCS Memo, 3/14/77.

17. Isbell, J.R., Function spaces and adjoints. Math. Scand. 36 (1975), 317 - 339.

18. Jackson, J.R., Spaces of mappings on topological products with applications to homotopy theory. Proc. A.M.S. 3 (1952), 327 - 333.

19. Lee, R.S., The category of uniform convergence spaces is cartesian closed. Bull. Austral. Math. Soc. 15 (1976), 461 - 465.

20. Lee, Rosalyn S., Function spaces in the category of uniform convergence spaces. Ph.D. Dissertation, Carnegie-Mellon University, 1978.

21. Michael, E.A., Topologies on spaces of subsets. Trans. A.M.S. 71 (1952), 152 - 182.

22. Nel, Louis D., Initially structured categories and cartesian closedness. Canad. J. Math. 27 (1975), 1361 - 1377.

23. Nel, Louis D., Cartesian closed topological categories. Categorical Topology — Mannheim 1975. Lecture Notes in Math. 540 (1976), 439 - 451.

24. Scott, Dana, Continuous lattices. Toposes, Algebraic Geometry and Logic. Lecture Notes in Math. 274 (1974), 93 - 136.

25. Steenrod, Norman E., A convenient category of topological spaces. Michigan Math. J. 14 (1967), 133 - 152.

26. Tashjian, Gloria, Cartesian-closed coreflective subcategories of Tychonoff spaces. Preprint, 1976.

27. Vietoris, Leopold. Bereiche zweiter Ordnung. Monatsh. für Math. und Physik 32 (1922), 258 - 280.

28. Vietoris, Leopold, Kontinua zweiter Ordnung. Monatsh. für Math. und Physik 33 (1923), 49 - 62.

29. Vogt, Rainer M., Convenient categories of topological spaces for homotopy theory. Archiv der Math. 22 (1971), 545 - 555.

30. Ward, A.J., Problem. Proceedings of the International Symposium on Topology and its Applications, Herceg-Novi 1968. Beograd, 1969.

31. Watson, P.D., On the limits of sequences of sets. Quart. J. Math. (Oxford) (2) 4 (1953), 1 - 3.

32. Wyler, Oswald, On the categories of general topology and topological algebra. Archiv der Math. 22 (1971), 7 - 17.

33. Wyler, Oswald, Filter space monads, regularitv, completions. TOPO 72 — General Topology and its Applications. Lecture Notes in Math. 378 (1974), 591 - 637.

34. Wyler, Oswald, Quotient maps. Gen. Topology Appl. 3 (1973), 149 - 160.

35. Wyler, Oswald, Convenient categories for topology. Gen. Topology Appl. 3 (1973), 225 - 242.

36. Wyler, Oswald, Are there topoi in topology? Categorical Topology -- Mannheim 1975. Lecture Notes in Math. 540 (1976), 699 - 719.

37. Wyler, Oswald, Algebraic theories of continuous lattices. Preprint, 1976.

Department of Mathematics
Carnegie-Mellon University
Pittsburgh, PA 15213
USA